Introduction to Particle Production in Hadron Physics

TO MAUREEN

— who during the course of this work increased our family from one to three by the addition of twins, and did it with so much good humour and so little complaint that my own labours were hardly disturbed.

Introduction to Particle Production in Hadron Physics

S. HUMBLE
*CERN 1211, Geneva 23
Switzerland*

1974

ACADEMIC PRESS
London and New York

A Subsidiary of Harcourt Brace Jovanovich, Publishers

ACADEMIC PRESS INC. (LONDON) LTD.
24/28 Oval Road.
London NW1

United States Edition published by
ACADEMIC PRESS INC.
111 Fifth Avenue
New York, New York 10003

Library of Congress Catalog Card Number: 73–19014
ISBN—0-12-361450-3

Set in 'Monophoto' Times and printed offset litho in Great Britain by

Page Bros (Norwich) Ltd, Norwich

Preface

With the current interest in particle production reactions in hadron physics this would seem an appropriate time to pull together the various threads of the subject and to catalogue them for those who would venture into the field. To this end I have tried to set out in a simple and understandable way the various regions of interest progressing from low energy single particle production to high energy many-particle reactions. The material follows the general lines of a series of lectures given at the Daresbury Nuclear Physics Laboratory in 1971, though of course, the present work has been greatly expanded and updated.

In writing a book of this kind one becomes aware that one is often treading on quick-sand. New experimental information is accumulating so quickly that some statements may be found to be in error, even before the manuscript is completed. To minimize this risk I have deliberately tried to avoid dependence in the text on specific experimental results, and confined myself as far as possible to mentioning trends in the current data in only a fairly general manner. I would refer the reader to the many excellent conference reviews for a more detailed discussion of the most recent experimental situation.

Since this work is intended to be an introduction to particle production I have, in places, abandoned absolute rigour in order to present a coherent if not completely watertight account of current ideas. In all cases I have tried to provide a very comprehensive guide to the original literature for those who wish to pursue a particular topic further. In order not to be completely overwhelmed by the references, however, at times I have had to be somewhat selective in my choice. I must apologise, therefore, to the many authors who have produced noteworthy and significant results and yet have not been cited here. I would assure them that this omission reflects more on my ignorance, judgement and lack of stamina than on the merits of their work.

S. Humble

February 1974

Acknowledgments

I am indebted to Professor A. Donnachie and my colleagues at DNPL for several interesting and informative conversations during the course of this work. I am particularly grateful to Drs. J. F. L. Hopkinson and G. D. Kaiser for their guidance on, and careful reading of, various parts of the manuscript.

Contents

Chapter 1   Introduction

Chapter 2   Single Particle Production

**Chapter 3   Analysis of Production Processes
at Low and Intermediate Energies**

Chapter 4   High Energy Exclusive Reactions

Chapter 5   Inclusive Reactions

Chapter 6   Hadrodynamics

1

Introduction

1.1. Problems of dimensionality

During the past few decades a great number of strongly interacting particles, hadrons, have been discovered and there has been considerable interest in determining the dynamics underlying the interactions of such particles. Most of this work has concentrated on the study of two body reactions in which two particles initially collide to produce a final state which also consists of just two hadrons. These may be the same as the initial particles in which case the collision is termed an elastic process. Alternatively they may differ from the initial pair by certain quantum numbers and then the collision is called an exchange process since certain conserved quantities such as charge or strangeness have been exchanged between the two colliding bodies.

Besides elastic and exchange collisions there are also production phenomena in which the final state consists not of two particles, but of many. However, nothing like the same systematic study has been attempted for production processes that has been carried out for two body reactions. Although this situation is regrettable the reasons for it are clear. In production processes one has to make more measurements to fully determine the configuration of produced particles. Also in order to determine the dynamical effects, reflecting the structure of the collision amplitude, one has to separate out the kinematical phase-space effects, which often is far from trivial. Furthermore, from a theoretical standpoint there has been the belief that the dynamics of two body reactions should be somewhat simpler to understand.

In fact after several years of intensive work it has to be admitted that two body reactions are still not understood very well. Because of this it might be argued that there seems little point in considering production phenomena at this stage. However, it should be stressed that this is a very negative and unrealistic point of view since, not only do production data exist, but production reactions are found to account for 80 % to 90 % of the total cross-section at high energies. Also there is reason to believe that an understanding of hadron dynamics will only come from a combined study of 2-body and n-body reactions. This is certainly the case if unitarity is an important ingredient in the theory; but more generally there is an increasing amount of evidence to suggest that the dynamics governing both types of reactions possess certain similarities. Thus it is probable that production reactions can provide further tests for the various assumptions and dynamical models that have been made for two particle scattering.

It would be misleading, however, to suggest that production phenomena should be considered *only* as a guide to two-body phenomenology. It is already clear that a detailed analysis of production data leads to features of hadron physics which are peculiar to production processes. For example, with several particles in the final state there are now several final state interactions to consider, many leading to possible resonance formations and reflections into other sub-systems. Also, it is found that produced particles have a pronounced tendency to travel in the same direction as the beam particle; a tendency which is apparently fairly independent of both their momenta and the momentum of the beam. Moreover, it is very notable that there are considerably fewer particles produced than we might have expected, and of those that are created there are generally two (leading particles) which carry away a substantial fraction of the available energy. No doubt as the technology for the identification and detailed measurement of produced particles improves and with the advent of very high energy accelerators, the subsequent data will show many more interesting features. For these reasons we would suggest that far from being the uninteresting by-product of particle collisions, production phenomena should be an important, if not dominant, part of the study of hadron dynamics. Fortunately in the past two or three years this point of view has become widely accepted and considerable effort is now given to understanding such phenomena.

The outstanding problem we face when discussing particle production is one of dimensionality. Simply, we have too many quantities to measure. This is easily seen if we consider a typical experiment with a target at rest bombarded by a beam particle travelling in the z direction. Each of the n final state particles will generally have components of momentum in the x, y and z directions. Hence each production event will be characterized by these $3n$ momentum components restricted only by the four overall energy-momentum constraints. We have therefore $3n - 4$ independent variables which we can take to be either momentum components or alternatively some other set of measurable quantities. For two body processes ($n = 2$) we have only two independent variables such as the incident particle momentum and the scattering angle. However, for three body reactions there are five independent variables and for four body reactions there are eight and so on. The problem then arises, how are we to present these measurements in a meaningful way? The usual graphical methods become inadequate since we cannot plot a distribution of the events over all the variables simultaneously. We could, of course, "publish" data on magnetic tapes, but this does not give us the same insight into the structure of the distribution that is possible with a graphical picture. The answer must be to choose one or two variables which we believe are significant and combine together all those events which give rise to specific values of these variables. In this way we obtain averaged distributions which we can again display graphically.

Unfortunately with $3n - 4$ variables to choose from the possible number of selections of one or two of these becomes very large. It is necessary therefore to have an indication beforehand which distributions in terms of which quantities will be most meaningful. Also since these averaged distributions combine together events coming from different regions of phase space we would like to have some idea that we are not averaging out some very meaningful features of the data. In other words we have to use what knowledge we have of particle dynamics to make a reasonable selection of variables. To do this it is often desirable to make mathematical models for the production amplitudes based on a few simple physical assumptions such as resonance formation, particle exchange, exponential cut off of transverse momenta, etc. Before this can be done, however, we have to derive such things as kinematics, physical region and phase space, as well as either a low energy partial wave decomposition or some high energy approximation. For two body reactions this is fairly straightforward and has been extensively covered in the literature. For n-body production, on the other hand, the complications arising from the increased dimensionality of the problem often intimidate both the casual onlooker and the student.

The object of this book therefore is two-fold. Firstly we shall try to show that while the production of extra particles in the final state does give rise to algebraic complications these are fairly straightforward if tackled in a systematic way. Secondly it is intended to provide a simple reference guide to some formalism, techniques and models in current use in particle production physics. In achieving these objectives we shall concentrate to some extent on three body reactions at least in the early chapters. Such reactions represent the simplest extension to two body scattering and yet indicate many of the difficulties associated with the increased dimensionality. At the same time we shall try to indicate most of the main areas of current interest in production dynamics.

1.2 Phase-space

If, as we have suggested, it is necessary to rely on models to solve the problem of dimensionality in production reactions we must be able to deduce the observable consequences from such models. This sounds simple enough, though as we shall see, in practice it is not always easy.

Suppose we have an initial state consisting of two particles with four momenta p_a, p_b producing a final state consisting of n particles with four momenta lying between q_f and $q_f + \mathrm{d}q_f$, $i = 1, 2, \ldots . n$. The cross-section element for this process can be written as

$$\mathrm{d}\sigma = \frac{(2\pi)^4}{4} \frac{|\bar{T}_{fi}|^2}{F_{ab}} \delta^4(p_a + p_b - q_1 - q_2 \cdots - q_n) \prod_{f=1}^{n} \frac{\mathrm{d}^3\mathbf{q}_f}{(2\pi)^3(2w_f)}, \quad (1.2.1)$$

where w_f is the energy of the fth final state particle; $|\bar{T}_{fi}|^2$ is the spin averaged amplitude squared, i.e.

$$|\bar{T}_{fi}|^2 = \frac{1}{(2s_a + 1)(2s_b + 1)} \sum_{\text{spins}} |T|^2, \quad (1.2.2)$$

F_{ab} is Møller's invariant flux

$$F_{ab} = \{(p_a \cdot p_b)^2 - m_a^2 . m_b^2\} \quad (1.2.3)$$

and we have chosen units such that $\hbar = c = 1$.

The derivation of (1.2.1) is already well covered in the literature and assumes a very standard normalization (see Chapters 1 and 2, H. Pilkuhn, 1967) which relates the amplitude T to the S matrix element by the symbolic expression

$$S = 1 + i(2\pi)^4\delta^4(P) \frac{T}{\prod_{j=1,f} (2w_j)}. \quad (1.2.4)$$

Since this is the normalization we shall use throughout this work (unless otherwise stated) it is perhaps helpful to note that eqn. (1.2.1) for two body processes becomes

$$\frac{\mathrm{d}\sigma}{\mathrm{d}\Omega} = \frac{1}{64\pi^2 s} \frac{q}{p} |\bar{T}_{fi}|^2,$$

where p, q are the magnitudes of the three-momenta of an initial and final state particle in the centre of mass (c.m.) system respectively and $\mathrm{d}\Omega$ is the element of solid angle subtended by $d^3\mathbf{q}$. In this case the optical theorem can be written as

$$\text{Im } T_{ii} = 2F_{ab}\sigma^{ab}_{\text{tot}}. \tag{1.2.6}$$

It should be noted that the r.h.s. of (1.2.1) must be divided by 2 if there are two identical particles in the *final* state. This is necessary since in obtaining a differential or total cross-section one considers all allowed values of the final momentum and would otherwise count the same final state twice. More generally when there are N groups of $n_1,\ldots,n_N$ identical particles (1.2.1) should be divided by $n_1! n_1! \ldots n_N!$. This solution causes confusion because the scattering amplitude is often renormalized so that these factors of $n_1!$ are apparently absent from the phase-space normalization. We would suggest that errors are less likely to be made if the factorials are inserted in the phase-space element.

Considering (1.2.1) in more detail we see that the momenta of the final state particles cannot be chosen at will since they are restricted by the four delta functions expressing energy–momentum conservation. For example, in the overall centre of mass system in which $\mathbf{p}_a = -\mathbf{p}_b$ we have

$$W = \sum_{f=1}^{n} w_f. \tag{1.2.7}$$

Together with the physical requirement that $w_f > m_f$, where m_f is the mass of the fth particle, (1.2.7) has the trivial implication that the sum of the masses of the produced particles must be less than the total energy of the system—a not unexpected result! However, when we integrate over some region of the allowed volume of phase-space to obtain various distributions to compare with experiment, other implications are not so trivial. It often happens that this energy-momentum conservation requirement already leads to distributions which have some apparent structure, i.e. so-called kinematical effects, and great care must be taken not to confuse such kinematical distortions with dynamical effects arising from the structure of the amplitude T.

In high energy, high multiplicity situations, by analogy with statistical mechanics (see Hagedorn, 1964) it may be possible to think of the cross-section distributions being largely kinematic in origin, while for two body reactions (cf. (1.2.5)) for the most part it is the dynamics which determine the structure of the cross-sections. In between these two extreme cases we have a mixture of kinematical and dynamical effects, and the problem once again is to distinguish between them.

To help us in this it is useful to be able to calculate what phase-space alone predicts for the differential cross-sections, i.e. with $|\overline{T}_{fi}|^2$ in (1.2.1) approximated by a constant. Unfortunately it turns out, because of the relationship

$$w_f = (\mathbf{q}_f^2 + m_f^2)^{\frac{1}{2}} \tag{1.2.8}$$

between the energy and magnitude of the three-momentum of each particle f, that the phase-space integrals cannot be performed analytically except for the special cases $n = 1, 2$ or 3. For $n > 3$, therefore, we must resort to numerical computational techniques to evaluate these integrals. This we shall investigate in more detail in Section 4.1. For the moment we shall concentrate on some of the general properties of the phase-space volume which will be required for our discussion of production phenomena in the following chapters.

Using the notation of Hagedorn (1964) let us denote the phase-space volume by R_n such that

$$R_n(P, m_1, m_2 \ldots m_n) = \int \ldots \int \prod_{f=1}^{n} \frac{\mathrm{d}^3\mathbf{q}_f}{w_f} \delta^4(P - q_1 - q_2 \ldots - q_n), \tag{1.2.9}$$

where $P = p_a + p_b$ is the total four-momentum of the system, and we have omitted certain constant factors such as $(2\pi)^3$ etc. The range of the integration in (1.2.9) depends on the particular distributions we are considering. For instance, if the integrals run over all allowed values of $\mathbf{q}_f$, R_n is proportional to the (phase-space) cross-section for n-particle production. On the other hand if we wish to investigate the momentum spectrum of some produced particle f, the range of integration is restricted to that region of phase-space where particle f has some momentum between $\mathbf{q}_f$ and $\mathbf{q}_f + \mathrm{d}^3\mathbf{q}_f$. In a similar manner other distributions can be defined. Notice, however, that provided the corresponding region of phase-space (i.e. restrictions on the range of integration) is invariant under Lorentz transformations, since each of the $\mathrm{d}^3\mathbf{q}_f/w_f$ factors as well as the δ^4 function is invariant, the phase-space volume R_n must also be relativistically invariant, This has the very nice feature that we can consider R_n in any Lorentz frame of reference we may choose in order to simplify its calculation. For instance, re-writing (1.2.9) as

$$R_n(P, m_1, \ldots, m_n) = \int \frac{\mathrm{d}^3\mathbf{q}_n}{w_n} R_{n-1}(P - q_n, m_1, \ldots, m_n) \tag{1.2.10}$$

we can calculate R_n in the overall c.m. system (with $\mathbf{P} = 0$, $P_0 = W$) in terms of R_{n-1} evaluated in its own c.m. system (i.e. where $\mathbf{q}_1 + \mathbf{q}_2 + \ldots + \mathbf{q}_{n-1} = \mathbf{P} - \mathbf{q}_n = 0$). The total energy W_n of this system is given in the overall c.m. system by

$$W_n^2 = (W - w_n)^2 - \mathbf{q}_n^2$$

$$= W^2 + m_n^2 - 2Ww_n,$$

so that

$$R_n(W, m_1, \ldots, m_n) = \int \frac{\mathrm{d}^3\mathbf{q}_n}{w_n} R_{n-1}[\sqrt{(W^2 + m_n^2 - 2Ww_n)}, m_1, \ldots, m_{n-1}].$$

$$(1.2.11)$$

Thus starting from the one particle phase-space R_1 given by (1.2.9) as

$$R_1(W, m_1) = \int \frac{\mathrm{d}^3\mathbf{q}_1}{w_1} \delta[W - \sqrt{(q_1^2 + m_1^2)}] \delta^3(\mathbf{q}_1)$$

$$= \delta(W - m_1)/m_1, \qquad (1.2.12)$$

the two particle phase-space volume can be determined from (1.2.11) as

$$R_2(W, m_1, m_2) = \pi q/W, \qquad (1.2.13)$$

where q is the momentum of either particle 1 or 2 in the c.m. system; and so on, by iteration, for all n.

In deriving (1.2.12) and (1.2.13) the integrations extended over all allowed regions of phase-space. However, from what we have already said the range of integrations in (1.2.9) and (1.2.11) can be restricted to some particular region of phase-space in order to determine certain kinematic distributions. For example, the momentum spectrum of particle n is obtained by dropping the integration over $\mathrm{d}^3\mathbf{q}_n$ in (1.2.11) and is therefore simply proportional to R_{n-1} divided by its energy w_n.

The spectrum of several particles (k say) may be computed by splitting the integral in (1.2.1) so that

$$R_n(P, m_1, \ldots, m_n) = \int \delta^4\left(P - \sum_{f=1}^{n} q_f\right) \left\{\prod_{f=1}^{k} \frac{\mathrm{d}^3\mathbf{q}_f}{w_f}\right\}\left\{\prod_{f=k+1}^{n} \frac{\mathrm{d}^3\mathbf{q}_f}{w_f}\right\}. \quad (1.2.14)$$

(Notice the numbering of the particles 1 to n is completely arbitrary so that (1.2.14) is quite general.) Because of the delta function relations

$$\delta^4\left(P - \sum_1^n q_f\right) = \int \mathrm{d}^4Q_k\delta^4\left(P - Q_k - \sum_{k+1}^n q_f\right)\delta^4\left(Q_k - \sum_1^k q_f\right) \qquad (1.2.15)$$

and

$$1 = \int_{\sum_{f=1}^{k} m_f}^{\infty} \delta(Q_k^2 - M^2)\,\mathrm{d}M^2$$

(1.2.14) can be rewritten as

$$R_n(P, m_1, \ldots, m_n) = \int^{\infty} \mathrm{d}M^2 \left\{\int \delta(Q_k^2 - M^2)\,\mathrm{d}^4Q_k\delta^4\right.$$

$$\times \left(P - Q_k - \sum_{k+1}^{n} q_f\right) \prod_{k+1}^{n} \frac{d^3\mathbf{q}_f}{\omega_f}\Bigg\} \Bigg\{\int \delta^4\left(Q_k - \sum_{1}^{k} q_f\right) \prod_{1}^{k} \frac{d^3\mathbf{q}_f}{w_f}\Bigg\}. \quad (1.2.16)$$

On performing the $d(Q_k)_0$ integration we find the $\delta(Q_k^2 - M^2)\,d^4Q_k$ term in the first curly bracket is equivalent to $d^3\mathbf{Q}_k/2(Q_k)_0$ where $(Q_k)_0^2 = \mathbf{Q}_k^2 + M^2$. Hence

$$R_n(P, m_1, \ldots, m_n) = \int^{\infty} dM^2 R_{n-k+1}(P, M, m_{k+1}, \ldots m_n) R_k(Q_k, m_1, \ldots, m_k),$$
$$(1.2.17)$$

where R_{n-k+1} describes the phase-space appropriate to $n - k$ particles of mass $m_{k+1}, \ldots, m_n$ plus some system (quasi particle) of mass M. The internal configuration of this system is described by the function R_k which again from relativistic invariance may be calculated for simplicity in its own c.m. system in which $(Q_k)_0 = M$. In this form we see that the distribution given by phase-space for the square of the k-particle invariant mass distribution

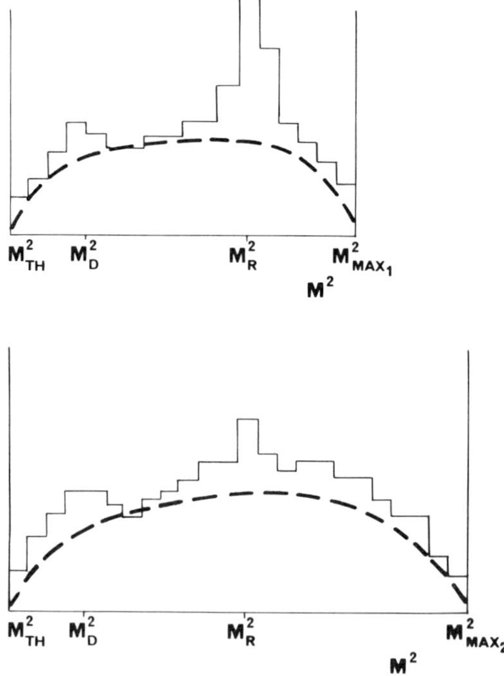

FIG. 1. Typical (idealized) histograms showing invariant mass-squared distribution at two different incident energies. M_R is the mass of a resonance and M_D indicates the position of a low mass (Deck) enhancement. The broken curve represents the pure phase-space distribution with an arbitrary normalization.

$d\sigma/dM^2$ is just proportional to

$$R_{n-k+1}(P, M, m_{k+1}, \ldots, m_n)R_k(M, m_1, \ldots, m_k). \qquad (1.2.18)$$

By comparing the experimentally observed distribution to this phase-space prediction (1.2.18) we should be able to obtain some indication of the dynamical effects which give rise to differences between the two distributions. For instance, if a resonance of mass M_R is produced which has a strong decay mode into the k particles $1 + 2 + \ldots + k$ we would expect an enhancement above the phase-space distribution at values of M close to M_R as shown in Fig. 1. Many resonances were first discovered in this way, a good example being the ω meson which decays predominantly into three pions (Maglic *et al.*, 1961). However, it is perhaps worthwhile adding a cautionary note that not all enhancements necessarily correspond to resonances. It is fairly easy to simulate bumps in these distributions which have nothing to do with second sheet resonance poles in the scattering amplitude. Some may be distinguished from resonances by considering the mass distribution at another widely separated incident energy. An enhancement corresponding to a resonance, of course, would occur at a fixed value of M irrespective of the total energy of the system, whereas bumps produced by other effects may well vary with energy. However, this need not always happen, as in the low mass enhancements produced by the Deck effect (Deck, 1964; Cohen *et al.*, 1972). Also a resonance peak may tend to disappear as the energy of the system changes depending on the way in which it is produced. Thus other tests are required for the confirmation of resonance phenomena such as analysing the angular distribution of decay products to see if it corresponds to some specific spin state. These will be considered later.

For the moment let us conclude our remarks concerning phase-space. Clearly it is going to be an important theme in our discussion of production processes in the following chapters. From now on, however, it will be more useful to consider it in more detail for particular situations as they arise in the text.

2

Single Particle Production

2.1. Invariants and kinematics

Let us begin this section by the introduction of our notation for the typical 3-body process $a + b \to c_1 + c_2 + c_3$ shown in Fig. 2a. From the corresponding four-momenta of these particles $p_a + p_b \to q_1 + q_2 + q_3$ (besides

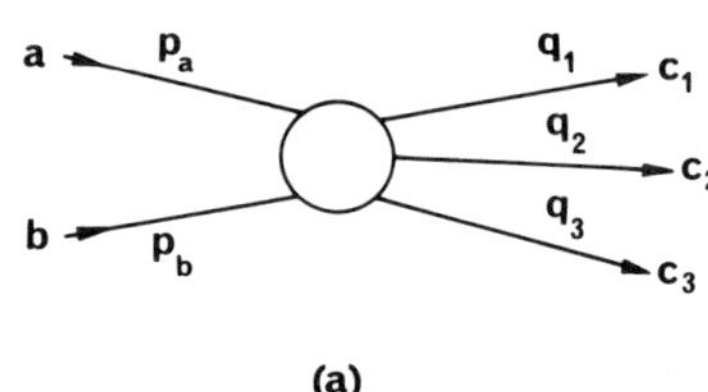

(a)

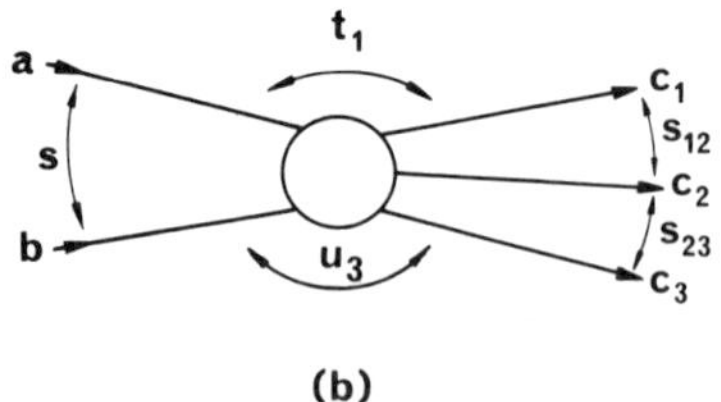

(b)

FIG. 2. The three-body process $a + b \to c_1 + c_2 + c_3$.

the masses m_a, m_b, m_1, m_2, m_3) we can construct the following ten scalar invariants

$$s = (p_a + p_b)^2$$
$$s_{ij} = (q_i + q_j)^2 \qquad i \neq j = 1, 2, 3$$
$$t_i = (p_a - q_i)^2 \qquad i = 1, 2, 3 \qquad (2.1.1)$$
$$u_i = (p_b - q_i)^2 \qquad i = 1, 2, 3.$$

Only five of these are independent, however, since the imposition of energy-momentum conservation demands the invariants satisfy certain linear relationships. For example, $s + t_1 + u_1$ can be written as

$$\begin{aligned}
s + t_1 + u_1 &= (q_1 + q_2 + q_3)^2 + (p_a - q_1)^2 + (p_b - q_1)^2 \\
&= 3m_1^2 + m_a^2 + m_b^2 + s_{23} - 2q_1(p_a + p_b - q_2 - q_3) \\
&= m_1^2 + m_a^2 + m_b^2 + m_b^2 + s_{23}.
\end{aligned} \qquad (2.1.2)$$

Similarly

$$s + t_2 + u_2 = m_2^2 + m_a^2 + m_b^2 + s_{13}$$

etc.

We shall take as one useful independent set $s, s_{12}, s_{23}, t_1, u_3$ as indicated in Fig. 2b. Then the remaining invariants are related to this set by the matrix equation†

$$
\begin{bmatrix}
t_3 - m_a^2 - m_b^2 - m_3^2 \\
u_1 - m_a^2 - m_b^2 - m_1^2 \\
t_2 - m_a^2 - m_1^2 - m_2^2 \\
u_2 - m_b^2 - m_2^2 - m_3^2 \\
s_{13} - m_1^2 - m_2^2 - m_3^2
\end{bmatrix}
=
\begin{bmatrix}
-1 & 1 & 0 & 0 & -1 \\
-1 & 0 & 1 & -1 & 0 \\
0 & -1 & 0 & -1 & 1 \\
0 & 0 & -1 & 1 & -1 \\
1 & -1 & -1 & 0 & 0
\end{bmatrix}
\begin{bmatrix}
s \\
s_{12} \\
s_{23} \\
t_1 \\
u_3
\end{bmatrix}
\qquad (2.1.3)
$$

which of course contains all the information of (2.1.2). The expression (2.1.3) can be thought of as the extension of the requirement $s + t + u = \Sigma m_i^2$ for 2-body scattering.

Having thus defined our invariants we can now use them to calculate any energy, three-momentum or relative angle of any particle in any frame of

---

† (Although not relevant to this chapter it is worthwhile pointing out that when we consider more than three particles in the final state we obtain quadratic relationships between the invariants as well as linear ones. This is particularly important when we consider the analytic structure of production amplitudes defined as functions of Lorentz scalars since the quadratic relations will in general give rise to *two* values of some scalars for *one* fixed set of values for the other).

reference. This is important if we wish to test some given model for a production amplitude against experiment. From the theorists' point of view it is frequently more desirable either to work in a frame independent way in terms of invariant amplitudes as functions of Lorentz scalars, or to choose some particular reference frame to simplify the mathematics. However, from a practical point of view we have, in the end, to relate to experimental data usually given in terms of energies and scattering angles measured in the laboratory, i.e. where one of the initial particles is at rest.

We give below the relationship between these quantities and our invariants, as well as the corresponding relationships in the overall centre of mass-system with $\mathbf{p}_a + \mathbf{p}_b = 0$ and the c.m. system of particles c_1 and c_2, i.e. where $\mathbf{q}_1 + \mathbf{q}_2 = 0$. These latter frames of reference will be useful for the definition of the physical region and the phase-space manipulations outlined in the next sections.

In determining these expressions it is helpful to consider initially a particle of mass $\sqrt{y}$ colliding with a particle of mass $\sqrt{z}$ in their c.m. system with total energy $\sqrt{x}$, i.e. where

$$\sqrt{x} = \sqrt{(p^2 + y)} + \sqrt{(p^2 + z)}.$$

Inverting this relation the magnitude of the momentum, p, of either particle is given by

$$p = P(x, y, z) = [(x^2 + y^2 + z^2 - 2xy - 2yz - 2zx)/4x]^{\frac{1}{2}} \qquad (2.1.4a)$$

and the energy, $\sqrt{(p^2 + y)}$, of particle of mass $\sqrt{y}$ is

$$E(x, y, z) = (x + y - z)/(2\sqrt{x}). \qquad (2.1.4b)$$

In this way it is a simple matter to determine the energy and momentum of all particles in our 3-body process in the overall c.m. system. In other systems these quantities can also be determined either in an analogous fashion to the above, or by considering specific Lorentz transformations from the c.m. system (e.g. Hagedorn, 1964). We leave this as an exercise for the reader and here merely state the results. (For further discussion regarding the definition of physical quantities in terms of Lorentz scalars see, for example, Ferrari and Selleri (1962) and Razmi (1964)).

(L) *In the lab. frame*, $\mathbf{p}_b = 0$: the magnitude of the three momenta $\mathbf{p}_a, \mathbf{q}_1, \mathbf{q}_2$ and $\mathbf{q}_3$ can be written as

$$p_a^L = \frac{\sqrt{s}}{m_b} P(s, m_a^2, m_b^2) \qquad (2.1.5a)$$

$$q_i^L = \frac{\sqrt{-u_i}}{m_b} P(-u_i, -m_b^2, -m_i^2), \qquad i = 1, 2, 3$$

and their kinetic energies as

$$T = \frac{1}{2m_b}\left[s - (m_a + m_b)^2\right] \tag{2.1.5b}$$

$$T_i = \frac{1}{2m_b}\left[-u_i + (m_b - m_i)^2\right], \qquad i = 1, 2, 3.$$

The scattering angles of $\mathbf{q}_1$, $\mathbf{q}_2$, $\mathbf{q}_3$ with respect to the incident particle momentum $\mathbf{p}_a$ are given by

$$2q_i^L P_a^L \cos\theta_i^L = 2T(T_i + m_i) + 2m_a T_i - (m_i - m_a)^2 + t_i \qquad i = 1, 2, 3 \tag{2.1.5c}$$

and F_{ab} of (1.2.3) is just $p_a^L m_b$.

(c) *In the overall c.m. frame,* $\mathbf{p}_a + \mathbf{p}_b = 0$: the three momenta of p_a, p_b, q_i are

$$p_a^c = p_b^c = P(s, m_a^2, m_b^2)$$

$$q_i^c = P(s, m_i^2, s_{jk}), \qquad i = 1, 2, 3 \tag{2.1.6a}$$

and have total energies

$$E_a^c = \sqrt{s} - E_b^c = E(s, m_a^2, m_b^2)$$

$$w_i^c = E(s, m_i^2, s_{jk}), \qquad i = 1, 2, 3 \tag{2.1.6b}$$

The scattering angles w.r.t. $\mathbf{p}_a$ are now

$$2q_i^c p_a^c \cos\theta_i^c = 2E_a^c w_i^c - m_a^2 - m_i^2 + t_i, \qquad i = 1, 2, 3, \tag{2.1.6c}$$

while the angle λ_k^c between $\mathbf{q}_i$ and $\mathbf{q}_j$ is given by

$$2q_i^c q_j^c \cos\lambda_k^c = 2w_i^c w_j^c = m_i^2 + m_j^2 - s_{ij} \qquad i \neq j \neq k = 1, 2, 3 \tag{2.1.6d}$$

and $F_{ab} = p_a^c \sqrt{s}$.

(s) *In the* $c_1 + c_2$ *frame,* $\mathbf{q}_1 + \mathbf{q}_2 = 0$: the three-momenta become

$$p_a^s = P(s_{12}, m_a^2, u_3)$$

$$p_b^s = P(s_{12}, m_b^2, t_3)$$

$$q_1^s = q_2^s = P(s_{12}, m_1^2, m_2^2) \tag{2.1.7a}$$

$$q_3^s = \sqrt{(s/s_{12})}\,P(s, m_3^2, s_{12})$$

and energies

$$E_a^s = E(s_{12}, m_a^2, u_3)$$

$$E_b^s = E(s_{12}, m_b^2, t_3)$$

$$w_1^s = E(s_{12}, m_1^2, m_2^2) \qquad\qquad (2.1.7b)$$

$$w_2^s = E(s_{12}, m_2^2, m_1^2)$$

$$w_3^s = \sqrt{(s/s_{12})}E(s, -m_3^2, s_{12}).$$

The scattering angles of $\mathbf{q}_1$ w.r.t. $\mathbf{p}_a$ and $\mathbf{p}_b$ are given by

$$2q_1^s p_a^s \cos \theta_1^s = 2E_a^s w_1^s - m_1^2 - m_a^2 + t_1$$

$$2q_1^s p_b^s \cos \lambda^s = 2E_b^s w_1^s - m_1^2 - m_b^2 + u_1 \qquad\qquad (2.1.7c)$$

and the angle β^s between $\mathbf{p}_a$ and $\mathbf{p}_b$ by

$$2p_a^s p_b^s \cos \beta^s = 2E_a^s E_b^s + m_a^2 + m_b^2 - s \qquad\qquad (2.1.7d)$$

By a permutation of the outgoing particle subscripts i, j, k it is a simple matter to write the analogous expressions in $c_2 + c_3$ and $c_1 + c_3$ c.m. systems.

2.2. Physical region

The definition of the various scattering angles has a very practical purpose. Just as in two-body scattering $(a + b \rightarrow c_1 + c_2)$ where the physical region is defined as those values of invariants which give positive three-momentum q, and physical scattering angle, (e.g. $|\cos \theta| < 1$ for elastic scattering gives the constraints $0 > t > -4q^2$) so in production processes it is the momentum and scattering angles which define the allowed value of the invariants. Clearly, we need only consider one of the frames c, L, s etc. for this purpose since they are related to each other by Lorentz transformations which leave the invariants unchanged. However, as we shall see, it is often much easier to obtain the physical boundaries by considering quantities defined in different frames of reference, although some care must be taken in this case to ensure that we have completely defined the physical limits.

As we might expect when we are dealing with a five dimensional space, the physical region is not simple but is the intersection of various hypersurfaces each of which represents a condition on some of the invariants. For example, (2.1.6c) can be written as

$$|\cos \theta_i^c| = |(2E_a^c w_i^c - m_a^2 - m_i^2 + t_i)/(2q_i^c p_a^c)| \leqslant 1$$

which from (2.1.6a), (2.1.6b) provides a relationship between s, s_{jk}, t_i while $|\cos \lambda_k^c| \leqslant 1$ in (2.1.6d) relates s, s_{jk}, s_{ik}.

The totality of these conditions can be expressed very neatly in a form similar to the Kibble function for two-body scattering (Kibble, 1960)† i.e.

---

† Note that this cannot be extended to cases where there are more than three particles in the final state since the corresponding determinant is identically zero. This occurs because the determinant is expressed in terms of four-vectors. It follows therefore that the rows or columns of the $n \times n$ matrix (with $n > 4$) cannot be independent.

$$\begin{vmatrix} p_a^2 & p_a q_1 & p_a q_2 & p_a q_3 \\ q_1 p_a & q_1^2 & q_1 q_2 & q_1 q_3 \\ q_2 p_a & q_2 q_1 & q_2^2 & q_2 q_3 \\ q_3 p_a & q_3 q_1 & q_3 q_2 & q_3^2 \end{vmatrix} \leqslant 0 \qquad (2.2.1)$$

which simultaneously gives the physical region boundaries for all crossing related processes involving a, b, c_1, c_2, c_3. However, for practical purposes it is often easier to look at the restraints provided by individual angles. The choice of angle, of course, will depend on which invariant one wishes to bound. To help with this we indicate in Table 1 which of the independent set

TABLE 1. Kinematic Variables as Functions of Invariants

Quantity	Defined by
$T, p_a^L, p_a^c, p_b^c, E_a^c, E_b^c$	s
$q_1^s, q_2^s, w_1^s, w_2^s$	s_{12}
T_3, q_3^L	u_3
$q_3^c, q_3^s, w_3^c, w_3^s$	s, s_{12}
q_1^c, w_1^c	s, s_{23}
p_a^s, E_a^s	s_{12}, u_3
$\cos\theta_3^L, \cos\theta_3^c, \cos\beta^s, p_b^s, E_b^s$	s, s_{12}, u_3
$\cos\theta_1^L, \cos\theta_1^c, T_1, q_1^L$	s, s_{23}, t_1
$\cos\lambda_i^c, q_2^c, w_2^c$	s, s_{23}, s_{12}
$\cos\theta_1^s$	s_{12}, u_3, t_1
T_2, q_2^L	s_{23}, t_1, u_3
$\cos\theta_2^L, \cos\theta_2^c, \cos\lambda^s$	$s, s_{12}, s_{23}, t_1, u_3$

of invariants are required to calculate the quantities defined in (2.1.5) (2.1.6) and (2.1.7).

Besides the angles defined above there is another set of azimuthal angles which, while not independent, often simplify the analysis (Ferrari and Selleri, 1962). A typical azimuthal angle ϕ_2 is defined in Fig. 3 and is related to θ_1, θ_3 and λ_2 by the expression

$$\cos\lambda_2 = \cos\theta_1 \cos\theta_3 + \sin\theta_1 \sin\theta_3 \cos\phi_2.$$

For physical values of ϕ_2 i.e. $|\cos\phi_2| \leqslant 1$ this implies the conditions

$$\cos(\theta_1 + \theta_3) \leqslant \cos\lambda_2 \leqslant \cos(\theta_1 - \theta_3)$$

on θ_1, θ_3 and λ_2 which can be written symmetrically as

$$1 - \cos^2\theta_1 - \cos^2\theta_3 - \cos^2\lambda_2 + 2\cos\theta_1 \cos\theta_3 \cos\lambda_2 \geqslant 0. \qquad (2.2.2)$$

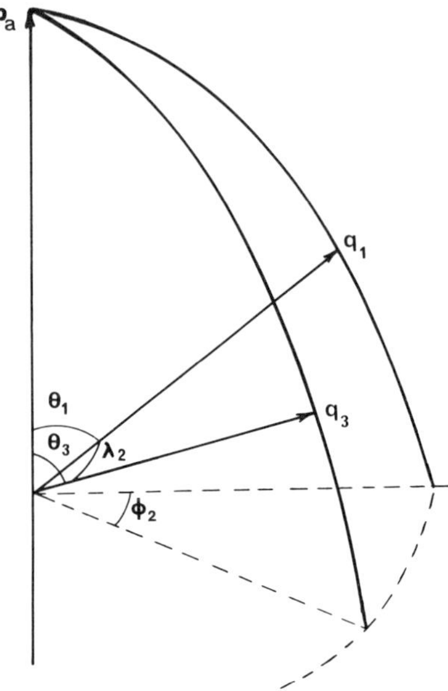

FIG. 3. Definition of the azimuthal angle ϕ_2.

The value of this result is that given four of our set or independent invariants it provides limits on the fifth, that is to say, it defines the extent of the physical region (cf. 2.2.1). Indeed, multiplying by $4q_1^2q_3^2p_a^2$ we find using (2.1.5) (2.1.6) or (2.1.7) depending on our arbitary choice of Lorentz frame that

$$4q_1^2q_3^2p_a^2(1 - \cos^2\theta_1 - \cos^2\theta_3 - \cos^2\lambda_2 + 2\cos\theta_1\cos\theta_3\cos\lambda_2) \geqslant 0 \quad (2.2.3)$$

reduces to a second degree polynomial in these invariants and as such is clearly very useful for Monte Carlo phase-space integrals (see Section 4.1).

2.2a. *High energy approximation.* At high energies the above physical conditions simplify somewhat, since many of the masses may be neglected, and the allowed regions for processes such as double Regge pole exchange can be readily obtained. As an illustration let us consider this particular example of double Regge exchange, where we assume s, $s_{ij}(i \neq j = 1, 2, 3)$ are all large while t_1 and u_3 are small, $0 > t_1, u_3 > -m^2$. (This choice of limit will be explained in Section (4.2)). Then, from (2.1.6) to leading order in s

$$\cos\theta_i^c \approx -1 - \frac{2u_i}{s - s_{jk}} \approx 1 + \frac{2t_i}{s - s_{jk}}$$

and

$$\cos \lambda_k^c \approx -1 + \frac{2(s_{ik}s_{jk} - sm_k^2)}{(s - s_{ik})(s - s_{jk})}, \qquad i \neq j \neq k = 1, 2, 3$$

which, since $s > s_{12}$ and $s > s_{23}$, for instance implies

$$s_{12}s_{23} > sm_2^2. \tag{2.2.4}$$

However, $s_{12}s_{23}$ cannot grow indefinitely. To see this we note that since θ_1^c, θ_3^c and λ_2^c are the angles subtended by $\mathbf{p}_a$ and $\mathbf{q}_1$, $\mathbf{p}_a$ and $\mathbf{q}_3$, $\mathbf{q}_1$ and $\mathbf{q}_3$ respectively it follows from Fig. 3 that

$$\cos \lambda_2^c = \cos \theta_1^c \cos \theta_3^c + \sin \theta_1^c \sin \theta_3^c \cos \phi_2^c$$

i.e.

$$\cos \lambda_2^c \leqslant \cos \theta_1^c \cos \theta_3^c + |\sin \theta_1^c||\sin \theta_3^c|$$

and

$$-1 + \frac{2(s_{12}s_{23} - sm_2^2)}{(s - s_{12})(s - s_{23})} \leqslant \left[1 + \frac{2t_1}{s - s_{12}}\right]\left[-1 - \frac{2u_3}{s - s_{23}}\right]$$
$$+ \left\{1 - \left[1 + \frac{2t_1}{s - s_{12}}\right]^2\right\}^{\frac{1}{2}}\left\{1 - \left[1 + \frac{2u_3}{s - s_{23}}\right]^2\right\}^{\frac{1}{2}}$$

Expanding in powers of t_1 and u_3 we obtain

$$\frac{s_{12}s_{23} - sm_2^2}{(s - s_{12})(s - s_{23})} \leqslant -\frac{t_1}{s - s_{12}} - \frac{u_3}{s - s_{23}} + 2\left[\frac{t_1 u_3}{(s - s_{12})(s - s_{23})}\right]^{\frac{1}{2}} + O(t_1 u_3)$$

where the remainder, $O(t_1 u_3)$, is less than zero.
    Thus

$$s_{12}s_{23} \leqslant s[m_2^2 - t_1 - u_3 + 2\sqrt{(t_1 u_3)}] \tag{2.2.5}$$

and while from (2.2.4) the product $s_{12}s_{23}$ is large for large s, s_{12} and s_{23} cannot both be of order s simultaneously. We shall see the importance of this result later.

2.3. Differential cross-sections

Having outlined how we may obtain the physically allowed range of variables we can now proceed to again discuss the problem of phase-space, and the manipulations necessary to obtain various particle distributions. As we have already noted the element of phase-space given in (1.2.9) i.e.

$$dR_3 = \frac{d^3\mathbf{q}_1}{w_1} \frac{d^3\mathbf{q}_2}{w_2} \frac{d^3\mathbf{q}_3}{w_3} \delta^4(p_a + p_b - q_1 - q_2 - q_3) \qquad (2.3.1)$$

has the very important property that each d^3q_i/w_i and the delta function are Lorentz invariant and so can be evaluated in any frame we please. Here let us consider, for example, the first two terms in the $\mathbf{q}_1 + \mathbf{q}_2$ c.m. system (system s). Instead of $q_{1\mu}q_{2\mu}$ we can choose $P_\mu = q_{1\mu} + q_{2\mu}$ and $Q_\mu = q_{1\mu} - q_{2\mu}$; but since $\mathbf{q}_1^s = -\mathbf{q}_2^s = \mathbf{q}_s$,

$$d^3P\,d^3Q = d^3P\,d^3q_s$$
$$= d^3P q_s^2\,dq_s\,d\Omega_s$$

where $d\Omega_s$ is the element of solid angle defined by $d^3\mathbf{q}_1$ in the s-frame. Also since

$$P_0 = (q_s^2 + m_1^2)^{\frac{1}{2}} + (q_s^2 + m_2^2)^{\frac{1}{2}}$$

we have

$$dP_0 = q_s\,dq_s[(q_s^2 + m_1^2)^{-\frac{1}{2}} + (q_s^2 + m_2^2)^{-\frac{1}{2}}]$$
$$= \frac{q_s\sqrt{s_{12}}}{w_1^s w_2^s}\,dq_s$$

and dR_3 becomes

$$dR_3 = \delta^4(p_a + p_b - P - q_3)\,d^4P\,d\cos\theta_1^s\,d\phi_1^s\,\frac{q_1^s}{\sqrt{s_{12}}}\,\frac{d^3q_3}{w_3} \qquad (2.3.2)$$

where $\theta_1^s\,\phi_1^s$ are the polar and azimuthal angles of $\mathbf{q}_1$ in the s system. The d^4P integration can be done immediately using the delta function while d^3q_3/w_3 can be written in the overall c.m. system (c) as

$$\frac{d^3q_3}{w_3} = (q_3^c)^2\,\frac{dq_3^c}{w_3^c}\,d\cos\theta_3^c\,d\phi_3^c.$$

For unpolarized targets we are free to choose the azimuthal angle ϕ_3^c arbitrarily and hence $\int d\phi_3^c = 2\pi$.

We now wish to express the phase space element

$$dR_3 = 2\pi\frac{q_1^s}{\sqrt{s_{12}}}\,d\cos\theta_1^s\,d\phi_1^s\,\frac{(q_3^c)^2}{w_3^c}\,dq_3^c\,d\cos\theta_3^c \qquad (2.3.3)$$

in terms of our invariants. Using the relation

$$\cos\lambda^s = \cos\beta^s\cos\theta_1^s + \sin\beta^s\sin\theta_1^s\,\cos\phi_1^s \qquad (2.3.4)$$

and noting from Table 1 that

$$q_3^c \text{ depends on } s_{12}$$

$$\cos \theta_3^c \text{ depends on } s_{12}, u_3$$

$$\cos \theta_1^s \text{ depends on } s_{12}, u_3, t_1$$

$$\text{while } \cos \beta^s \text{ depends on } s_{12} \text{ and } u_3$$

$$\text{and } \cos \lambda^s \text{ depends on } s_{12}, u_3, t_1, s_{23}$$

the Jacobian describing the change of variables can be written as

$$
J = \begin{vmatrix}
\dfrac{\partial q_3^c}{\partial s_{12}} & 0 & 0 & 0 \\[2ex]
\dfrac{\partial \cos \theta_3^c}{\partial s_{12}} & \dfrac{\partial \cos \theta_3^c}{\partial u_3} & 0 & 0 \\[2ex]
\dfrac{\partial \cos \theta_1^s}{\partial s_{12}} & \dfrac{\partial \cos \theta_1^s}{\partial u_3} & \dfrac{\partial \cos \theta_1^s}{\partial t_1} & 0 \\[2ex]
\dfrac{\partial \phi_1^s}{\partial s_{12}} & \dfrac{\partial \phi_1^s}{\partial u_3} & \dfrac{\partial \phi_1^s}{\partial t_1} & \dfrac{\partial \phi_1^s}{\partial s_{23}}
\end{vmatrix}
$$

Because of the triangular nature of this determinant only the diagonal elements contribute so that

$$
dq_3^c \, d\cos \theta_3^c \, d\cos \theta_1^s \, d\phi_1^s = \left[\frac{\partial q_3^c}{\partial s_{12}}\right]\left[\frac{\partial \cos \theta_3^c}{\partial u_3}\right]\left[\frac{\partial \cos \theta_3^c}{\partial t_1}\right]\left[\frac{\partial \, d\phi_1}{\partial s_{23}}\right]
$$

$$\times \, ds_{12} \, du_3 \, ds_{23} \, dt_1 \qquad (2.3.5)$$

and from (2.1.6), (2.1.7) a little simple algebra gives

$$
dq_3^c \, d\cos \theta_3^c \, d\cos \theta_1^s \, d\phi_1^s = \frac{w_3^c}{2q_3^c \sqrt{s}}
$$

$$
\frac{ds_{12} \, du_3 \, ds_{23} \, dt_1}{(2q_3^c p_b^c)(2q_1^s p_a^s)(2q_1^s p_b^s) \sin \beta^s \sin \theta_1^s \sin \phi_1^s}. \qquad (2.3.6)
$$

Combining (2.3.6) (2.3.3) and (1.2.1) we obtain

$$
\frac{d\sigma}{ds_{12} \, du_3 \, ds_{23} \, dt_1} = \frac{1}{(8\pi)^4}
$$

$$
\times \frac{|\overline{T}|^2}{s\sqrt{s_{12} p_a^c p_b^c}(2q_1^s p_a^s p_b^s \sin \theta_1^s \sin \phi_1^s \sin \beta^s)} \qquad (2.3.7)
$$

where we have included the correct factors of 2π etc. which were omitted from dR_3. Note that the term in brackets in the denominator of (2.3.7) is the square root of the boundary function given in (2.3.3) defined in terms of the s-frame variables. However, it should also be pointed out that since the left hand side of (2.3.7) and $|\bar{T}|^2$ are invariant, the denominator of the right hand side of (2.3.7) must also be invariant. Thus we would have obtained the same result by initially considering instead of the s-frame any other two particle subsystem, e.g. the $\mathbf{q}_1 + \mathbf{q}_3$ c.m. system.

From (2.3.7) we can now integrate over some or all of the four invariants s_{12}, u_3, s_{23}, t_1 to obtain the desired differential or total cross-section. We list below a few typical distributions. In each case we relate it to a distribution defined in the laboratory frame since, as we have already pointed out, our aim is to be able to compare our models for the amplitude T to the experimental situation. For example,

$$\frac{1}{4m_b q_3^L p_a^L} \cdot \frac{\mathrm{d}^2\sigma}{\mathrm{d}T_3\,\mathrm{dcos}\,\theta_3^L} = \frac{\mathrm{d}\sigma}{\mathrm{d}s_{12}\,\mathrm{d}u_3}$$

$$= \int_{t_1^-}^{t_1^+} \mathrm{d}t_1 \int_{s_{23}^-}^{s_{23}^+} \mathrm{d}s_{23}\, \frac{\mathrm{d}\sigma}{\mathrm{d}s_{12}\,\mathrm{d}u_3\,\mathrm{d}s_{23}\,\mathrm{d}t_1} \qquad (2.3.8)$$

where the range of integration is that region where $\sin\phi_1^s$, $\sin\theta_1^s$ in (2.3.7) is real and non-zero, i.e.

$$t_1^{\pm} = m_1^2 + m_a^2 - 2E_a^s w_1^s \pm 2q_1^s p_a^s$$

$$s_{23}^{\pm} = s + t_1 + m_a^2 - 2E_b^s w_1^s + 2q_1^s p_b^s(\cos\beta^s \cos\theta_1^s \pm \sin\beta^s \sin\theta_1^s).$$

Also

$$\frac{1}{2m_b}\frac{\mathrm{d}\sigma}{\mathrm{d}T_3} = \frac{\mathrm{d}\sigma}{\mathrm{d}u_3} = \int_{(m_1+m_2)^2}^{s_{12}^+} \mathrm{d}s_{12}\, \frac{\mathrm{d}\sigma}{\mathrm{d}s_{12}\,\mathrm{d}u_3}. \qquad (2.3.9)$$

In this case although s_{12}^+ can be given by a detailed examination of the condition (2.2.3) it is simpler to consider (2.1.5) (2.1.6) and (2.1.7) with the aid of Table 1 and to obtain from the definition of $\cos\theta_3^L$ in (2.1.5)

$$s_{12}^+ = u_3 + s - m_b^2 - 2(T + m_a)(T_3 + m_3) + 2q_3^L p_a^L.$$

In the same way we find

$$\frac{1}{2q_3^L p_a^L}\frac{\mathrm{d}\sigma}{\mathrm{d}\cos\theta_3^L} = \frac{\mathrm{d}\sigma}{\mathrm{d}s_{12}} = \int_{u_3^-}^{u_3^+} \mathrm{d}u_3\, \frac{\mathrm{d}\sigma}{\mathrm{d}s_{12}\,\mathrm{d}u_3}, \qquad (2.3.10)$$

where

$$u_3^{\pm} = -2E_b^c w_3^c + m_3^2 + m_b^2 \pm 2p_b^c q_3^c$$

so that

$$\sigma = \int_{(m_1+m_2)^2}^{(\sqrt{s}-m_3)^2} \mathrm{d}s_{12}\, \frac{\mathrm{d}\sigma}{\mathrm{d}s_{12}}. \qquad (2.3.11)$$

2.3a *High energy approximation.* As in Section 2.2 it is again worth noting that in the high energy limit many of the limits of integration in the above expressions simplify quite considerably. As an example of this let us assume we have a matrix element of the form

$$T = c(t_1, u_3) s_{12}^{\alpha_1(t_1)} s_{23}^{\alpha_2(u_3)}$$

(which as we shall see in Section (4.2) is suggested by a double Regge pole model) and calculate the Muirhead distribution $d^2\sigma/dt_1 du_3$. Then taking s, s_{12}, s_{23} large and t_1, u_3 small we can rewrite (2.3.7) to leading order in s, s_{12}, s_{23} as

$$\frac{d^2\sigma}{dt_1 \, du_3} \approx \frac{1}{s^3} \int_{(m_1+m_2)^2}^{(\sqrt{s}-m_3)^2} ds_{12} s_{12}^{2\alpha_1} \int_a^b ds_{23} \frac{s_{23}^{2\alpha_2} g(t_1, u_3)}{\left[\left(\frac{s_{23}}{a} - 1\right)\left(1 - \frac{s_{23}}{b}\right)\right]^{\frac{1}{2}}} \qquad (2.3.12)$$

where the limits a, b of s_{23} come from (2.2.4) and (2.2.5) i.e.

$$a = sm_2^2/s_{12} \leqslant s_{23} \leqslant sm_2^2 f/s_{12} \equiv b$$

and

$$f = 1 - [t_1 + u_3 - 2\sqrt{(t_1 u_3)}]/m_2^2.$$

The denominator of the integrand in (2.3.12) may be obtained by a detailed study of the expression $2q_1^s p_a^s p_b^s \sin\theta_1^s \sin\phi_1^s \sin\beta^s$ found in our definition of the fully differential cross-section (2.3.7). Alternatively we can again note that this factor is the boundary function (2.3.3) given in terms of the s-frame variables. Therefore just as the limits (2.2.4) and (2.2.5) were derived from (2.3.3) so must the form of the denominator in (2.3.12) be given in terms of these same limits a, b.

Putting $y = (2s_{23} - a - b)/(b - a)$ we find the integral over s_{23} in (2.3.12) becomes

$$\left(\frac{s}{s_{12}}\right)^{2\alpha_2+1} \cdot h(t_1, u_3)$$

where

$$h(t_1, u_3) = \frac{2f^{\frac{1}{2}}}{(f-1)}\left[\frac{(f-1)m_2^2}{2}\right]^{2\alpha_2+1} \int_{-1}^{1} dy \left[y + \frac{f+1}{f-1}\right]^{2\alpha_2} (1-y^2)^{-\frac{1}{2}}$$

and therefore

$$\frac{d^2\sigma}{dt_1 \, du_3} \approx s^{2\alpha_2-2} k(t_1 u_3) \int_{(m_1+m_2)^2}^{(\sqrt{s}-m_3)^2} ds_{12} (s_{12})^{2\alpha_1-2\alpha_2-1}$$

$$\approx k_1(t_1, u_3) s^{2\alpha_1-2} + k_2(t_1, u_3) s^{2\alpha_2-2} \qquad (2.3.13)$$

with k_1 and k_2 known functions of t_1 and u_3.

This is a most interesting result since it implies that even though the amplitude was assumed to be a product of Regge pole exchanges the cross-section (to leading order in s) depends on the sum of such exchanges. This is a direct consequence of our result (2.2.5) that both sub-energies cannot be of the order of s simultaneously. Thus if *either* of the trajectories α_1 or α_2 is the Pomeron with unit intercept ($\alpha(0) = 1$) the $d^2\sigma/dt_1 du_3$ distribution for small t_1, u_3, and hence also the integrated cross-section will have little if any energy dependence just as in two body reactions.

2.3b *Dalitz plot.* One differential cross-section which deserves special mention here is the distribution over the Dalitz plot $d^2\sigma/ds_{12}\, ds_{23}$. This can be obtained by integrating (2.3.7) over t_1 and u_3 so that

$$\frac{d^2\sigma}{ds_{12}\, ds_{23}} = \frac{1}{4s}\frac{d^2\sigma}{dw_1^c\, dw_3^c} = \int_{t_1^-}^{t_1^+} dt_1 \int_{u_3^-}^{u_3^+} du_3 \frac{d\sigma}{ds_{12}\, ds_{23}\, dt_1\, du_3}, \qquad (2.3.14)$$

where

$$t_1^\pm = -2E_a^c w_1^c + m_1^2 + m_a^2 \pm 2q_1^c p_a^c$$

and

$$u_3^\pm = t_1 - s_{23} + m_3^2 + 2E_b^c w_2^c + 2q_2^c p_b^c(\cos\theta_1^c \cos\lambda_3^c \pm \sin\theta_1^c \sin\lambda_3^c)$$

However, since the Dalitz plot has played such a significant role in the study of production phenomena, let us go back to our definition of the three-body phase-space factor R_3 and consider the derivation of $d^2\sigma/ds_{12}\, ds_{23}$ (or equivalently $d^2\sigma/dw_1^c\, dw_3^c$) in more detail.

In the overall c.m. system

$$R_3 = \int \frac{d^3 \mathbf{q}_1^c}{2w_1^c}\frac{d^3 \mathbf{q}_2^c}{2w_2^c}\frac{d^3 \mathbf{q}_3^c}{2w_3^c} \delta(W - w_1 - w_2 - w_3)\delta^3(-\mathbf{q}_1 - \mathbf{q}_2 - \mathbf{q}_3) \quad (2.3.15)$$

and using the δ^3 functions, the $d^3\mathbf{q}_2^c$ integrations can be performed immediately to give

$$R_3 = \int \frac{d^3 \mathbf{q}_1^c}{2w_1^c}\frac{d^3 \mathbf{q}_3^c}{2w_3^c} \delta[W - w_1 - w_3 - \sqrt{(|\mathbf{q}_1 + \mathbf{q}_3|^2 + m_2^2)}].$$

Now writing

$$\frac{d^3 q_1}{2w_1} = \frac{\tfrac{1}{2}q_1\, dq_1^2\, d\Omega_1}{2w_1}$$

$$\frac{d^3 q_3}{2w_3} = \frac{\tfrac{1}{2}q_3\, dq_3^2\, d\Omega_3}{2w_3},$$

where $d\Omega_i = d\cos\theta_i\, d\phi_i$, $i = 1, 3$ and choosing axes so that $\mathbf{q}_1$ lies in the

z-direction, and $\cos \theta_3$ is the angle between $\mathbf{q}_1$ and $\mathbf{q}_3$ we can integrate over $d\Omega_1\, d\phi_3$ to give

$$R_3 = 2 \int q_1 q_3 \, d\cos\theta_3 \, \frac{dq_1^2 \, dq_3^2}{2w_1 \, 2w_3}$$

$$\times \, \pi^2 \delta(W - w_1 - w_3 - \sqrt{(m_2^2 + q_1^2 + q_3^2 + 2q_1 q_3 \cos\theta_3)}).$$

The Dalitz plot density is obtained by noting from $w_i^2 = q_i^2 + m_i^2$ that

$$\frac{dq_i^2}{2w_i} = dw_i, \qquad i = 1, 3$$

so that the $\cos\theta_3$ integration leads to the result

$$R_3 = \int dw_1 \, dw_3 \pi^2 \theta(4q_1^2 q_3^2 - [(W - w_1 - w_3)^2 - m_2^2 - q_1^2 - q_3^2]^2) \quad (2.3.16)$$

The derivation of the Dalitz plot density in this form has two practical advantages. Firstly it indicates the allowed region of w_1, w_3 (and therefore also s_{12}, s_{23}) for any c.m. energy W and masses m_1, m_2, m_3. This region is bounded by a closed contour in the w_1, w_3 plane given by the θ-function in (2.3.16) i.e.

$$4(W_1^2 - m_1^2)(w_3^2 - m_3^2)$$

$$= [(W - w_1 - w_3)^2 + m_1^2 - m_2^2 + m_3^2 - w_1^2 - w_3^2]^2 \quad (2.3.17)$$

to which we add the subsidiary conditions

$$w_i^2 \geqslant m_i^2$$

$$w \geqslant w_1 + w_3.$$

Secondly from (2.3.16) we can see that if the matrix element is assumed to be constant then there is no dependence in the $d^2\sigma/dw_1 \, dw_3$ distribution on w_1 and ω_3, i.e. the distribution has a uniform density proportional to π^2. It is worth noting that this is one of the very few distributions to have this nice feature, and means that we can immediately conclude that any deviation from a flat distribution must be a dynamical effect, arising from the matrix element T.

This has been an extremely effective technique in the discovery of resonances. By plotting each event of a 3-body production experiment as a point in the w_1, w_3 plane it is possible to identify a resonance-like structure in either the $1 + 2$ or $2 + 3$ subsystem by looking for an accumulation of points along a line of constant w_3 (which is also a line of constant s_{12}) or along a line of constant w_1 (s_{23}). At the same time because of the relation (2.1.3) a resonance in the $1 + 3$ system can also be indicated by an accumulation of points along a line of constants $s_{13} = s + m_1^2 + m_2^2 + m_3^2 - (s_{12} + s_{23})$.

A typical example of a Dalitz plot distribution is shown in Fig. 4 for the process $K^+ p \to K^0_1 \pi^+_2 p_3$. The accumulation of points near $s_{12} = 0{\cdot}8$ GeV^2 and $s_{23} = 1{\cdot}5$ GeV^2 indicate the presence of the K^* (892 MeV) and Δ (1236 MeV) resonances in the $K^0_1 \pi^+_2$ and $\pi^+_2 p_3$ systems respectively. Notice that by projecting this Dalitz distribution on to the s_{12}, s_{23} axes we obtain

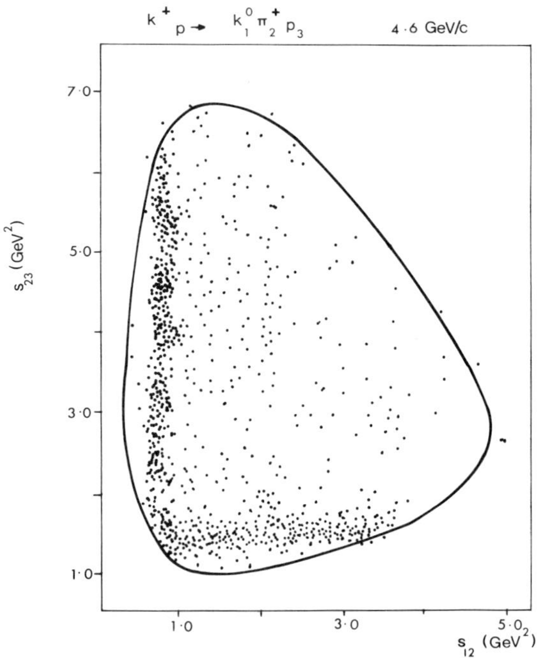

FIG. 4. Dalitz plot for the process $k^+ p \to k^0_1 \pi^+_2 p_3$ (G. Dehm 1973).

the (mass)^2 distribution $d^2\sigma/ds_{ij}$ which we considered in Section 1.2. However, it is clear that the Dalitz plot contains much more information than both of the single distributions $d\sigma/ds_{12}$ $d\sigma/ds_{23}$ put together.

2.3c *One particle exchange and the Chew–Low extrapolation.* Before leaving our discussion of differential production cross-sections we should mention that in certain circumstances it can be more convenient to work with the phase-space element defined as in (2.3.2) rather than (2.3.7). For instance, in the case of one particle exchange from the bc_3 vertex, as in Fig. 5, the $c_1 + c_2$

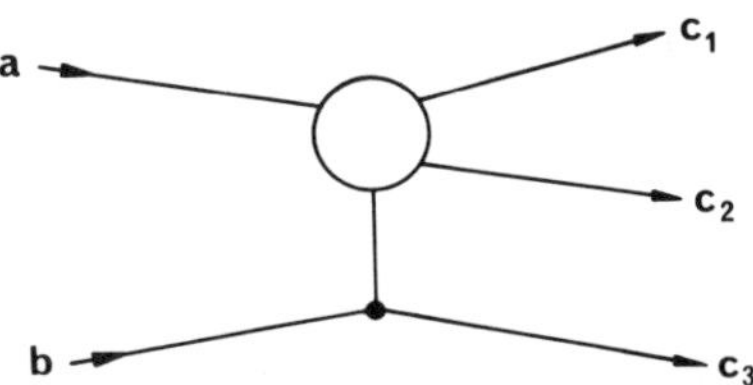

FIG. 5. Single particle exchange diagram.

system factors from the rest of the production amplitude. As an illustration let us consider the process $\pi N \to \pi\pi N$ in the single pion exchange approximation (i.e. the peripheral model (Ferrari and Selleri (1962)) such that

$$T_{fi} = gF(u_3)\bar{u}_f(q_3)\frac{\gamma_5 A_{\pi\pi}(s_{12}, \cos\theta_1^s; u_3)}{u_3 - m_\pi^2}u_i(p_b) \qquad (2.3.18)$$

where g and F are the πNN coupling constant and form factor, $\bar{u}, u$ are Dirac four-spinors and $A_{\pi\pi}(s_{12}, \cos\theta_1^s; u_3)$ is the invariant $\pi\pi$ scattering amplitude where one of the pions has a virtual mass $\sqrt{u_3}$. Inserting (2.3.18) and (2.3.2) into (1.2.1) we obtain

$$d\sigma = \frac{1}{2\pi F_{ab}}\frac{g^2}{4\pi}F^2(u_3)\frac{-u_3}{(u_3 - m_\pi^2)^2} \qquad \frac{d\sigma_{\pi\pi}}{d\Omega_s}(s_{12}, \cos\theta_1^s; u_3)$$

$$\times\, d\cos\theta_1^s\, d\phi_1^s\, \frac{(q_3^c)^2}{w_3^c}\, dq_3^c\, d\cos\theta_3^c, \qquad (2.3.19)$$

where the differential cross-section $d\sigma_{\pi\pi}(s_{12}, \cos\theta_1^s; u_3)/d\Omega_s$ for off-mass shell $\pi\pi$ scattering is the only quantity dependent on θ_1^s and ϕ_1^s. Integrating over $d\theta_1^s$ and $d\phi_1^s$ is therefore straightforward and gives

$$\frac{d^2\sigma}{dq_3^c\, d\cos\theta_3^c} = \frac{1}{2\pi F_{ab}}\frac{g^2}{4\pi}F^2(u_3)\frac{(-u_3)}{(u_3 - m_\pi^2)^2}\sigma_{\pi\pi}(s_{12}; u_3)\frac{(q_3^c)^2}{w_3^c}. \qquad (2.3.20)$$

It is pertinent to mention that if the exchanged particle carried spin (unlike the pion considered here) or if more than one particle is exchanged the corresponding differential cross-section (2.3.19) would not be isotropic in the angle ϕ_1^s. The ϕ_1^s distribution therefore can provide a sensitive test of single pseudo-scalar meson exchange. This was first noted by Treiman and Yang (1962). More precisely they suggested studying the distribution in the angle ϕ where ϕ is defined as the angle between the plane containing the two particles c_1 and c_2 and the production plane (containing particles a, b and c_3) as seen in the rest frame of particle a. However, it has been shown by Jackson (1964) that ϕ is equivalent to ϕ_1^s where ϕ_1^s is defined in terms of a

z axis which lies along the direction of particle a viewed from the $c_1 + c_2$ rest frame, and the y axis is taken as the normal to the production plane.

Returning now to (2.3.20) and noting from (2.1.5) and (2.1.6) that

$$\frac{\mathrm{d}^2\sigma}{\mathrm{d}T_3 \, \mathrm{d}\cos\theta_3^L} = \frac{\mathrm{d}^2\sigma}{\mathrm{d}q_3^c \, \mathrm{d}\cos\theta_3^c} \frac{w_3^c q_3^L}{(q_3^c)^2}$$

we see that by measuring this differential distribution $\mathrm{d}^2\sigma/\mathrm{d}T_3 \, \mathrm{d}\cos\theta_3^L$ and extrapolating to the pole position $u_3 = m_\pi^2$ we should be able to obtain information about the $\pi\pi$ cross-section. This is the well-known Chew–Low extrapolation (Chew and Low, 1959) (see also Goebel, 1958). Unfortunately, in practice it is not as easy as it sounds, since the single pion exchange mechanism shown in Fig. 5 is only one of many possible diagrams which contribute to the process $\pi N \to \pi\pi N$. In order to extract the $\pi\pi$ cross-section from the $\mathrm{d}^2\sigma/\mathrm{d}T_3\phi\cos\theta_3^L$ distribution therefore it is necessary to select events for which this pion exchange graph is dominant. This means we must restrict ourselves to fairly high incident pion momenta so that the possibility of one of the pions interacting strongly with the nucleon (perhaps even producing an N^* resonance) is reduced. Furthermore, because of the smallness of the pion mass m_π such events should correspond to small values of the momentum transfer u_3 where the propagator $1/(u_3 - m_\pi^2)^2$ in (2.3.20) is extremely large. However, on the other hand the factor $-u_3$ in the numerator of (2.3.20) arising from the pseudo-scalar coupling $\bar{u}\gamma_5 u$ in (2.3.18) implies the cross-section will be strongly suppressed at *very* small momentum transfers. Therefore the extrapolation of $(u_3 - m_\pi^2)^2 \, \mathrm{d}^2\sigma/\mathrm{d}T_3 \, \mathrm{d}\cos\theta_3^L$ to the point $u_3 - m_\pi^2$ has to be performed from a region where the number of events is expected to be low, and the statistical errors are therefore likely to be large.

One way round this problem is to consider the process $\pi^\pm p \to \pi^\pm \pi^- \pi^+ p$ and select events corresponding to the formation of the Δ^{++} resonance in the final state π^+ system. The single pion exchange graph for the pseudo three-body processes $\pi^\pm p \to \pi^\pm \pi^- \Delta^{++}$ can be written down in a similar fashion to (2.3.18). However, in this case the factor of $-u_3$ is absent because of the different spin and parity of the Δ resonance. Thus the extrapolation to positive values of u_3 is somewhat easier. Even so it should be mentioned that the choice of the process $\pi N \to \pi\pi\Delta$ over $\pi N \to \pi\pi N$ is not without its drawbacks. Because of the larger mass of the Δ, the smallest physically allowed value of $-u_3$ is larger than in the corresponding $\pi N \to \pi\pi N$ reaction, i.e.

$$(-u_3)_{\min} \underset{s\to\infty}{=} (s_{12} - m_\pi^2)(m_\Delta^2 - m_N^2)/s \quad \text{for} \quad \pi N \to \pi\pi\Delta$$

compared with

$$(-u_3)_{\min} \underset{s \to \infty}{=} (s_{12} - m_\pi^2)m_N^2/s^2 \quad \text{for} \quad \pi N \to \pi\pi N$$

Therefore at any given value of s, the Chew–Low extrapolation to the pion pole has to be performed over a longer range in the case of Δ production. Also since the Δ has a finite width the value of $(-u_3)_{\min}$ changes as one considers events coming from different regions of the resonance peak.

Unfortunately in both reactions $\pi N \to \pi\pi N$ and $\pi N \to \pi\pi\Delta$ it is found (e.g. by examining the u_3 and ϕ_1^s dependence) that the single pion exchange graph is always modified to some extent by background effects in all regions of phase-space. These background effects, in the region of interest, are most likely to arise from final state interactions between the pions and the nucleon even when we choose a kinematic region where no prominent low-mass N^* resonance can be produced. These interactions in the final state (as well as possibly in the initial state) i.e. so-called "absorption effects", have been found to be important features of two-body phenomenology and various ways of parameterizing these effects have been suggested. See for example, Gottfried and Jackson (1964), Durand and Chiu (1965), Bander and Shaw (1965) and Ross *et al.* (1970). A detailed discussion of these techniques lies outside the scope of the present work, being already very well publicised in several books on two-body phenomenology. Nevertheless, as a prelude to our next section on the *full* partial wave analysis of single particle production amplitudes, it is relevant to introduce here the density matrix elements which form the basis for most discussion on absorption models.

Suppose, for example, in the process $\pi N \to \pi\pi N$, the $\pi\pi$ system was in a pure $j = 1$ ($I = 1$) state, with the mass of the rho-meson. In this case instead of the full five point amplitude we could consider the two-body $\to$ quasi-two-body amplitude $f_{0\lambda_b;\lambda\lambda_3}^{j=1}(s, u_3; s_{12} = m_\rho^2)$ for the process $\pi N \to \rho N$, where λ_b, λ_3 are the helicities of the initial and final state nucleons and λ is the helicity of the spin-one ρ-meson. However, it is clear that generally the $\pi\pi$ system is not just in a $j = 1$ state and has a variable $(\text{mass})^2 = s_{12}$. Therefore the $\pi N \to \pi\pi N$ amplitude must be considered as a sum over all possible angular-momentum states, i.e.

$$T_{\lambda_b\lambda_3} = \sum_{j=0}^{\infty} \sum_{\lambda=-j}^{j} \sqrt{(2j + 1)}\, \mathrm{d}_{\lambda,0}^{j}(\theta_1) \exp(i\lambda\phi_1) f_{0\lambda_b,\lambda\lambda_3}^{j}(s, u_3; s_{12}) \qquad (2.3.21)$$

where θ_1 and ϕ_1 describe the two-pion decay. This incidentally is the first step in the Wick three-body partial wave analysis presented in the next section. The spin-averaged $(\text{amplitude})^2$ $|\overline{T}|^2$ is given from (2.3.21) as

$$|\overline{T}|^2 = \tfrac{1}{2} \sum_{j,j'} \sum_{\lambda,\lambda'} \sqrt{(2j + 1)}\, \sqrt{(2j' + 1)}\, \mathrm{d}_{\lambda,0}^{j}(\theta_1)\, \mathrm{d}_{\lambda',0}^{j'}(\theta_1)$$

$$\times \cos(\lambda - \lambda')\phi\, \mathrm{Re}\,\{\rho_{\lambda\lambda'}^{jj'}\} \qquad (2.3.22)$$

where the density matrix elements $\rho^{jj'}_{\lambda\lambda'}$ are bilinear products of the helicity amplitudes $f^{j}_{0\lambda_b,\,\lambda\lambda_3}$. (Sometimes certain kinematic functions are included in the definition of $\rho^{jj'}_{\lambda\lambda'}$ in order to provide a simple relationship between them and the differential cross-section.) Now by examining the θ_1, ϕ_1 structure of the differential cross-section

$$\frac{\mathrm{d}^4\sigma}{\mathrm{d}s_{12}\,\mathrm{d}u_3\,\mathrm{d}\cos\theta_1\,\mathrm{d}\phi_1}, \quad \text{or equivalently} \quad \frac{\mathrm{d}^4\sigma}{\mathrm{d}q_3^c\,\mathrm{d}\cos\theta_3^c\,\mathrm{d}\cos\theta_1\,\mathrm{d}\phi_1},$$

it is possible to deduce which of the density matrix elements are dominant in the production reaction.

The advantage of this approach is that as yet we have made no assumption about the production mechanism, so that we are not restricted to the single pion exchange approximation of (2.3.18). Conversely by examining the behaviour of the $\rho^{jj'}_{\lambda\lambda'}$ we can determine which helicity amplitudes in (2.3.21) must include other effects besides this single-particle exchange mechanism.

In conclusion let us mention that as always in the definition of helicity amplitudes the helicities can be defined in various reference frames, and it is purely a matter of taste which frame one chooses. Here let us mention only two, the so-called s-channel frame and the t-channel frame. In both cases θ_1 and ϕ_1 refer to the decay of the $\pi\pi$ system viewed in its own c.m. system. However, in the former case θ_1 is the polar angle measured in a coordinate frame in which oz is the incoming beam direction in the overall c.m. system and ϕ_1 represents the angle between the production and decay planes. This is the more "usual" frame for the definition of helicities and is particularly useful for discussing absorption effects. In the t-channel helicity frame on the other hand the z-axis is the incident beam direction now viewed in the $\pi\pi$ c.m. system and the x-axis lies in the production plane. In this frame (often called the Jackson frame) the angles θ_1, ϕ_1 correspond to θ_1^s and the Treiman–Yang angle ϕ_1^s and therefore as we have already noted is useful for isolating single particle exchange contributions. These two definitions of the helicity frames, of course, lead to different density matrix elements. However, it should be clear that they can be simply connected by Wigner rotations. For further discussion on this see, for instance, (Cohen-Tannoudji et al., 1968).

2.4. Partial wave decomposition

In two body scattering the analysis of low energy data has been considered almost exclusively in terms of a partial wave decomposition for the scattering amplitude. That is to say, the scattering amplitude is written as

$$T_{\lambda,\mu}(s,\cos\theta) = \sum_{J} f^{J}_{\lambda,\mu}(s)D^{J}_{\lambda,\mu}(0,\theta\,\psi) \tag{2.4.1}$$

where μ and λ are the relative helicities of the initial and final states respectively and $D^J_{\lambda,\mu}$ are the well-known relation matrices

$$D^J_{\lambda,\mu}(\phi, \theta, \psi) = \exp(-i\lambda\phi)\, d^J_{\lambda,\mu}(\theta)\exp(-i\mu\psi).$$

ψ is a rotation about the z-axis and can be omitted except in polarization measurements. For the scattering of spinless particles, (2.4.1) reduces to

$$T(s, \cos\theta) = \sum_J (2J + 1) f^J(s) P_J(\cos\theta)$$

where $P_J(\cos\theta)$ are Legendre polynomials.

These partial wave expansions are useful because close to threshold the partial waves with orbital angular momentum l are proportional to $(pq)^l$ where p and q are the relative momenta of the initial and final state particles. So long as the energy is not too large it is plausible that this l-dependent factor (centrifugal barrier) requires dominance of the scattering amplitude by just these low partial waves. Thus it may be a good approximation to truncate the infinite sum in (2.4.1) at some finite value of $J = J_{max}$, in which case, instead of having to determine a function of two variables s and $\cos\theta$ the problem has now been reduced to finding $J_{max} + 1$ functions $f^J_{\lambda,\mu}(s)$ of one variable s, suitably weighted by known polynomials of $\cos\theta$. Moreover, it may be very plausible that of these $J_{max} + 1$ partial waves only those which contain resonance poles will give a sizeable contribution to the partial wave sum.

Let us now try to apply the same reasoning to low energy production

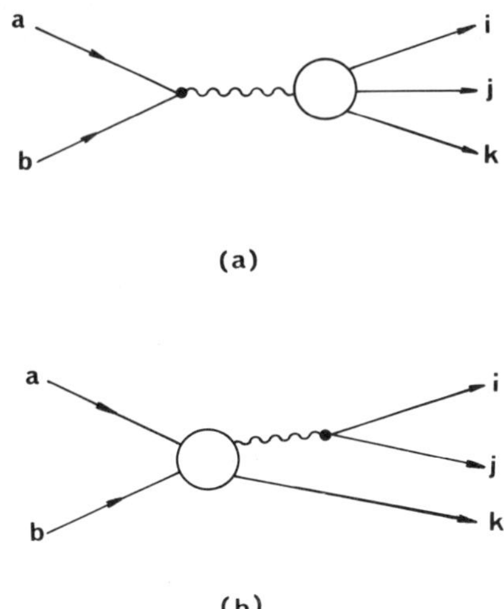

FIG. 6. Typical resonance contributions to the three-body amplitude.

amplitudes. Here, unfortunately, the problem is considerably more complex. For instance, in general, not only can the final state be achieved via the decay of an intermediate resonance as in Fig. 6a but also by the formation and decay of resonances in the final state as in Fig. 6b. Also, as the number of particles involved in a reaction increases so does the number of ways in which the production amplitude can be decomposed into partial waves. Even for our particular concern in this chapter of 3-body reactions this represents quite a considerable complication.

For example, one way in which the 3-body amplitude may be decomposed into states of definite angular momentum, suggested by Branson, Landshoff and Taylor (1963), is to consider the rotations of the plane containing the momenta of the final state particles with respect to some axes fixed in space as shown in Fig. 7. Such a "rigid body" interpretation of the final state has the

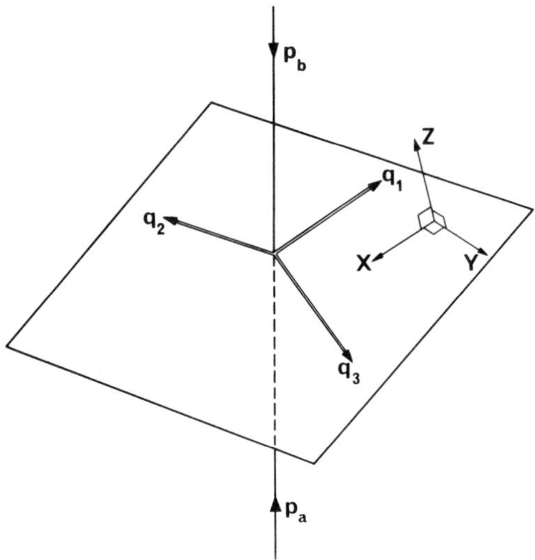

FIG. 7. Illustration of the rigid body frame of reference for a three-particle final state.

beauty that it leads to the relatively simple expansion for the production amplitude

$$T^{\mu_a, \mu_b}_{\lambda_1\lambda_2\lambda_3}(s, s_{ij}; \Phi\Theta\Psi) = \sum_{J, \Lambda} \sqrt{\left(\frac{2j + 1}{4\pi}\right)} D^J_{\Lambda, \mu_a - \mu_b}(\Phi\Theta\Psi) T^{J; \mu_a\mu_b}_{\lambda_1\lambda_2\lambda_3}(s, s_{ij}) \quad (2.4.2)$$

where Θ, Φ are the polar and azimuthal angles of the initial c.m. momentum with respect to the axes $OXYZ$ fixed relative to the final state momenta.

The angle Ψ corresponds to a rotation of the initial state about the initial momentum. It occurs throughout only in the phase-factor $\exp[-i(\mu_a - \mu_b)\Psi]$ and is therefore only relevant in polarization experiments.

The decomposition (2.4.2) is clearly the closest analogue to the two-body expansion (2.4.1) that can be achieved, and with its inverse, i.e.

$$T^{J;\,\mu_a,\,\mu_b}_{\lambda_1\lambda_2\lambda_3}(s, s_{ij}) = \sqrt{\frac{2j+1}{4\pi}} \int D^{J*}_{\Lambda,\,\mu}(\Phi\Theta\Psi) T^{\mu_a,\,\mu_b}_{\lambda_1\lambda_2\lambda_3}(s, s_{ij}; \Phi\Theta\Psi) \, \mathrm{d}\cos\Theta \, \mathrm{d}\Phi \quad (2.4.3)$$

is very useful for certain analyses of the production data. For instance, choosing $\Psi = 0$ (which we may do for unpolarized particles) so that the rotation matrices $D^J_{\lambda,\,\mu}$ are related to the normalized spherical harmonics by

$$D^J_{\Lambda,\,\mu}(\Phi\Theta\Psi) = \sqrt{\left(\frac{4\pi}{2J+1}\right)} Y^{J*}_{\Lambda}(\Theta\Phi)$$

and using the relation

$$Y^{J_1}_{\Lambda_2} Y^{J_1}_{\Lambda_2} = \sum_{|\Lambda_1-\Lambda_2|\,\leqslant\,\Lambda_3\,\leqslant\,\Lambda_1+\Lambda_2} \sqrt{\frac{[(2J_1+1)(2J_2+1)]}{4\pi(2J_3+1)}} C^{J_1J_2J_3}_{000} C^{J_1J_2J_3}_{\Lambda_1\Lambda_2\Lambda_3} Y^{J_3}_{\Lambda_3} \quad (2.4.4)$$

where the C^{ABC}_{abc} are Clebsch Gordon coefficients, we see that the fully differential cross-section can also be expanded in a series of spherical harmonics, i.e.,

$$\frac{\mathrm{d}^4\sigma}{\mathrm{d}s_{12}\mathrm{d}s_{23}\mathrm{d}\cos\Theta\,\mathrm{d}\Phi} = \sum_{J,\Lambda} \sqrt{\left(\frac{2J+1}{4\pi}\right)} W^J_{\Lambda}(s, s_{ij}) Y^{J*}_{\Lambda}(\Theta\Phi) \quad (2.4.5)$$

Therefore by examining the angular dependence of the differential cross-section one can deduce which of the weighting factors W^J_{Λ} are dominant in a low energy body process and hence via (2.4.4) and (2.4.2) we can infer which are the important partial wave amplitudes. However, for a detailed dynamical study it will also be necessary to describe the important final state resonance production effects shown in Fig. 6b. To do this we must be able to take account of specific two body angular momentum sub-systems in the definition of the partial wave amplitudes $T^{J,\,\mu_a,\,\mu_b}_{\lambda_1\lambda_2\lambda_3}(s, s_{ij})$ in (2.4.2). This has been considered extensively in the literature, though sometimes in a fairly approximate or rather restrictive way. A partial list of references includes Lindenbaum and Sternheimer (1957a, b) (1958), Bergia *et al.* (1960), Olsson and Yodh (1963a, b), Macfarlane (1963) and Berman and Jacob (1965).

One of the more systematic treatments of combining two-body final-state partial waves within a three-body partial wave analysis was suggested by Wick (1962). The detailed mathematics is somewhat lengthy and is very well covered in the literature (Wick (1962); Namyslowski *et al.*, 1967). Also later in this section we shall be describing the decomposition of the particularly useful case $a^0 + b^{\frac{1}{2}} \to c_1^0 + c_2^0 + c_3^{\frac{1}{2}}$, (where the superscripts 0, $\frac{1}{2}$ refer to the spin of the particle) in the essentially equivalent formalism of Deler and Valladas. Here, therefore, let us merely indicate the essential steps in the Wick approach.

The expansion of a two particle helicity state in its own c.m. system into angular momentum states is well understood and has been expounded at length by Jacob and Wick (1959). Some particular two particle sub-system (e.g. $c_i + c_j$) is initially expanded in this manner in its own c.m. system (i.e. where $\mathbf{q}_i + \mathbf{q}_j = 0$). Then a Lorentz transformation is applied to give the two particle system its corresponding momentum $\mathbf{q}_i + \mathbf{q}_j$ in the overall c.m. frame where $\mathbf{q}_i + \mathbf{q}_j = -\mathbf{q}_k$. The resulting $(c_i + c_j) + c_k$ pseudo two-body state can now again be easily expanded in states of total angular momentum. The end product of these manipulations is that the matrices describing the transformations from plane wave states to states of definite J and j_k (i.e. total angular momentum and angular momentum of the $i + j$ system) are given by

$$\langle q_1\lambda_1, q_2\lambda_2, q_3\lambda_3 | Q, JM; s_{ij}j_k m_k; \mu_i\mu_j\mu_k \rangle$$

$$= \delta^4(q_i + q_j + q_k - Q)\delta((q_i + q_j)^2 - s_{ij})\,\delta_{\lambda_k\mu_k}N_J N_{j_k} 4(ss_{ij})^{\frac{1}{4}}(q_k k_k)^{-\frac{1}{2}}$$

$$\times\, d^{s_i}_{\lambda_i,\mu_i}(\beta_{ij})\, d^{s_j}_{\lambda_j,\mu_j}(\beta_{ji})\, d^{j_k}_{m_k,\mu_{ij}}(\theta_k)D^{J*}_{M,m_k-\mu_k}(\sigma_k), \qquad (2.4.6)$$

where the normalization of Wick (1962) has been assumed. (This normalization absorbs the factors of $(2\pi)^3$ in our standard definition of the cross-section element.) The angles β_{ij}, θ_k are given by†

$$\cos\beta_{ij} = \frac{1}{4k_k q_i\sqrt{(ss_{ij})}}\{(s_{ij}+m_i^2-m_j^2)(s-s_{jk}+m_i^2)-2m_i^2(s+s_{ij}+m_k^2)\}$$
$$(2.4.7)$$

$$\cos\theta_i = \frac{1}{4k_i q_i\sqrt{(ss_{jk})}}\{(s-m_i^2)(m_k^2-m_j^2)-s_{jk}(s_{ik}-s_{ij})\}, \qquad i,j,k = 1, 2, 3$$

and σ_k represents the three Eulerian angles $(\phi_k^c, \theta_k^c, \psi_k)$. θ_k^c, ϕ_k^c are the polar and azimuthal angles defined in (2.1.6) and the equation leading to (2.2.2), and ψ_k is the azimuthal angle of particle k in the (ij) c.m. system. The normalization factor $N_J = [(2J + 1)/4\pi]^{\frac{1}{2}}$, and superscript s_i and subscripts μ_i, λ_i denote the spins and helicities of initial and final state particles i, while $\mu_{ij} = \mu_i - \mu_j$. Also q_i represents the momentum of particle i in the overall system (i.e. $q_i = q_i^c$) and k_i is the momentum of particles j and k in their own c.m. system (e.g. $k_3 \equiv q_1^s = q_2^s$ of (2.1.7a).)

The corresponding transformation matrix for a two particle state is

$$\langle p_a\lambda_a, p_b\lambda_b | P, JM, \mu_a, \mu_b \rangle$$

$$= \delta^4(p_a + p_b - P)\delta_{\lambda_a\mu_a}\delta_{\lambda_b\mu_b}N_J\left[\frac{4\sqrt{s}}{p_a^c}\right]^{\frac{1}{2}} D^{J*}_{M\mu_{ab}}(\hat{n}_a) \qquad (2.4.8)$$

† There is an error in the definition of $\cos\theta_i$ given by Namyslowski $et\ al.$ (1967).

where $\hat{n}_a$ represents a unit vector in the direction of $\mathbf{p}_a$. When $\mathbf{p}_a$ lies along the axis of quantization we just have

$$D^J_{M,\mu_{ab}}(\hat{n}_1) = \delta_{M,\mu_{ab}} \qquad (2.4.9)$$

Thus for a partial wave analysis of the T-matrix element defined now as

$$\delta^4(p_a + p_b - q_1 - q_2 - q_3)T = \langle q_1\lambda_1, q_2\lambda_2, q_3\lambda_3 | T | p_a\mu_a, p_b\mu_b\rangle$$

we have only to insert the three and two particle unit operators

$$\sum_{\substack{JM, j_km_k \\ \lambda_i\lambda_j, \lambda_k}} \int d^4P\, | PJM, s_{ij}j_km_k\lambda_i\lambda_j; \lambda_k\rangle\langle PJM, s_{ij}j_km_k\lambda_i\lambda_j; \lambda_k |$$

$$\sum_{\substack{JM \\ \mu_a\mu_b}} \int d^4P\, | PJM\mu_a\mu_b\rangle\langle PJM\mu_a\mu_b |$$

to obtain the following angular momentum decomposition

$$T = \sum_{\substack{J, j_km_k \\ v_iv_j}} 8\left[\frac{s\sqrt{(s_{ij})}}{p_aq_kk_k}\right]^{\frac{1}{2}} N_J^2 N_{jk} \exp(-i\pi s_j)$$
$$\times\; d^{s_i}_{\lambda_iv_i}(\beta_{ij})\, d^{s_j}_{\lambda_jv_j}(\beta_{ji})\, d^{j_k}_{m_kv_{ij}}(\theta_k) D^{J*}_{\mu_{ab},\,m_k-\lambda_k}(\sigma_k)\mathcal{T}^{Jj_km_k,\,v_iv_j,\,\lambda_k}_{\mu_a\mu_b} \qquad (2.4.10)$$

The $\mathcal{T}^{Jj_km_k,\,v_iv_j,\,\lambda_k}_{\mu_a\mu_b}$ are the corresponding partial wave amplitudes and are functions of the total energy and final state sub-energies. Each one denotes the contribution of a particular two-body angular momentum state j_k coupling to a state of total angular momentum J, and as such is clearly relevant for discussing resonance formation in the $i + j$ sub-system. We shall consider some approximate forms for $\mathcal{T}^J$ in the next chapter. However, now the problem arises how to also describe possible final state interactions between particles $c_i + c_k$ or $c_j + c_k$. To do this we should expand the production amplitude in terms of the angular momentum of the (ik) or (jk) two particle systems. But, since (2.4.10) has been derived in terms of a complete set of states such expansions must be equivalent and can be transformed into one another. The necessary matrices have also been determined by Wick (1962) and Namyslowski $et\ al.$ (1967) and can be written as

$$\langle PJM, s_{jk}j_im_i, \mu_j\mu_k, \mu_i | P'J'M's_{ik}j_jm_j, \lambda_k\lambda_i; \lambda_j\rangle$$
$$= \delta^4(P - P')\delta_{JJ'}\delta_{MM'}U^{j_im_i\mu_j\mu_k;\,\mu_i}_{j_jm_j\lambda_k\lambda_i;\,\lambda_j}(s_{jk}, s_{ik}), \qquad (2.4.11)$$

where

$$U^{j_im_i\mu_j\mu_k;\,\mu_i}_{j_jm_j\lambda_k\lambda_i;\,\lambda_j}(s_{jk}, s_{ik}) = \frac{1}{4}\left[\frac{\sqrt{(s_{jk}s_{ik})}}{sq_ik_iq_jk_j}\right]^{\frac{1}{2}} [(2j_i + 1)(2j_j + 1)]^{\frac{1}{2}}(-1)^{s_i-\mu_i+s_k+\mu_k}$$

$$\times\; d^J_{\Lambda_i\Lambda_j}(\chi_k)\, d^{j_j}_{m_j\lambda_j}(\theta_j)\, d^{j_i}_{m_i\mu_i}(\theta_i)\, d^{s_i}_{\mu_i\lambda_i}(\beta_{ij})\, d^{s_j}_{\mu_j\lambda_j}(-\beta_{kj})\, d^{s_k}_{\mu_k\lambda_k}(\beta_{ji} + \beta_{ij})$$

$$\times\, \delta(\cos\theta_i - (4k_iq_i\sqrt{(ss_{jk})})^{-1}[(s - m_i^2)(m_k^2 - m_j^2) - s_{jk}(s_{jk} + s_{ij})]) \qquad (2.4.12)$$

with

$$\Lambda_i = m_i - \lambda_i$$

and†

$$\cos \chi_i = \frac{1}{4sq_j q_k}\left[s_{ik}s_{ij} + (s - m_j^2)s_{ij} + (s - m_k^2)s_{ik} - (s + 2m_i^2 - m_j^2 - m_k^2)s + m_j^2 m_k^2\right]$$

such that

$$\mathcal{T}^{Jj_i m_i \lambda_j \lambda_k;\, \lambda_i}_{\mu_a \mu_b} = \sum_{\substack{j_k m_k \\ \nu_i \nu_j \nu_k}} \int \mathrm{d}s_{ij}\, U^{j_i m_i \lambda_j \lambda_k;\, \lambda_i}_{j_k m_k \nu_i \nu_j;\, \nu_k}\, \mathcal{T}^{Jj_k m_k \nu_i \nu_j;\, \nu_k}_{\mu_a \mu_b}. \tag{2.4.13}$$

Notice, if we assume there is a final state resonance of spin j_k in the $c_i + c_j$ system which couples to the total angular momentum J, (2.4.13) gives the effect of this resonance in the $c_j + c_k$ system with angular momentum j_i. That is to say, (2.4.13) gives the background in one sub-energy channel arising from a resonance in another. Thus if we wish to consider generally the interference of resonances in all sub-energies, for example

$$\pi N \to \rho N \to (\pi_1 \pi_2)N$$
$$\to \pi_1 N_2^* \to \pi_1(\pi_2 N)$$
$$\to \pi_2 N_1^* \to \pi_2(\pi_1 N)$$

we must consider all three particle bases. The above transformation matrices can then be used to determine the relative contribution of each resonance production mechanism to *one* particular set of states (e.g. the $|(\pi\pi)N\rangle$ system). It has been stressed by Namyslowski, Razmi and Roberts (1967) that this is the way one should make a systematic study of the momentum and angular distributions in a three-body process which takes account of interference effects and in principle, at least, avoids double counting. However, in practical applications the infinite sums over J and j_k etc. must be truncated to a manageable number of partial waves. Therefore we are no longer working with a complete set of states and hence the probability of performing some kind of double counting error again arises. The size of the error is a phenomenological question depending on the assumed truncation of the partial wave series as well as the dynamics of the two-body to three-body process. Nevertheless it is clear that such an error must always exist to some extent.

Finally in this brief outline of the Wick decomposition let us recollect that at low energies the centrifugal barrier tends to depress contributions with high orbital angular momentum L. Because of this it is often desirable to be

---

† We hope the use of m_i for both a mass and the third component of angular momentum does not cause confusion.

able to express the helicity representation given above in terms of states of definite orbital angular momentum, i.e. in terms of the so-called JLS system or canonical basis. The required transformation matrices have been determined by McKerrell (1964). For the two particle state we have

$$|PJM\lambda_1\lambda_2\rangle = \frac{1}{(2\sqrt{s})^{\frac{1}{2}}} \sum_{L,S} \sqrt{\left(\frac{2L+1}{2J+1}\right)}\, C^{LSJ}_{0\lambda_{12}\lambda_{12}}\, C^{s_1 s_2 S}_{\lambda_1 - \lambda_2 \lambda_{12}} |PJLS\rangle \quad (2.4.14)$$

where $|PJLS\rangle$ denotes the state of orbital angular momentum L and total spin S, and $\lambda_{12} = \lambda_1 - \lambda_2$. The corresponding formula for the three particle system is

$$|JM, s_{jk} j_i \lambda_k; \lambda_i\rangle = \frac{1}{[2\sqrt{(s_{jk})}]^{\frac{1}{2}}} \sum_{\substack{l_i \sigma_i \\ LS}} \sqrt{\left(\frac{(2l_i+1)}{(2j_i+1)}\frac{(2L+1)}{(2J+1)}\right)}$$

$$\times\, C^{LSJ}_{0,\, m_i - \lambda_i,\, m_i - \lambda_i}\, C^{j_i s_i S}_{m_i,\, -\lambda_i,\, m_i - \lambda_i}\, C^{l_i \sigma_i j_i}_{0\lambda_{jk}\lambda_{jk}}\, C^{s_j s_k \sigma_i}_{\lambda_j - \lambda_k \lambda_{jk}} |JMLS, l_i \sigma_i\rangle \quad (2.4.15)$$

where S is the total spin of the three particle system, L is the relative orbital momentum of the particle i and the (jk) system and $l_i \sigma_i$ are the relative orbital angular momentum and total spin of this (jk) system.

The partial wave decomposition (2.4.10) is quite general and is applicable for any 2-body $\rightarrow$ 3-body process where the particles have any spin. Although it looks rather different from the relatively simple form (2.4.2) it can be related to (2.4.2) by a rotation to the $OXYZ$ axes of Fig. 7. However, rather than discussing this in any generality here it is probably more instructive to consider the important practical case of a system with two spinless and one spin-$\frac{1}{2}$ particles (e.g. $\pi N \rightarrow \pi\pi N$, $k^- p \rightarrow \pi\pi\Lambda$) and derive the corresponding Deler, Valladas decomposition. In doing so we shall repeat some of the steps already outlined above and follow quite closely the progression given in the original paper of Deler and Valladas. In view of the complexity of the problem we believe this repetition is desirable in order to give a coherent treatment of the subject.

2.4a. *Deler-Valladas Formalism.* To make our discussion more concrete let us specify the particular process $\pi_a N_b \rightarrow \pi_1 \pi_2 N_3$ and consider the effects of a $\pi_2 N$ final state interaction with total and orbital angular momentum j, l. We do this simply to identify the particles. Any spin zero mesons or spin $\frac{1}{2}$ baryons would do instead of the π's and N's considered here. Then with respect to a set of axes $oxyz$ where oz is in the direction of the incident pion and ox lies in the production plane (i.e. the plane containing $\mathbf{p}_a$ and $\mathbf{q}_1$) we have†

---

† We use $\simeq$ rather than $=$ sign because for each event we have to make a rotation about the z axis so that the x axis lies in the plane of the three produced particles. However, for an unpolarized nucleon target the cross-section has no dependence on the azimuthal angle which describes the rotation.

$$\langle \pi_1(\pi_2 N_3)|T|\pi N\rangle \simeq \langle q_1 0; k_1 \Sigma|T|p_a 0; p_b \lambda_b\rangle$$

$$\simeq \sum_{JLL'} \langle q_1 0; k_1 \Sigma|JM; L'j\rangle T_{jl}^{JLL'}(s, s_{23})$$

$$\times \langle JM, L\tfrac{1}{2}|p_a 0; p_b \lambda_b\rangle, \qquad (2.4.16)$$

where

$$T_{jl}^{JLL'}(s, s_{23}) = \langle JML'j|T|JML\tfrac{1}{2}\rangle$$

k_1 is again the momentum of the π_2 or N in the $(\pi_2 N_3)$ c.m. system, and Σ is the third component of the spin of N_3 projected on to a z^* axis defined as the direction of the $\pi_2 N_3$ sub-system in the overall c.m. system. $z^* = -\,\mathbf{q}_1/|\mathbf{q}_1|$, ($x^*$ lies in the production plane as shown in Fig. 8).

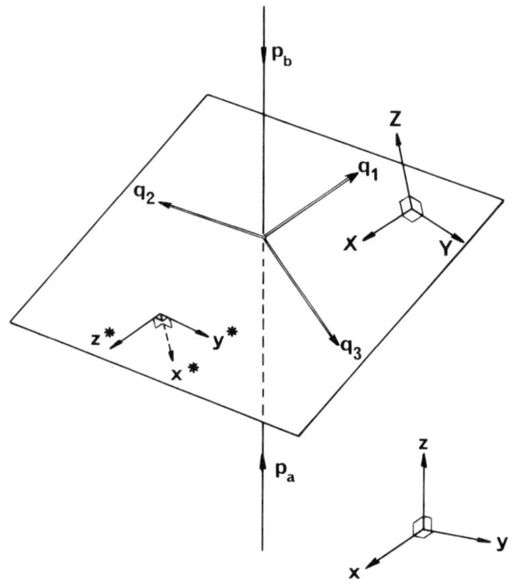

FIG. 8. Illustration of the three reference frames $oxyz$, $ox^*y^*z^*$, $OXYZ$.

In (2.4.16) we have already written our matrix element in terms of orbital angular momentum $L'L$ and l. In order to project our eigenstates of total angular momentum J and helicities λ_b, λ (the helicity of the $\pi_2 N$ system) let us rewrite (2.4.16) as

$$\langle \pi_1(\pi_2 N_3)|T|\pi N\rangle \simeq \sum_{JLL'} \sum_{\lambda, \mu_b} \langle q_1 0; k_1 \Sigma|JMj\lambda\rangle\langle JMj\lambda|JML'j\rangle$$

$$\times T_{jl}^{JLL'}\langle JM; L\tfrac{1}{2}|JM\mu_b\rangle\langle JM\mu_b|q_a 0; p_b \lambda_b\rangle$$

We can now make the substitutions (see (2.4.14), (2.4.15) and Jacob and Wick (1959))

$$\langle JMj\lambda | JML'j\rangle = (-1)^{j+\lambda} C^{JjL'}_{-\lambda\lambda 0}$$

$$\langle JM, L\tfrac{1}{2} | JM\mu_b\rangle = (-1)^{\frac{1}{2}+\mu_b} C^{J\frac{1}{2}L}_{-\mu_b\mu_b 0} \qquad (2.4.17)$$

$$\langle JM\mu_b | p_a 0; p_b\lambda_b\rangle = \left[\frac{4\sqrt{s}}{p_a}\right]^{\frac{1}{2}} \sqrt{\left(\frac{2J+1}{4\pi}\right)} \delta_{M\lambda_b}\delta_{\mu_b\lambda_b}$$

and note that $\langle q_1 0; k_1\Sigma | JMj\lambda\rangle$ is the product of two wave functions, one in the overall c.m. system, the other in the $\pi_2 N_3$ rest frame such that

$$\langle q_1 0; k_1\Sigma | JMj\lambda\rangle = \langle k_1\Sigma | jl\lambda\rangle\langle q_1 0; q_{23}\lambda | JM\lambda\rangle$$

$$= \left\{ \left[\frac{4\sqrt{s_{23}}}{k_1}\right]^{\frac{1}{2}} C^{l\frac{1}{2}j}_{m\Sigma\lambda} Y^l_m(\theta^*_1\phi^*_1) \right\}$$

$$\times \left\{ \sqrt{\left(\frac{2J+1}{4\pi}\right)} \left[\frac{4\sqrt{s}}{q_1}\right]^{\frac{1}{2}} D^J_{m\lambda}(0\theta_1 0) \right\}$$

where θ_1 is the production angle of the $(\pi_2 N_3)$ sub-system $(\theta_1 = -\theta^c_1)$ and $\theta^*_1\phi^*_1$ are the polar angles of the nucleon in the $\pi_2 N_3$ rest frame. Thus (2.4.16) becomes

$$\langle \pi_1(\pi_2 N) | T | \pi N\rangle = \frac{1}{2\pi} \sum_{JLL'} \frac{4\sqrt{(4s\sqrt{s_{23}})}}{\sqrt{(p_a q_1 k_1)}} \sqrt{\left(\frac{2J+1}{2}\right)} T^{JLL'}_{jl}(s, s_{23}) f^{JLL'}_{\lambda_b\Sigma} \qquad (2.4.19)$$

where

$$f^{JLL'}_{\lambda_b,\Sigma} = \sqrt{\left(\frac{2J+1}{2}\right)} \sum_\lambda (-1)^{\frac{1}{2}+\lambda_b} (-1)^{j+\lambda} C^{J\frac{1}{2}L}_{-\lambda_b\lambda_b 0} C^{l\frac{1}{2}j}_{m\Sigma\lambda} Y^l_m(\theta^*_1\phi^*_1)\, d^J_{\lambda_b\lambda}(\theta_1)$$

So far we have written (2.4.16) in terms of quantities defined in two reference frames, i.e. the $oxyz$ and the $Ox^*y^*z^*$ frames. However, our aim in this sub-section is specifically to refer the T-matrix elements to the "rigid body" axes $OXYZ$ where OZ is perpendicular to the production plane and OX is in the direction of the $\pi_2 N_3$ sub-system. This can be achieved by writing symbolically

$$\langle \mu_f | T | \mu_i\rangle \simeq \sum_{\Sigma,\lambda_b} \langle \mu_f | \Sigma\rangle\langle \Sigma | T | \lambda_b\rangle\langle \lambda_b | \mu_i\rangle$$

where μ_i, μ_f are the third components relative to the initial and final spin states of the nucleon in this reference frame. From Fig. 9 we see that this implies the corresponding function $f^{JLL'}_{\mu_i\mu_f}$ is obtained from $f^{JLL'}_{\lambda_b\Sigma}$ of (2.4.19) by the expression

$$f_{\mu_i,\mu_f} = \sum_{\Sigma,\lambda_b} D^{\frac{1}{2}}_{\mu_f,\Sigma}(0, \pi/2, \pi/2 - \phi^*_1) f^{JLL'}_{\lambda_b,\Sigma} D^{\frac{1}{2}}_{\lambda_b,\mu_i}(0, -\Theta - \Phi) \qquad (2.4.20)$$

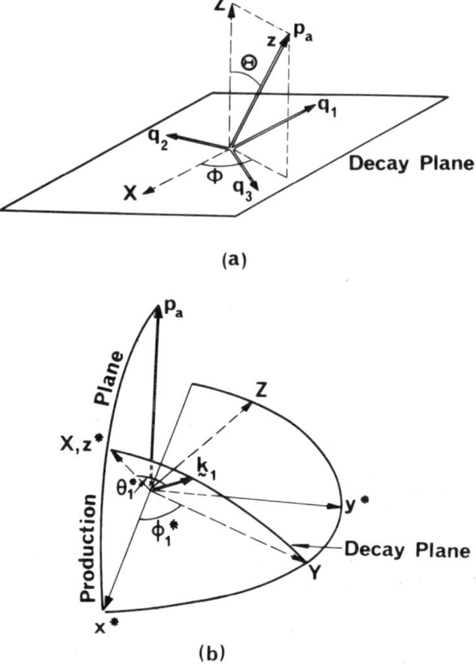

FIG. 9. Definition of, and relationship between angles Θ, Φ; $\theta_1^* \phi_1^*$.

where $(0, -\Theta, -\Phi)$ and $(0, \pi/2, \pi/2 - \phi_1^*)$ are the Euler angles connecting the $oxyz$ and $OXYZ$ frames and the $OXYZ$, $ox^*y^*z^*$ frames. If we note that the sums over λ and Σ in (2.4.19) and (2.4.20) respectively are restricted by the C–G coefficient such that $m + \Sigma = \lambda$, and also that

$$\exp(i\lambda\phi_1^*) \, \mathrm{d}^J_{\lambda_b\lambda}(\theta_1) = (-1)^{\lambda_b - \lambda} D^J_{-\lambda_b - \lambda}(0, \theta_1, \phi_1^*)$$

$$\simeq \sum_{m'} G^J_{-\lambda_b m'}(0, -\Theta, -\Phi) D^J_{m' - \lambda}(0\pi/2, \pi/2)$$

where θ_1 is the angle between oz^* and OZ some simple though tedious manipulations of the C–G coefficients and rotation matrices lead to the result (see Deler and Valladas, 1966)

$$f^{JLL'(jl)}_{\mu_i\mu_f} = \sqrt{\left(\frac{2J+1}{2L+1}\right)} \sum_{\nu} \sum_{\mathscr{L}} \sum_{\Sigma} \sum_{\lambda>0} (-1)^{\frac{1}{2}+j} C^{JjL'}_{-\lambda\lambda 0}[1 + (-1)^{L+\nu+1}]$$

$$C^{l\frac{1}{2}j}_{m\Sigma\lambda} C^{\frac{1}{2}J\mathscr{L}}_{\Sigma-\lambda-m} C^{\frac{1}{2}J\mathscr{L}}_{\mu_f\nu-\mu_f\nu} C^{J\frac{1}{2}L}_{\nu-\mu_f\mu_i\nu-\mu_f+\mu_i}$$

$$\times \, \mathrm{d}^{\mathscr{L}}_{\nu,-m}(\pi/2)(-i)^m P^m_l(\theta_1^*) Y^L_{\nu+\mu_i-\mu_f}(\Theta\Phi) \qquad (2.4.21)$$

where $\mathscr{L} = J \pm \frac{1}{2}$ and P_l^m are associated Legendre functions. This expression (2.4.21) together with (2.4.19) is our required expansion of the T-matrix element in terms of spherical harmonics $Y_\Lambda^L(\Theta\Phi)$ for a given two-body final state sub-system with total and orbital angular momentum j and l respectively.

We might mention that parity conservation indicates that only those values of $L'L$ and l connected by the relation

$$L + 2n = L' + l + 1 \qquad (n \text{ an integer})$$

contribute to the expression (2.4.21). Moreover, it can be shown that only two of the four functions $f_{\pm\frac{1}{2},\pm\frac{1}{2}}^{JLL'}$ need be calculated explicitly since we have

$$f_{-\mu_i-\mu_f}^{JLL'} = (-1)^{\mu_i+\mu_f}(f_{\mu_i\mu_f}^{JLL'})^* \qquad (2.4.22)$$

Finally in this section let us note that a similar expansion to (2.4.21) could have been obtained for the $(\pi_1 N_3)\pi_2$ system by suitably redefining the $OXYZ$ reference frame so that the x axis now lies along the $\pi_1 N_3$ direction and the z axis is inverted. The only formal difference to (2.4.21) in this case is that the angle θ_1^* is replaced by $-\theta_2^*$, where θ_2^* is the polar angle of the nucleon in the $\pi_1 N$ rest frame and $(-i)^m$ becomes $(+i)^m$ because of the z axis inversion. However, if as often happens, we wish to consider the *simultaneous* effects of interactions in two sub-systems, e.g. the $(\pi_1 N_3)$ and $(\pi_2 N_3)$ systems it is more convenient to rotate the $OXYZ$ axes so that OX is now the bisector of the angle θ_{12} between the π_1 and π_2 directions. This brings in an extra factor of $\exp(\pm iv\theta_{12}/2)$ into the respective definitions of $f_{\mu_i\mu_f}^{JLL'}(jl)$. Moreover, since these functions as they stand provide a relativistically correct description of the nucleon spin states only in the rest frame of the $(\pi_i N_3)$ sub-system our new choice of axes means that the $f_{\mu_i\mu_f}^{JLL'}$ have to undergo Lorentz rotations in their final spin indices. The effect of these rotations is to introduce further factors of $\exp[i\Omega_1\mu_f]$ and $\exp[-i\Omega_2\mu_f]$ respectively into the definition of $f_{\mu_i\mu_f}^{JLL'}$ for the $\pi_1(\pi_2 N_3)$ and $(\pi_1 N_3)\pi_2$ decompositions. The angles Ω_k have been determined by Stapp (1957) as

$$\sin|\Omega_k| = \frac{(V_k^* \times V_k)(1 + \gamma_k^* + \gamma_k + \gamma_N)}{(1 - \gamma_k^*)(1 - \gamma_k)(1 - \gamma_N)}$$

where

$$\gamma_k = \sqrt{(V_k^2 + 1)}, \quad V_k^* = \mathbf{k}_k^*/m_k, \quad V_k = \mathbf{q}_k/\omega_k, \quad V_N = \mathbf{q}_3/m_3,$$

The above rotations play the same role as do the transformation matrices defined in (2.4.12) using the Wick helicity convention. They allow us to consider in a relativistically correct manner the $\pi_1(\pi_2 N_3)$ and $(\pi_1 N_3)\pi_1$ final state effects referred to one complete set of states. Although we do not consider it here the $(\pi_1\pi_2)N_3$ system can also be given in the Deler–Valladas formalism.

The differences therefore between the two approaches of Namyslowski *et al.* and Deler and Valladas are purely formal: the corresponding partial wave decompositions merely being defined in terms of different sets of axes and spin bases. The former approach, as we have considered it, is quite general and can be applied to 2-body $\rightarrow$ 3-body processes with particles of any spin. It can also be extended to processes with any number of particles in the final state, though the algebra quickly becomes unmanageable. The latter approach was specifically derived for the $a^0 b^{\frac{1}{2}} \rightarrow c_1^0 c_2^0 c_3^{\frac{1}{2}}$ system and is written so that its dependence on the spherical harmonics $Y_\Lambda^L(\Theta\Phi)$ is explicit. This form, as we have already pointed out, is particularly useful for comparison with data.

Unfortunately, whichever decomposition one uses the corresponding expressions, by the very nature of the problem, tend to look rather complicated at first sight. However, if one understands the main steps in the derivation of these partial wave decompositions, the complications are seen to be of a purely algebraic nature. Moreover, as so often happens, even these difficulties tend to disappear when one considers specific cases. As a final illustration of this let us consider in the Deler and Valladas notation the function $f_{\mu_i \mu_f}^{JLL',jl}$ for the particular example of $\pi + \Delta\,(j = \frac{3}{2}, l = 1)$ production with total angular momentum J and orbital angular momentum L' from an initial πN state with orbital angular momentum L. From (2.4.21) we can deduce for values of $L = J + \frac{1}{2}, L' = J - \frac{3}{2}$ that

$$
f_{+\frac{1}{2},\,\pm\frac{1}{2}}^{JLL',\,\frac{3}{2},\,1}(\theta^*) = \frac{(-1)^L \sqrt{3}}{(2L - 1)\sqrt{[L(L-1)(2L+1)]}} \sum_m (\mathrm{im}\,\sin\theta^* - (L-1)\cos\theta^*)
$$

$$
\times\, P_m^{L-1}(0) \times \left\{ \begin{array}{l} \sqrt{[(L+m)(L-m)]}\,Y_m^L(\Theta\Phi) \\ \sqrt{[(L-m+1)(L-m)]}\,Y_{m-1}^L(\Theta\Phi) \end{array} \right\} \left. \begin{array}{l} (\mu_f = +\frac{1}{2}) \\ (\mu_f = -\frac{1}{2}) \end{array} \right\} \quad (2.4.23)
$$

Many other useful examples are given in Tables I, II and III of Deler and Valladas' paper, to which we would refer the interested reader. We would only remark that we chose the particular case $L = J + \frac{1}{2}, L' = J - \frac{3}{2}$ given above because there is a misprint in the corresponding entry in Table III. Hence it must always be a good idea in a practical calculation to derive as much of the formalism as one can for oneself, rather than relying on even the most detailed texts.

2.5. Isospin

The isospin structure of a production amplitude is fairly straightforward and does not require much elaboration here. As one may imagine, the isospin decomposition follows the same lines as the angular momentum analysis though is considerably less complex. Firstly we combined two of the final state particles i and j (say) and determine $\mathscr{I}_{ij}$, $\hat{\mathscr{I}}_{ij}$ the total isospin of the (ij) system and its third component. i.e.

$$|\mathscr{I}_{ij} \cdot \hat{\mathscr{I}}_{ij}\rangle = C^{I_i I_j \cdot \mathscr{I}_{ij}}_{i_i i_j \mathscr{I}_{ij}} |I_i i_i\rangle |I_j i_j\rangle$$

where I_i, i_i represent the total isospin of the ith particle and its third component, and C is the relevant Clebsch–Gordan coefficient. We then combined these isospin states with the third particle k and again expand in isospin eigenstates. In this way we find the production amplitudes can be written in terms of isospin amplitudes as

$$\langle I_1 i_1, I_2 i_2, I_3 i_3 | T | I_a i_a, I_b i_b\rangle = \sum_{I, \mathscr{I}_{ij}} G^{i_1 i_2 i_3; i_a i_b}_{I \mathscr{I}_{ij}} T^{I; \mathscr{I}_{ij} I_k}_{I_a, I_b} \tag{2.5.1}$$

where I is the total isospin of the system and G is given by the

$$G = C^{I_i I_j \mathscr{I}_{ij}}_{i_j i_j \cdot \mathscr{I}_{ij}} C^{I_{ij} I_k I}_{\mathscr{I}_{ij} i_k i} C^{I_a I_b I}_{i_a i_b i} \tag{2.5.2}$$

In the same manner we could have first considered the isospin of the (j, k) system and then combined it with the third particle i. The resulting isospin amplitudes for this case are related to those given above by (Barut and Leung, 1965)

$$T^{I, \mathscr{I}_{jk}, I_i}_{I_a, b} = \sum_{ij} V^{I: \mathscr{I}_{jk}, I_i}_{\mathscr{I}_{ij}, I_k} T^{I. \mathscr{I}_{ij}, I_k}_{I_a I_b} \tag{2.5.3}$$

where the recoupling coefficient V is given in terms of the $6-j$ symbol $\begin{Bmatrix} A & B & C \\ a & b & c \end{Bmatrix}$ as

$$V = (-1)^{2I + I_j + I_k - \mathscr{I}_{jk}} \left[(2\mathscr{I}_{ij} + 1)(2\mathscr{I}_{jk} + 1)\right]^{\frac{1}{2}} \begin{Bmatrix} I_i I_j \mathscr{I}_{ij} \\ I_k I \mathscr{I}_{jk} \end{Bmatrix} \tag{2.5.4}$$

Although these equations may look rather complicated they are really fairly trivial. To reassure those who find the numerous superscripts and subscripts intimidating we list below the isospin amplitudes for the particular cases of $\pi N \to \pi_1 \pi_2 N_3$ and $NN \to \pi_1 N_2 N_3$. (Of course, so far as the isospin structure is concerned this covers many other possible processes in which the pions and nucleons may be replaced by any other particles with isospin 1 or $\frac{1}{2}$ respectively). We would recommend that the reader spend ten minutes with a table of C–G coefficients to verify some of the coefficients listed in Tables 2 and 3, and convince himself that it is not too difficult.

The isospin amplitudes corresponding to an initial decomposition with respect to $\mathscr{I}_{13}$, i.e. the $(\pi_1 N_3)$ isospin component for example, can be read off from Table 2a by interchanging π_1 and π_2. Note also that the recoupling

TABLE 2a. $\pi N \to \pi_1(\pi_2 K_3)$ Isospin Amplitude Coefficients. (Each entry has a square root sign implied. For instance $\pi^+ p \to \pi^0 \pi^+ p = \sqrt{(\tfrac{3}{5})}T$. The minus signs, of course, are taken outside the square root.)

$T^I_{\mathscr{I}_{23}(\pi_2 N_3)}$	$T^{3/2}_{3/2}$	$T^{3/2}_{1/2}$	$T^{1/2}_{3/2}$	$T^{1/2}_{1/2}$
$\pi^+ p \to \pi^+ \pi^+ n$	$-2/15$	$2/3$		
$\pi^+ p \to \pi^+ \pi^0 p$	$-4/15$	$-1/3$		
$\pi^+ p \to \pi^0 \pi^+ p$	$3/5$			
$\pi^- p \to \pi^+ \pi^- n$	$-2/15$		$-1/3$	
$\pi^- p \to \pi^- \pi^+ n$	$8/135$	$2/27$	$-1/27$	$-8/27$
$\pi^- p \to \pi^0 \pi^0 n$	$-2/135$	$2/27$	$4/27$	$2/27$
$\pi^- p \to \pi^- \pi^0 p$	$16/135$	$-1/27$	$-2/27$	$4/27$
$\pi^- p \to \pi^0 \pi^- p$	$-1/135$	$-4/27$	$2/27$	$-4/27$

TABLE 2b. $\pi N \to (\pi_1 \pi_2) N_3$ Isospin Amplitude Coefficients. (Each entry has a square root sign implied, as in Table 2a.)

$T^I_{\mathscr{I}_{12}(\pi_1 \pi_2)}$	$T^{3/2}_{2}$	$T^{3/2}_{1}$	$T^{1/2}_{1}$	$T^{1/2}_{0}$
$\pi^+ p \to \pi^+ \pi^+ n$	$4/5$			
$\pi^+ p \to \pi^+ \pi^0 p$	$-1/10$	$1/2$		
$\pi^+ p \to \pi^0 \pi^+ p$	$-1/10$	$-1/2$		
$\pi^- p \to \pi^+ \pi^- n$	$1/45$	$1/9$	$-1/9$	$-2/9$
$\pi^- p \to \pi^- \pi^+ n$	$1/45$	$-1/9$	$1/9$	$-2/9$
$\pi^- p \to \pi^0 \pi^0 n$	$4/45$			$2/9$
$\pi^- p \to \pi^- \pi^0 p$	$-1/10$	$-1/18$	$-2/9$	
$\pi^- p \to \pi^0 \pi^- p$	$-1/10$	$1/18$	$2/9$	

matrices V can be simply derived by considering specific charge states in Tables 2 and 3. For instance, from the process $\pi^+ p \to \pi^0 \pi^+ p$ in Tables 2a, 2b we obtain

$$T^{3/2}_{\frac{3}{2}(\pi_2 N_3)} = \sqrt{\tfrac{5}{3}} \left\{ -\sqrt{(\tfrac{1}{10})}\, T^{3/2}_{2(\pi_1 \pi_2)} - \sqrt{(\tfrac{1}{2})}\, T^{3/2}_{1(\pi_1 \pi_2)} \right\}$$

i.e.

$$T^{3/2}_{\frac{3}{2}} = -\frac{1}{\sqrt{6}}\, T^{3/2}_{2} - \sqrt{(\tfrac{5}{6})}\, T^{3/2}_{1}$$

and the coefficients $-\sqrt{\tfrac{1}{6}}$ and $-\sqrt{\tfrac{5}{6}}$ are the two elements of $V^{\frac{3}{2}\frac{3}{2}1}_{f_{12}f_{12}}$ with $\mathscr{I}_{12} = 2,1$ respectively. Clearly the remaining elements of V can be determined in a similar manner.

TABLE 3a. $NN \to \pi_1(N_2 N_3)$ Isospin Amplitude Coefficients.
(Each entry has a square root implied)

$T^I_{\mathscr{I}_{23}}$	T^1_1	T^1_0	T^0_1
$pp \to \pi^0 pp$	1/2		
$pp \to \pi^+ pn$	$-1/4$	1/2	
$pp \to \pi^+ np$	$-1/4$	$-1/2$	
$pn \to \pi^0 pn$		1/4	$-1/12$
$pn \to \pi^0 np$		$-1/4$	1/12
$pn \to \pi^- pp$	$+1/4$		1/6

TABLE 3b. $NN \to (\pi_1 N_2)N_3$ Isospin Amplitude Coefficients.
(Each entry has a square root implied.)

$T^I_{\mathscr{I}_{12}}$	$T^1_{\frac{3}{2}}$	$T^1_{\frac{1}{2}}$	$T^0_{\frac{1}{2}}$
$pp \to \pi^0 pp$	$-1/6$	$-1/3$	
$pp \to \pi^+ pn$	3/4		
$pp \to \pi^+ np$	$-1/12$	2/3	
$pn \to \pi^0 pn$	1/6	$-1/12$	$-1/12$
$pn \to \pi^0 np$	$-1/16$	1/12	$-1/12$
$pn \to \pi^- pp$	$-1/12$	$-1/6$	$+1/6$

3

Analysis of Production Processes at Low and Intermediate Energies

In previous chapters we have been concerned with the dimensionality problem in particle production insofar as it affects the kinematics of the situation. That is to say, we have only discussed the rather formal difficulties associated with various phase-space manipulations and the definition of the physical region. The production amplitude on the other hand, has so far only been considered in the general terms of partial wave decompositions. However, if we wish to investigate the dynamical mechanisms which govern particle production reactions it is clear that we must now consider how to determine the amplitudes from the experimental data. From (1.2.1) we see that the modulus of the amplitude $|T|$ for the elastic scattering of two *spinless* particles can be calculated from the distribution $d\sigma/d\cos\theta$; while for a three body final state $|T|$ can be determined from the fully differential distribution i.e. $d^4\sigma/ds_{12}dt_1ds_{23}du_3$ of (2.3.7), again assuming all the particles have zero spin. However, in practical laboratory situations the target particle is a spin $\frac{1}{2}$ baryon so that the fully differential cross-sections of (1.2.1) determine only the modulus of the spin averaged amplitudes $|\overline{T}|$. In order to do better than this and determine the individual amplitudes corresponding to the different spin states of the particles it is also necessary to perform experiments which explicitly measure these spin orientations. Such measurements are only just becoming accurate enough in the case of elastic πN scattering to allow a determination of these individual spin amplitudes. For production processes, however, the requirement of fully differential distributions plus the measurement of the various spin states rules out a similar derivation of the production spin amplitudes in the foreseeable future.

In this case, therefore, it is necessary to assume, *a priori*, some form for these amplitudes, i.e. so-called model amplitudes, based upon the few physical ideas we have about particle production, and then to test these model guesses against whatever data is available at present. Of course, since these data for the most part represent only averages over the spin variables and are often only partially differential cross-sections, it is possible for two quite different models of the amplitudes to fit this same limited data. In practice, however, what more often happens is that when a model is generalized to try to describe several reactions simultaneously, it is found that it cannot do so without the addition of further rather ad hoc parameters. Such additions may then allow a good fit to the data, but often the original motivation for the model is lost in the process.

These two failings of model amplitudes, i.e. their lack of uniqueness and frequently ad hoc nature are well known. We repeat them here only because they are sometimes overlooked by their users on the one hand, or quoted by their critics on the other hand to discredit all model calculations. Nevertheless, if used carefully they can be an extremely powerful tool in the understanding of particle dynamics. In order to avoid some of the more obvious pitfalls of model calculations let us try to list a few simple points which might help in the phenomenological study of production reactions.

(i) Look carefully at the current data and decide what are the significant features. Part of this, of course, is to try to distinguish which are purely kinematical effects, i.e. phase-space, and which are dynamical in origin.

(ii) Construct a model which appears to be consistent with these features as well as the physical ideas to be tested, such as resonance formation, Regge pole exchange, transverse momentum cut-off, etc.

(iii) Deduce the further consequences of the model. From (i) and (ii) it should already be apparent that the model by its very construction, can describe at least some of the available data. However, one of the results of making any model is that it brings with it some further features which must also be tested. Clearly it is important to consider which is the best way of identifying such features in the data. [A good example is a model containing a final state resonance. This will normally give an enhancement in the appropriate mass distribution. However, it also suggests a particular form for the distribution in an angle variable; (a form determined by the spin and parity of the resonance) and the model can only be classed as satisfactory if it adequately describes both distributions].

(iv) Compute. By this stage we should have a good idea of the results one will obtain, and the computation is only for a more precise determination of parameters. If a fit to the data requires extremely careful adjustment of several parameters it is likely that the model is not a good one, since its generalization or extension to other reactions and other data will be very difficult.

With this preface let us now proceed to consider some of the existing models for particle production. In the present chapter we start by restricting our discussion to low energies and concentrate on three-body final states. However, with the introduction of the concept of duality which relates the high energy behaviour of the amplitude to its low energy structure, the classification of models by arbitrary energy ranges is no longer very satisfactory. Therefore, when we come to talk about the extension of Veneziano's dual model amplitude (Veneziano, 1968) to describe single particle production processes, (i.e. two-body to three-body amplitudes) we will necessarily have to consider both its low energy and high energy limits. Nevertheless, on the grounds that many particle production becomes more likely as the energy increases we shall leave the discussion of dual models for n-particle production to the next chapter.

3.1. Isobar model

When one looks at the integrated cross-section for any particular process, whether it is a two-body or n-body reaction, it is found that the data tend to display two fairly well defined regimes. The first, corresponding to low c.m. energies (e.g. $< 2.5\,\text{GeV}$), often exhibit a number of peaks which die away with increasing energy. Above about $2 \cdot 5\,\text{GeV}$, the data enter the "high energy" regime which is characterized by rather smooth cross-sections. It should be noticed that the low energy peaks tend to occur near the position of known resonances, and are therefore probably attributable, in the main, to a superposition of resonances in the various partial wave amplitudes. A similar low energy resonance regime can also be discerned in the final state mass distributions which are considered in (2.3.10) as well as in the Dalitz plot distributions of (2.3.14). For this reason it would seem likely that in order to construct a sensible model for single particle production processes at low energies we must be able to include resonance poles not only in the total energy squared variable s, but also in the final state sub-energies s_{ij}. Moreover, these resonances must be included with the correct angular functions corresponding to the various spins of these resonances. To this end we should first perform a partial wave decomposition of the production amplitude as in Section 2.4. Then the question arises how should we parameterize the resulting partial wave amplitudes $T_{jl}^{JLL'}(s, s_{ij})$ in order to describe the contribution of resonances produced in the $i + j$ final state partial wave with total and orbital angular momentum j and l?

One parameterization which comes to mind most readily is that usually known as the *isobar model*. Although this name covers several slightly different models, the basic assumptions which it implies can be summarized as follows:

(i) The production amplitude can be written as a sum of terms each of which corresponds to a specific two-body final state interaction, i.e.

$$T = \sum_{i \neq j = 1, 2, 3} T_{ij}, \qquad (3.1.1)$$

where T_{ij} represents the totality of diagrams in which the ij particles interact last. T_{ij} of course, is itself a sum over the various partial wave amplitudes $T_{jl}^{JLL'}(s, s_{ij})$ multiplied by the angular functions derived in Section 2.4.

(ii) Each $T_{jl}^{JLL'}$ can be written as a product of three terms. F^R, a formation amplitude for the reaction $a + b \rightarrow R_{ij} + c_k$ where R_{ij} is a resonance in the $i + j$ system; P^R the propagator for the resonance R_{ij}; and D_{ij}^R the form factor describing the decay of the resonance into particles i and j.

(iii) The formation amplitude F^R is simple.

These assumptions are shown diagrammatically in Fig. 10. Clearly we can always write the production amplitude in the form suggested by (i) plus perhaps some genuine three-body interaction which we shall ignore. Such a separation of the amplitude is the starting point of most dynamical theories of production processes; e.g. the non-relativistic Faddeev equations (Faddeev,

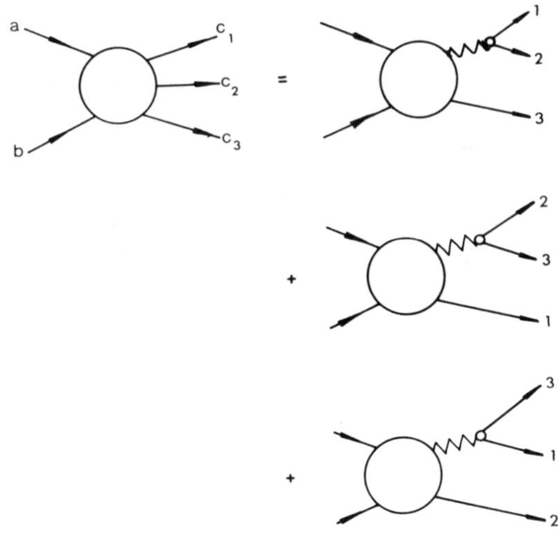

FIG. 10. Diagrammatic form of the isobar model.

1961, 1963; Lovelace, 1963); and indeed, we have already anticipated this separation by considering, in the last section, partial wave decompositions with respect to each of the possible two-particle final states.

As an example of (ii) let us consider a resonance in the $c_i + c_j$ system which is produced with orbital angular momentum L' relative to particle c_k from an initial state with orbital angular momentum L. Then provided the c.m. energy is not too large it may be plausible to assume that the formation amplitude F^R exhibits the centrifugal barrier factors, i.e.

$$F^R = \left(\frac{p_1^c}{E_1^c}\right)^L \left(\frac{q_k^c}{w_k^c}\right)^{L'} \cdot f^R, \tag{3.1.2}$$

where the energy factors are included to provide a scale. Similarly, for low final state sub-energies let us also write the decay amplitude as

$$D^R = \left(\frac{k_i}{w_i}\right)^l d^R, \tag{3.1.3}$$

where w_i, k_i are the energy and momentum of particle i in the $c_i + c_j$ c.m. system, and the reduced amplitude d^R of course will be proportional to the partial width for the decay of R into $c_i + c_j$. For the propagator P^R we take the form

$$P^R = \frac{1}{s_R - s_{ij} - i\rho(s_{ij})\Gamma(s_{ij})} \tag{3.1.4}$$

where $\rho(s_{ij})$ is just the kinematic factor

$$\rho(s_{ij}) = k_i/\sqrt{s_{ij}}$$

and $\sqrt{s_R}$, Γ are the mass and width of the resonance. Notice that in (3.1.4) we have assumed some energy dependence for this width $\Gamma(s_{ij})$ since empirically it is found that the corresponding resonance peak in the $d\sigma/ds_{ij}$ distribution, for example, turns out to be rather asymmetric in shape. In particular, when the resonance is rather broad, the energy dependence of $\Gamma(s_{ij})$ apparently distorts the shape so that the maximum of the peak falls somewhat below the position $s_{ij} = s_R$, and the shape of the peak is rather flatter above this maximum than below it. This situation is shown diagrammatically in Fig. 11.

It is fairly easy to see how these distortions and displacements can be produced, particularly for low mass resonances, if $\Gamma(s_{ij})$ again contains threshold factors such that

$$\Gamma(s_{ij}) \approx \Gamma_0 \left(\frac{k_i}{k_R}\right)^{2l}, \tag{3.1.5}$$

where k_R is the value of k_i at the point $s_{ij} = s_R$. Inserting (3.1.5) into (3.1.4)

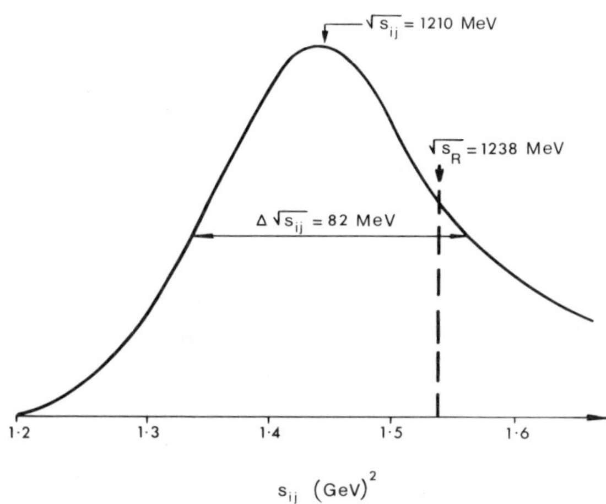

FIG. 11. The asymmetric shape of the Δ (1238) resonance produced in the reaction $K^+ p \to K^0 \pi^+ p$ at $1\cdot14\,\text{GeV}/c$ incident K^+ momentum. Notice the peak in this distribution is at 1210 MeV rather than the resonance position of 1238 MeV.

this means that the resulting peak maximum is located not at s_R but rather at the point

$$s_{ij} = s_R - (2l + 1)\Gamma_0^2 \left[\frac{s_R^2 - (m_i^2 - m_j^2)^2}{2s_R^2 k_R^2} \right]. \qquad (3.1.6)$$

We see therefore that if s_R is close to the $c_i + c_j$ threshold, the momentum k_R is small so that the second term in (3.1.6) can be quite large and the peak is consequently shifted a significant distance.

So far we have discussed the case when particles c_i and c_j are in a resonance state. However, it can also happen that we wish to describe a two-particle final state interaction which, while not perhaps resonant, could play a significant role in the structure of the production amplitude. Such a case is the $I = 0$, $J = 0$ $\pi\pi$ interaction which may or may not possess the so-called "epsilon" resonance. Even if it is, in fact, a resonance, its width is so large that no simple propagator such as that in (3.1.4) can adequately describe its effect. For this situation instead of (3.1.4) and (3.1.5) one might use an "effective" propagator $\exp(i^{\delta(s_{ij})})$ where δ is the appropriate partial wave phase, while the decay form factor D^R might be represented by $|f(s_{ij})|^{\frac{1}{2}}$ where $f(s_{ij})$, is the corresponding partial wave amplitude of elastic $c_i + c_j$ scattering. The square root of this amplitude is taken because in some sense it represents only half of the scattering process, i.e. the decay of the state into $c_i + c_j$ rather than the full amplitude.

It should be mentioned that these two-particle final state factors $P^R D^R$ either in the form of (3.1.3), (3.1.4) or in the form of the partial wave amplitude have been used for many years as enhancement factors which modify the n-body phase-space elements, i.e.

$$\frac{d\sigma}{ds_{ij}} \propto |P^R D^R|^2 \sigma(s_{ij}) \tag{3.1.7}$$

where $\sigma(s_{ij})$ is the (phase-space) cross-section for the production of n-2 particles plus a pseudo particle (resonance) of mass s_{ij}. See, for instance, Lindenbaum and Sternheimer (1957) and Jackson (1964). However, this procedure takes no account of the spin or angular momentum of the two-particle final state, to say nothing of its production mechanism from the initial state. To do this it is necessary to follow the procedure suggested here, i.e. assume some form for the formation amplitude F^R and use the product of all three factors $F^R P^R D^R$ in the partial wave decomposition of Section 2.4. In this way we ensure that the correct angular dependence is built into the two-body to three-body production amplitude corresponding to the particular angular momentum states in which the resonance is produced and sub-sequently decays. As we have previously pointed out, by this method it is also possible to consider interference effects between several produced resonances of final state interactions. The size of the interference, of course, is strongly dependent on the angular momentum states involved, and unless one uses a rigorous formalism such as a partial wave expansion to study these effects, it is very possible that large errors could be created in the resultant form of the production amplitude.

This brings us to a consideration of the formation amplitudes F^R and the third assumption of the isobar model, namely, that F^R is simple. Besides being rather ill-defined, this assumption is also the weak point of the model as we shall now try to explain. Generally speaking, F^R can be a complex function of all the energies of the system, i.e. s, s_{12}, s_{23} and therefore also s_{13}, because the partial wave expansion only separates out the dependence on the angles $\Theta \, \Phi$ or equivalently the dependence on the momentum transfers t_1 and u_3. Nevertheless, purely on the grounds of simplicity F^R is taken to be a function just of the overall c.m. energy, i.e. $F^R = F^R(s)$. (See, for example, Lindenbaum and Sternheimer (1961) Olsson and Yodh (1966), Morgan (1968), M. de Beer et al. (1969)). Unfortunately, even though this assumption gives F^R sufficient freedom to include the possible intermediate resonance poles in the s variable, which we have already suggested should be a feature of the low energy amplitude, such an extreme simplification is perhaps not justified except in certain particular examples. For instance, it should be noted that taking $F^R = F^R(s)$, the expansion (3.1.1) would imply the amplitude is constructed by considering each two-body final state effect without reference to the interactions that may occur in other sub-energy channels. For instance, for the reaction $\pi N \to \pi_1 \pi_2 N_3$ this prescription says we should consider each of the following resonance amplitudes separately,

$$\pi N \to \pi_1 N_1^* \to \pi_1 (\pi_2 N_3)$$

$$\pi N \to \pi_2 N_2^* \to \pi_2 (\pi_1 N_3)$$

$$\pi N \to \rho N_3 \to (\pi_1 \pi_2) N_3$$

and calculate, for example, the formation and decay of the ρ resonance in exactly the same way whether the values of s_{23} and s_{13} are such that $N_{1,2}^*$ represent either the $N^*(1470)$ or $N^*(1780)$. Of course, whether the amplitude is constructed this way or not is a question of dynamics. However, we can imagine situations where, say, the $\pi_2 N_3$ system decays from an N_1^* resonance and the nucleon N_3 subsequently interacts with the other pion, π_1. More complicated configurations readily come to mind. Such multiple scattering effects may turn out to be negligibly small, as they almost certainly will if there is only one two-body final state in which there is a strong interaction. (This is the basis of Watson's theorem for final state interactions (Watson, 1952)). Nevertheless, in principle it is clear they can give rise to correlations between the different final-state production mechanisms. Thus there is at least one reason why we might expect the formation amplitude to indeed be a function of the sub-energies s_{ij} as well as s.

Moreover, it should also be noticed that the isobar model assumptions imply each T_{ij} of (3.1.1) carries most of the information about the $c_i + c_j$ final state. However, Amado (1967) has suggested by a study of the non-relativistic Faddeev equations that this should not be the case and, indeed, a factor containing the features of the $c_i + c_j$ final state interaction multiplies the *entire* amplitude. We are therefore led to the possibility that while the expansion (3.1.1) is mathematically sound, it is only a good approximation scheme for phenomenological analyses provided we allow F^R to be a function of all the sub-energies. To put this in other words, there may well exist *strong energy dependent correlations* between T_{12}, T_{23} and T_{13}.

Again let us stress that to resolve these difficulties we must fully understand the dynamics of the situation. It may turn out that the isobar model does represent the physical production mechanism after all, in which case it is important to realise just how selective nature has been. So far, in fact, the model has been quite successful in fitting low-energy production data, but this may be due to the large number of parameters available compared to the fairly small number of experimental measurements. Also, since the isobar model has been used chiefly with comparatively narrow resonances (e.g. ρ and Δ) only the value of the formation amplitude near the resonance position will be important in data fitting. Thus any correlations can easily be overlooked. In this context it will be instructive to consider in the next section the dual resonance model of Bardakci and Ruegg (1968) where such correlations can be explicitly displayed.

3.2 B_5 dual model

In 1968 Veneziano showed, by means of an explicit model, that the high energy Regge behaviour of the $\pi\pi \to \pi\omega$ amplitude could be achieved by the summation of an infinite number of resonance poles. This is to say, his model demonstrated the scattering amplitude could be described *either* by the exchange of Regge poles in the cross channel *or* by a series of direct channel resonances; and therefore that these descriptions may in some sense be *equivalent* or *dual* to each other.

Specifically the Veneziano model for the invariant amplitude of the completely symmetric reaction $\pi\pi \to \pi\omega$ (which is therefore a particularly simple case to consider) is given by

$$A(s, t, u) = B_4(s, t) + B_4(t, u) + B_4(u, s),$$

where

$$B_4(s, t) = g\frac{\Gamma(1 - \alpha(s))\,\Gamma(1 - \alpha(t))}{\Gamma(1 - \alpha(s) - \alpha(t))}, \tag{3.2.1}$$

α is the ρ trajectory taken to have the linear form

$$\alpha(x) = \alpha_0 = \alpha' x \tag{3.2.2}$$

and s, t, u are the usual Mandelstam invariants for two-body scattering. For physical values of s and t, i.e. $s(t) > (m_\pi + m_\omega)^2$ the gamma functions in the numerator of (3.2.1) give rise to idealized (real!) resonance poles at values of $\alpha(s)$ or $\alpha(t) = n$, any positive integer. The extra gamma function in the denominator ensures that double poles arising at $\alpha(s) = n_1$ and $\alpha(t) = n_2$ are reduced to single poles. Thus (3.2.1) can be written explicitly as an infinite sum of single poles in the form

$$B_4 = g\sum_{n=1}^{\infty} \frac{\Gamma(n + \alpha(t))}{\Gamma(\alpha(t))\,(n - 1)!}\frac{1}{\alpha(s) - n}. \tag{3.2.3}$$

Alternatively, using Stirling's approximation for the gamma function, i.e.

$$\Gamma(z) \underset{|z|\to\infty}{=} (2\pi)^{\frac{1}{2}} \exp(-z)^{z-\frac{1}{2}} \qquad |\arg z| < \pi \tag{3.2.4}$$

it can be seen that $B_4(s, t)$ for large s has the form

$$B_4(s, t) \underset{s\to\infty(1+iE)}{=} g\frac{\pi(\alpha' s)^{\alpha(t)}}{\Gamma(\alpha(t))\sin \pi\alpha(t)}\exp(-i\pi\alpha(t)), \tag{3.2.5}$$

which is the high energy behaviour for the scattering amplitude compatible with the exchange of the ρ trajectory in the t-channel. Notice that the scale factor, i.e. α' in this case, which controls the rate at which this asymptotic limit is reached is a prediction of the model rather than an arbitrary parameter as in Regge phenomenology.

However, in spite of these nice features, the representation (3.2.1) is not without some rather serious problems. For example, the "resonance" poles are situated on the real axis rather than on the second Reimann sheet, while from the conditions on z in (3.2.4) it is seen that Regge asymptotic behaviour only holds off the positive real s axis. These difficulties are no doubt associated with the fact that the correct analytic structure and unitarity requirements of the amplitude have not been imposed. (These problems will be discussed later.) Even so, it would be wrong to minimize the achievement of Veneziano's work and the profound effect the concept of duality has had on the world of particle physics since 1968.

The question now comes to mind, what is the generalization of the Veneziano model to production reactions and in particular to the two-body to three-body process, i.e. the five point amplitude? To answer this question let us note that $B_4(s, t)$ of (3.2.1) is proportional to the Euler beta function $B(x, y)$, i.e.

$$B_4(s, t) = g(1 - \alpha(s) - \alpha(t))\, B(1 - \alpha(s), 1 - \alpha(t)),$$

which has the integral representation

$$B(x, y) = \int_0^1 du\, u^{x-1}(1 - u)^{y-1}. \tag{3.2.6}$$

Starting from this integral representation, Bardakci and Ruegg (1968) derived the following representation for the five amplitude, which is now normally called B_5, i.e.

$$B_5(s, t_1, s_{12}, s_{23}, u_3)$$

$$= \int_0^1 du_1 \int_0^1 du_4\, u_1^{x_1}(1 - u_1)^{x_2}(1 - u_4)^{x_3} u_4^{x_4}(1 - u_1 u_4)^{x_5 - x_2 - x_3 - 1}, \tag{3.2.7}$$

where $x_i = -1 - \alpha(v_i)$ and v_i, $i = 1$ to 5 are the five independent Lorentz scalars describing the amplitude, e.g. $s, t_1, s_{12}, s_{23}, u_3$ as above.

Since one of the features of dual models is that they treat all channels on the same footing it will be appropriate to formally consider all five particles a, b, c_1, c_2, c_3 of the process to be ingoing and re-label them as c_1, c_2, c_3, c_4, c_5. Then the Lorentz scalars can be denoted in a more symmetric fashion as shown in Fig. 12 as

$$s_{ii+i} = (p_i + p_{i+1})^2, \qquad i = 1, \ldots, 5,$$

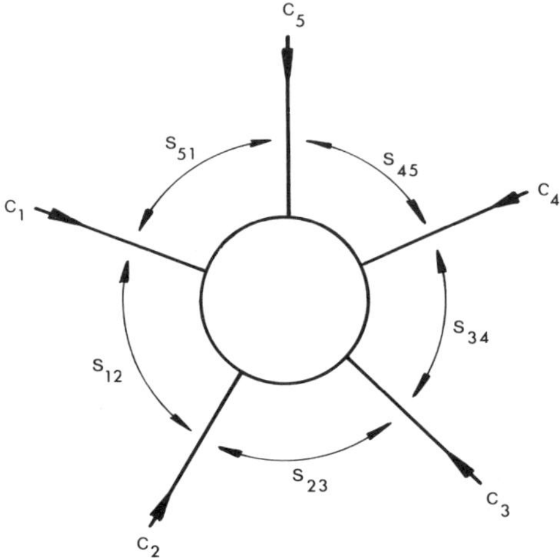

Fig. 12. Definition of the Lorentz scalars s_{ij} for the five-point amplitude.

where 6 and 1 are taken to be equivalent. With this slight change of notation we see that B_5 can be written in a cyclically symmetric form as

$$B_5 = \int_0^1 du_i \int_0^1 \frac{du_j}{1 - u_i u_j} u_1^{-\alpha_{12}-1} u_2^{-\alpha_{23}-1} u_3^{-\alpha_{34}-1} u_4^{-\alpha_{45}-1} u_5^{-\alpha_{51}-1}, \qquad (3.2.8)$$

where $\alpha_{i,i+1} = \alpha_0 + \alpha' s_{i,i+1}$ etc.†, i and j are any two non-successive integers from 1 to 5, and the variables u_k are chosen to satisfy the constraints

$$u_i = 1 - u_{i-1} u_{i+1} \qquad i = 1,\ldots,5, \qquad u_6 \equiv u_1, \qquad (3.2.9)$$

However, because the model does possess this symmetry under the cyclic interchange of indices 1 to 5, it will be sufficient for the moment to consider the particular case of $i = 1, j = 4$ and write (3.2.7) in the form of (3.2.8) as

$$B_5 = \int_0^1 du_1 \int_0^1 du_4 u_1^{-\alpha_{12}-1}(1 - u_1)^{-\alpha_{23}-1}(1 - u_4)^{-\alpha_{34}-1} u_4^{-\alpha_{45}-1}$$

$$\times (1 - u_1 u_4)^{\alpha_{23}+\alpha_{34}-\alpha_{51}}. \qquad (3.2.10)$$

The integrals in (3.2.10) are convergent when all the exponents have a negative real part and hence B_5 is analytic in this region. On the other hand, when $\mathrm{Re}(\alpha_{12})$, say, becomes positive the u_1 integral diverges at the lower end of integration and gives rise to poles in the s_{12} variable. Furthermore, if

† Although we have written the trajectory intercept as a universal constant it can be different from one channel to the next.

$\mathrm{Re}(\alpha_{45})$ also becomes positive the u_4 integral diverges at $u_4 = 0$ and therefore B_5 in this case develops poles in both variables s_{12} and s_{45}. Notice, however, that for $\mathrm{Re}(\alpha_{23}) > 0$ the u_1 integration diverges not at $u_1 = 0$ but at $u_1 = 1$, and hence B_5 cannot have poles in s_{12} and s_{23} at the same time. This is a consequence of the constraints (3.2.9) which imply that u_i and u_{i+1} cannot vanish simultaneously. More generally, therefore, it can be seen that the model does not simultaneously develop poles in s_{ij} and s_{kl} if any of i, j, k and l are equal. Conversely, since non-adjacent variables u_i and u_j are allowed to approach their limits of integration together, there may exist simultaneous poles in the corresponding variables $s_{i,i+1}s_{j,j+1}$. As we have already seen, this occurs for s_{12} and s_{45}, for example.

In order to express the dual model amplitude as an infinite series of these poles, it is useful to split up each of the integrals in (3.2.10) into two regions, i.e. from 0 to Λ and Λ to 1 where $0 < \Lambda < 1$.

Then writing

$$B_5 = \left(\int_0^\Lambda \int_0^\Lambda + \int_0^\Lambda \int_\Lambda^1 + \int_\Lambda^1 \int_0^\Lambda + \int_\Lambda^1 \int_\Lambda^1 \right) du_1\, du_4\, \{\text{integrand}\} \quad (3.2.11)$$

we see that only the first term leads to poles in both s_{12} and s_{45} while the second and third terms give poles in s_{12} and s_{45} respectively but not in both, and the fourth term is analytic in both s_{12} and s_{45}. Expanding the three terms in the integrand of (3.2.10),

$$(1 - u_1)^{-\alpha_{23} - 1}(1 - u_4)^{-\alpha_{34} - 1}(1 - u_1 u_4)^{\alpha_{23} + \alpha_{34} - \alpha_{51}}$$

in a double Taylor series about $u_1, u_4 = 0$ we obtain the Bardakci and Ruegg double pole expansion for (3.2.11)

$$B_5 = \sum_{k,l,m=0}^{\infty} \frac{\Lambda^{k+m-\alpha_{12}}}{k + m - \alpha_{12}} \frac{\Lambda^{l+m-\alpha_{45}}}{l + m - \alpha_{45}} \quad (3.2.12)$$

$$\times \{(-\alpha_{23} - 1)(-\alpha_{23} - 2)\ldots(-\alpha_{23} - k)\}$$

$$\times \{(-\alpha_{34} - 1)(-\alpha_{34} - 2)\ldots(-\alpha_{34} - l)\}$$

$$\times \{(\alpha_{23} + \alpha_{34} - \alpha_{51})(\alpha_{23} + \alpha_{34} - \alpha_{51} - 1)\ldots(\alpha_{23} + \alpha_{34} - \alpha_{51} - m + 1)$$

$+$ terms without multiple poles in s_{12} and s_{45}.

The leading order terms of the polynomial in curly brackets in (3.2.12), i.e.

$$B_5 \approx \sum_{j,j'} \frac{1}{(\alpha(s_{12}) - j)(\alpha(s_{45}) - j')} \sum_{k=0}^{k=\min(j,j')} c_k(s_{51})^k(s_{23})^{j-k}(s_{34})^{j'-k} \quad (3.2.13)$$

where c_k are constants, give rise to a finite number of degenerate poles at $\alpha(s_{12}) = k + m = j$, say, and $\alpha(s_{45}) = l + m = j'$. For instance, when s_{12}, s_{45}

are interpreted as energy variables then s_{51} and s_{23} represent momentum transfers so that (3.2.13) implies, by comparison with the partial wave decompositions of the last chapter,† that at each of the pole positions there exist only those resonances with spin from 0 to j or j' just as in the two-body case. Alternatively, if the term in brackets in (3.2.12) had not been a polynomial in s_{51} and s_{23} then there would be in general an infinite number of resonances at each pole position with spin not only from 0 to j but also from $j + 1$ to ∞ which are called ancestors. Hence the polynomial nature of the pole residues is a crucial feature of the model. It is worth stressing that if the constraints (3.2.9) were such as to allow poles in adjacent sub-energy variables this would, in fact, lead to ancestors since a pole clearly cannot be expressed as a finite polynomial.

Although we have chosen s_{12} and s_{45} as our two energy variables and expanded u_1 and u_4 in the integrand of (3.2.11) about zero we could equally well have chosen to expand u_1 about zero and u_4 about unity and therefore obtained the corresponding expansion for double poles in s_{12} and s_{34}. This is given simply from (3.2.12) with the interchange of indices 3 and 5. These two equivalent expansions of the B_5 amplitude are shown diagrammatically in Fig. 13, from which it should be clear there is a close connection between both of these expansions and the isobar amplitudes T_{ij} of Fig. 10. However, in the B_5 model we find there is indeed a very strong correlation between the formation amplitudes of the $3 + (4, 5)$ final state, (i.e. the residues of the s_{45} poles in the expansion (3.2.12) which are functions not only of s_{23}, s_{51} but also of s_{34} and s_{12}) and the formation amplitudes in the $(3. 4) + 5$ final state —the s_{34} pole expansion. The correlation ensures that when we sum up all the s_{45} poles we can re-expand it as a sum over s_{23} poles. This reinforces our view that the isobar model assumption of the last section may not be adequate to describe the physical situation except in a rather approximate way which averages over the functional dependence of the formation amplitudes.

This is perhaps as far as we need to go in discussing the Bardakci and Ruegg dual amplitude in the context of the isobar model. However, let us now expand our discussion and consider some other properties of B_5 which, while not always strictly relevant to particle reactions at low and intermediate energies, will provide a bridge to the following chapters. To do this it is convenient to note that B_5 defined by the double integral of (3.2.10) can alternatively be rewritten in the Zakrzewski form (Bialas and Pokorski, 1969; see also Hopkinson and Plahte, 1969)

$$B_5 = B(-\alpha(s_{45}), -\alpha(s_{51})) B(-\alpha(s_{34}), -\alpha(s_{23}))$$
$$\times {}_3F_2(\alpha(s_{12}) - \alpha(s_{45}) - \alpha(s_{34}), -\alpha(s_{51}), -\alpha(s_{23}); -\alpha(s_{45}) - \alpha(s_{51}),$$
$$-\alpha(s_{34}) - \alpha(s_{23}); \quad 1), \qquad (3.2.14)$$

---

† Notice that we have assumed that the external particles have no spin.

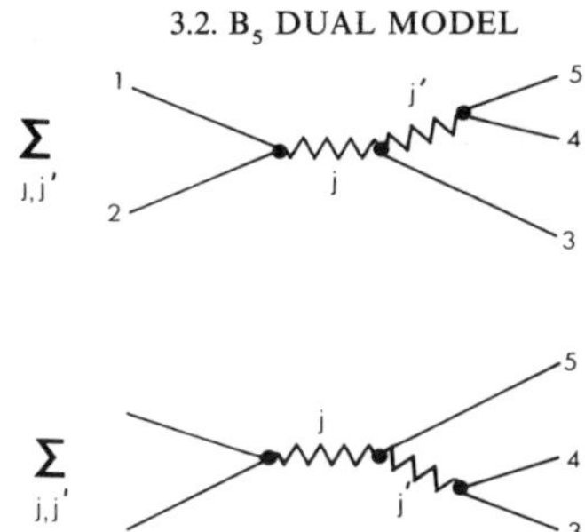

FIG. 13. The two equivalent double pole expansions of the B_5 function.

where $B(x, y)$ are the Euler beta functions defined in (3.2.6) and $_3F_2$ is a generalized hypergeometric function defined by

$$_3F_2(a, b, c; d, e; z) = \sum_{n=0}^{\infty} \frac{\Gamma(a + n)}{\Gamma(a)} \frac{\Gamma(b + n)}{\Gamma(b)} \frac{\Gamma(c + n)}{\Gamma(c)} \frac{\Gamma(d)}{\Gamma(d + n)} \frac{\Gamma(e)}{\Gamma(e + n)} \frac{z^n}{n!}$$

(3.2.15)

The second and third terms in (3.2.14) can be expanded as an infinite number of pole terms in the s_{34} variable, for example, such that

$$B_5 = \sum_{m=0}^{\infty} B(-\alpha(s_{12}), -\alpha(s_{51})) P_m D_m(s_{12}, s_{23}, s_{45}, s_{51})$$

(3.2.16)

where

$$P_m D_m = \sum_{m=0}^{\infty} \frac{(-1)^m}{m!} \frac{\Gamma(-\alpha(s_{23}))}{\Gamma(-\alpha(s_{23}) - m)} \cdot \frac{1}{m - \alpha(s_{34})}$$

$$_3F_2(\alpha(s_{45}) - \alpha(s_{12}) - \alpha(s_{23}), -\alpha(s_{51}), -m;$$
$$-\alpha(s_{51}) - \alpha(s_{12}), -\alpha(s_{23}) - m; 1)$$

(3.2.17)

Formula (3.2.16) exhibits clearly the factorization of the model: the amplitude factorizes into the two-body amplitude—a B_4 function, a vertex part—which itself is given by the Veneziano B_4 function, cf. (3.2.3); and intermediate "resonance" propagators which define the spins of the resonances as well as the masses.

We could also decompose (3.2.16) into a sum over poles in s_{12} as well as s_{34} and obtain the exact double pole expansion instead of the approximate form given in (3.2.12). It is probably more useful at this point, however, to derive the high energy behaviour of the B_5 amplitude. To do this let us take s_{12} as the total energy variable of the system and consider the limit in which s_{12} and one of the sub-energy variables s_{45} (say) are large, though another sub-energy, e.g. s_{34} is held fixed. (We shall also assume at least two of the momentum transfer variables, e.g. s_{51} and s_{23} are finite, otherwise it is possible to show the model has an exponentially vanishing limit). Now re-writing the Bardakci-Ruegg function (3.2.14) in yet another way (see Bialas and Pokorski, 1969) as

$$B_5 = B_4(-\alpha(s_{12}), -\alpha(s_{51}))B_4(-\alpha(s_{34}), \alpha(s_{51}) - \alpha(s_{23}))$$

$$\times \,_3F_2(-\alpha(s_{51}), -\alpha(s_{34}), \alpha(s_{23}) - \alpha(s_{45}) - \alpha(s_{51});$$

$$1 + \alpha(s_{23}) - \alpha(s_{51}), -\alpha(s_{12}) - \alpha(s_{51}); 1)$$

$$+ B(-\alpha(s_{12}), -\alpha(s_{23})) \, B(-\alpha(s_{45}), \alpha(s_{23}) - \alpha(s_{51}))$$

$$\times \,_3F_2(\alpha(s_{23}), -\alpha(s_{45}), \alpha(s_{51}) - \alpha(s_{34}) - \alpha(s_{23});$$

$$1 + \alpha(s_{51}) - \alpha(s_{23}), -\alpha(s_{12}) - \alpha(s_{23}); 1) \qquad (3.2.18)$$

the corresponding asymptotic limit is obtained once the asymptotic form for the above $_3F_2$ function is known. This can easily be deduced from the definition of the hypergeometric function (3.2.15) to be

$$\,_3F_2(a, b, c; d, e; z) \underset{c,\,e \to \infty}{=} F(a, b; d; cz/e) \qquad (3.2.19)$$

where $F(p, q, r, x)$ is a Gauss hypergeometric function. Inserting this limit in (3.2.18) therefore we obtain

$$B_5 \to B(-\alpha(s_{12}), -\alpha(s_{51})) \, B(-\alpha(s_{34}), \alpha(s_{51}) - \alpha(s_{23}))$$

$$\times \,_3F_2\left(-\alpha(s_{51}), -\alpha(s_{34}); 1 + \alpha(s_{23}) - \alpha(s_{51}); \frac{\alpha(s_{45})}{\alpha(s_{12})}\right)$$

$$+ B(-\alpha(s_{12}), -\alpha(s_{23})) \, B(-\alpha(s_{45}), \alpha(s_{23}) - \alpha(s_{51}))$$

$$\times F\left(-\alpha(s_{23}), \alpha(s_{51}) - \alpha(s_{34}) - \alpha(s_{23}); 1 + \alpha(s_{51}) - \alpha(s_{23}); \frac{\alpha(s_{45})}{\alpha(s_{12})}\right)$$

$$(3.2.20)$$

which can be recombined into one term as

$$B_5 \to B(-\alpha(s_{12}), -\alpha(s_{51}) \, B - \alpha(s_{34}), -\alpha(s_{23}))$$

$$\times F\left(-\alpha(s_{51}), -\alpha(s_{34}); -\alpha(s_{23}) - \alpha(s_{34}); 1 - \frac{\alpha(s_{45})}{\alpha(s_{12})}\right) \qquad (3.2.21)$$

[(3.2.21) follows from (3.2.20) because of the properties of the Gauss hypergeometric function. Specifically the above relation is given by Slater (1966), eqn (1.8.10)]. Also because of the assumption that all the trajectories are linear and have the same slope α', the argument $1 - \alpha(s_{45})/\alpha(s_{12})$ can be approximated for large s_{12}, s_{45} (and therefore large s_{35}) as $\alpha(s_{35})/\alpha(s_{12})$.

As we now let s_{12} and s_{45} tend to infinity only the first B_4 function in (3.2.21) is affected appreciably, going over to its asymptotic form (3.2.5) consistent with single Regge pole exchange. Thus (3.2.21) represents the "single Regge limit" of the B_5 function depicted in Fig. 14a. The "blob" at the

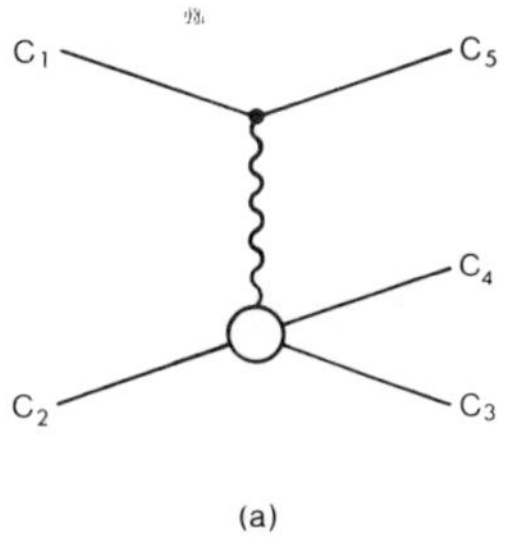

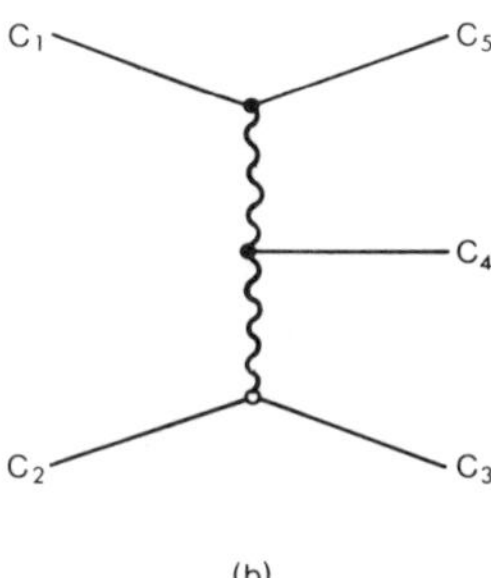

FIG. 14. (a) Single Regge pole limit of the five-point amplitude. (b) Double Regge pole limit of the five-point amplitude.

lower vertex, corresponding to the $c_2 + \text{Reggeon} \to c_3 + c_4$ amplitude is given by the other two terms in (3.2.21). Using the expansion (3.2.3) for the B_4 function, therefore, we can expand this amplitude also into a sum of resonances, i.e.

$$B_4(-\alpha(s_{34}), -\alpha(s_{23})) F\left(-\alpha(s_{51}), -\alpha(s_{34}); -\alpha(s_{23}) - \alpha(s_{34}); z \approx \frac{\alpha(s_{35})}{\alpha(s_{12})}\right)$$

$$= \sum_{m=0}^{\infty} \frac{1}{m - \alpha(s_{34})} \frac{(-1)^m}{m!} \frac{\Gamma(-\alpha(s_{23}))}{\Gamma(-\alpha(s_{23}) - m)} F(-\alpha(s_{51}), -m;$$

$$-\alpha(s_{23}) - m; z) \qquad (3.2.22)$$

Notice that the residues of these resonance poles depend on the momentum transfer variable s_{51}. In fact, comparing (3.2.22) with (3.2.3) we see that the only modification in this case is the introduction of the Gauss hypergeometric function which represents the effects of the off-mass-shell nature of the exchanged Reggeon. Thus the B_5 model would imply that these off-mass-shell corrections are of a simple polynomial nature, at least for small values of s_{51} where the single Regge pole exchange term is applicable.

The production of resonances from Regge pole exchanges as suggested by (3.2.22) is also frequently given the title of *isobar model*. In view of this it is worth pointing out that the first two assumptions of the isobar model, which we stated in the last section, also apply to this version. Of course, it is now a higher energy model, and we must consider the complete amplitude rather than the partial wave projections. However we can see that the third assumption concerning the simplicity of the resonance (isobar) formation amplitude is only partially applicable in the context of dual models. Although (3.2.22) represents a "simple" Regge pole exchange, the residues of the poles in s_{34} depend on *all* the other four independent Lorentz scalars in a very particular way.

It must be stressed that in the framework of dual models, the five point amplitude is given not only by $B_5(s_{12}, s_{23}, s_{34}, s_{45}, s_{51})$ but by the linear combination of *twelve* B_5 functions corresponding to all non-equivalent permutations of the particles. These twelve functions can be classified in two groups of six. The first contains all diagrams in which particles c_1 and c_2 are adjacent (called direct diagrams) and the second group contains those diagrams in which particles c_1 and c_2 have one particle between them (crossed diagrams) as shown in Fig. 15. Therefore, before any statement can be made about the simplicity of the resonance production amplitudes all twelve diagrams must be considered. We shall see the importance of this in the next section on the phenomenological uses of the B_5 model. For the present though let us merely comment that of the twleve contributions numbers 1, 2, 7 and 10 give both a non-zero asymptotic limit as $s_{12}, s_{45} \to \infty$ as well as a series of resonance poles in the s_{34} channel. Numbers 8 and 9 also provide a non-zero limit but do not possess resonance structure in the s_{34} variable. Thus, of the twleve diagrams there are six possible B_5 contributions to the asymptotic limit. Of course, some of these are closely related. For example 1 and 7 differ asymptotically only by Regge phase $\exp\left[-i\pi\alpha(s_{51})\right]$ so that by choosing their relative signs we can impose the correct signature on the exchanged trajectory. However, 8 and 9 are clearly different since they do not lead to resonances in the s_{34} channel. Nevertheless, they will provide non-negligible interference effects, or "smooth background", to the resonance production terms and must be considered.

In the above discussion we expanded the $c_2 + \text{Reggeon} \to c_3 + c_4$

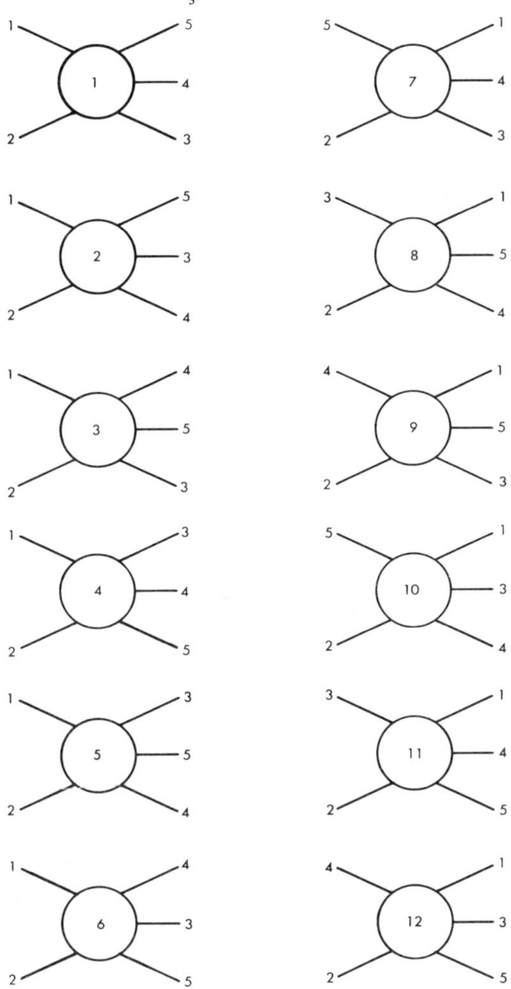

FIG. 15. The twelve non-equivalent configurations of the five-point amplitude.

amplitude as a sum over resonance poles. However, the duality hypothesis would suggest that for large values of s_{34} this resonance sum should be equivalent to Regge pole exchanges in the s_{23} channel. Although we do not discuss it in detail until the next chapter this leads us naturally to the situation shown in Fig. 14b—the *double Regge pole model*. In order to derive the functional dependence of the model from the B_5 amplitude let us return to the Bialas–Pokorski form (3.2.18) and consider the limit in which s_{12}, s_{34},

$s_{45} \to \infty$ with s_{51}, s_{23} finite. The B_4 functions go over to their asymptotic forms and the $_3F_2$ functions are easily shown to have the limit

$$_3F_2(a, b, c; d, e; z) \underset{b, c, e \to \infty}{=} \phi\left(a; d; \frac{bcz}{e}\right), \qquad (3.2.23)$$

where ϕ is a confluent hypergeometric function. Therefore, the double Regge limit of the first B_5 function in Fig. 15, i.e., $B_5(s_{12}, s_{23}, s_{34}, s_{45}, s_{51})$, is given by

$$B_5 \to B(-\alpha(s_{12}), -\alpha(s_{51})) \, B(-\alpha(s_{34}), \alpha(s_{51}) - \alpha(s_{23}))$$

$$\times \phi(-\alpha(s_{51}); 1 + \alpha(s_{23}) - \alpha(s_{51}); K)$$

$$+ \, B(-\alpha(s_{12}), -\alpha(s_{23})) \, B(-\alpha(s_{45}), \alpha(s_{23}) - \alpha(s_{51}))$$

$$\times \phi(-\alpha(s_{23}); 1 + \alpha(s_{51}) - \alpha(s_{23}); K), \qquad (3.2.24)$$

where

$$K = -\frac{\alpha(s_{45})\,\alpha(s_{34})}{\alpha(s_{12})}. \qquad (3.2.25)$$

The only other diagrams which contribute in this limit are 7, 8 and 9 which are just the terms necessary to provide the signature factors for the two Regge pole exchanges.

Notice that the double Regge limit of the B_5 function as we have written it looks more like the sum of two single Regge exchange diagrams rather than the direct product of two Regge poles. (cf. the models of Berger (1969) and Chan *et al.* (1967)). This is not quite the case, of course, since each of the two terms does indeed give rise to a double Regge contribution in the appropriate limit, s_{12}, s_{34}, $s_{45} \to \infty$. Nevertheless, it is true to say that B_5 does not give the naive product

$$B(-\alpha(s_{45}), -\alpha(s_{51})) \, B(-\alpha(s_{34}), +\alpha(s_{23})) \qquad (3.2.26)$$

except in the extreme limit $K \to -\infty$. We have already seen, however, in Section 2.2 that at high energies $-K \approx (s_{34} \cdot s_{45})/s_{12}$ must be bounded by the expression $m_{c_4}^2 + [\sqrt{(-s_{51})} + \sqrt{(-s_{23})}]^2$. See (2.2.5). For other values of K (3.2.24) contains corrections arising from the off-mass-shell nature of the exchanged Reggeons. Even in the limit $K \to 0$ when the central vertex functions ϕ tend to unity, this off-mass-shell dependence is still present in the arguments of the B_4 functions.

3.2a. **B$_5$ *Phenomenology*.** So far we have considered the B_5 function purely as a *mathematical* model of the five point amplitude, which indeed, is exactly what it is. As we have seen the model has been constructed so that it incorporates possible resonance poles, not only in the initial channel, but also in all three final state sub-systems. In doing so it specifies their interference effects as well as possible background contributions. Also, it is capable (at least mathematically) of describing the Regge asymptotic behaviour of the amplitude and allows a description of high-energy quasi-two-body reactions again with respect to possible background contributions. Moreover, since all reaction channels are considered on the same footing in terms of the contributing Regge trajectories, B_5 has also the important feature of crossing-symmetry. That is to say, the same set of B_5 functions in principle should describe all possible 2-body to 3-body reactions involving the same five particles. In view of these many attractive features it is evident that B_5 has a very special role as a *theoretical laboratory* for studying many of the unique problems associated with production phenomena. Precisely because it has these features, however, it is also tempting to investigate how closely it can provide a *quantitative description* of these production reactions. Unfortunately, before it can be used in a phenomenological manner. several defects of the model must be disguised if not eliminated. We list below some of the more obvious problems in identifying the mathematical dual model with a physical amplitude.

(i) *Resonance poles must be moved off the real axis.* In phenomenological fits this has normally been done simply by giving the trajectories in the energy channels imaginary parts. The size of these imaginary terms is determined by fitting to the known widths of the leading resonances on the trajectories (see, for example Petersson and Törnqvist (1969)). However, this is a rather arbitrary procedure. If done in an analytic way, so that all trajectories in all channels are complex above threshold, i.e. when $s_{ij} > (m_i + m_j)^2$, then the residues of the resonance poles will no longer be simple polynomials and ancestors are introduced. On the other hand, if ancestors are avoided by modifying only those trajectories in the energy channels then the model is no longer crossing-symmetric.

From a phenomenological point of view a more serious difficulty is that this method of moving resonance poles off the real axis means that all resonances in a resonance tower are given the same total width. Consequently it is found that daughter states in the model play a much too dominant role in the description of the five point amplitude. This point has been stressed by Berger (1971).

The problem of moving the poles off the real axis and imposing the required analytic structure to the amplitude is clearly a central difficulty for dual models whether considered theoretically or phenomenologically. Because of its importance, therefore, we shall return to this problem again in the next section and report on two possible solutions which have recently been suggested. (In the end, however, a complete solution must involve the construction of dual amplitudes—probably using the dual resonance models as "Born terms"—which ensure the S-matrix is unitary. This is a very difficult and ambitious programme which has received a great deal of attention recently. For the current status of the programme we would refer the interested reader to the review by Alessandrini (1971). From a phenomenological point of view it is clear that much work still needs to be done to make the scheme a realistic model of the physical situation.

(ii) *Duality and the Pomeron and pion-trajectories.* It is well known that the Pomeron trajectory is hard to incorporate into any dual model scheme. For instance, if the Pomeron were dual to resonances in the same way as other trajectories are considered to be, then since it does not have an exchange degenerate partner, its presence would suggest the existence of exotic resonances. Here "exotic" implies those resonances which cannot be constructed either out of three quarks in the case of a baryon, or a quark and an anti-quark in the case of a boson. To date there is very little evidence for such states. In view of this it has been suggested by Schmid (1968), Harari (1968), and Freund (1968) that Pomeron exchange, rather than be dual to resonances, must be dual to background effects. This conjecture could be consistent with a duality between the Pomeron and so-called "dual loops" or Regge–Regge cuts as suggested by Lovelace (1971). Nevertheless, in terms of simple dual *resonance* models such as B_5, it is clearly difficult to see how to incorporate these features in a phenomenological calculation. We shall, however, describe one or two attempts to do this in the next chapter.

A further point with regards to the Pomeron is that it was invented purely to account for diffractive effects in high energy scattering reactions and unlike ordinary trajectories, it has no correspondence with the exchange of known particles. That is to say, as yet we know of no particle which unequivocally lies on the Pomeron trajectory. This is not a major worry, since the established particles and resonances to date are very likely only a fraction of those actually present in nature. However, it does encourage the view that the Pomeron is a rather special and different object.

We have also included the pion trajectory into the list of phenomenological difficulties with B_5 because it too appears to be somewhat different from "normal" trajectories. Of course, in principle, the pion trajectory can be included in the B_5 framework in such a way as to provide a proper description for pion exchange both in the Regge limit and also near threshold. Nevertheless, it is different in the sense that duality is normally understood to relate the *imaginary* parts of Regge exchanges to resonances, and pion exchange gives a predominantly real exchange amplitude. In this sense it is not clear which resonances the pion should be dual to, or how they should be dual to it. Also there is little conclusive evidence for any recurrences on the pion trajectory which would establish that it has the same "universal" slope as other trajectories.

One further difficulty with pion exchange in dual models is the strong dependence of its coupling to the various spin states of external baryons. This brings us to the third phenomenological problem of B_5:

(iii) *The inclusion of external particle spins.* Since each B_5 function is a Lorentz invariant function, it would seem reasonable to equate the B_5 function, or rather some linear combination of B_5 functions, to each invariant spinor amplitude for the five point amplitude. This would mean, however, a large increase in the number of functions involved in a fit, together with a corresponding increase in the number of parameters to be adjusted. These extra parameters would indicate the size of the couplings of the Regge trajectories to the various spin-states of the external particles. Unfortunately,

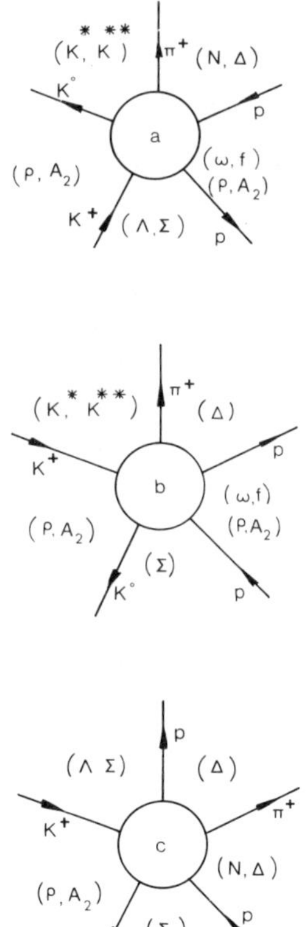

FIG. 16. The three B_5 functions for the $k^+p \to k^0\pi^+p$ system.

it is probably fair to say, even if it is theoretically plausible, such couplings can only be deduced accurately from the data if detailed information on the polarization parameters is available. In practice, however, the model is fitted only to the spin-averaged amplitudes so that such spin effects of the external particles are often masked. In view of this, in phenomenological applications, the external particle spin for the most part has been ignored, i.e. baryons are treated as if they were spin zero mesons. One overall kinematic factor is introduced to control the angular distributions of the decay particles from the first resonance on each trajectory.

(iv) *Multiplicity of terms.* Even with this simplified treatment of spin there is still, generally, a problem with the number of B_5 functions one must include to describe the production amplitude. Besides the spin states of the external particles there are also the isospin states to take into account. Chan and Paton (1969) have shown how this can be done in a systematic way; but in practice authors tend to restrict their discussion to specific charge states. Hence there is often an overlap between different isospin trajectories in the same channel, e.g. the isospin $I = \frac{3}{2}$, Δ trajectory and the $I = \frac{1}{2}$, N_α trajectory. Moreover, it is clear from two-body reactions that frequently more than one trajectory, or pair of exchange degenerate trajectories, is required to describe the data. Since each new trajectory included implies a new set of B_5 diagrams, one can see how the number of terms can rapidly become unmanageable. In this case it is necessary to make some drastic and perhaps unwarranted assumptions as to which trajectories are dominant in the various channels. Even then it is desirable to choose reactions which contain several exotic channels, so that not all of the twelve non-equivalent permutation diagrams of Fig. 15 contribute to the production amplitude.

(v) *Further difficulties.* In this category we must also include the property of dual models that require baryon trajectories to be parity doubled. Such a feature is not supported by the empirical evidence, and authors such as Green and Heimann (1969) and Bardakci and Halpern (1970) have suggested possible modifications to eliminate this doubling at least on the leading trajectory. However, such modifications are of a mathematical rather than a phenomenological nature, and lose the simplicity of the original model. Moreover, they still cannot eliminate from the model many of the daughter states which are not seen in the data, or tend to predict that some resonance states have negative widths, i.e. are ghosts. To some extent this can be over-come by the addition of satellite terms, i.e. terms defined in terms of daughter rather than parent trajectories. There is nothing in the model which says this should not be done, but it raises the rather philosophical question of where should one stop. Clearly one of the beautiful features of the dual model scheme is its relative simplicity. If now we have to keep adding more and more terms to obtain even a partial description of the physical amplitude there comes a point where we must decide to abandon the model, at least as a phenomenological tool and look for a new one.

In spite of these difficulties there have been countless analyses of two-body to three-body scattering data in terms of the Bardakci-Ruegg model over the last few years. It would be impossible to provide a complete list of references though such a list would include Peterson and Törnqvist (1969), Törnqvist (1970), Hoyer *et al.* (1970), Chan *et al.* (1970), Adjei *et al.* (1971), Hoyer *et al.* (1971), Hirshfeld and Schmidt (1970), Ross and Lyons (1970).

As an example of this work let us briefly outline the simultaneous analysis of the processes $K^+ p \to K^0 \pi^+ p$, $\pi^- p \to K^0 K^- p$ and $K^- p \to \overline{K}^0 \pi^- p$ undertaken by Chan *et al.* (1970), which is perhaps the most frequently quoted reference on B_5 phenomenology. The $Kp \to K\pi p$ system is a particularly nice system to study since only three of the twelve topologically different B_5 graph of Fig. 15 contribute. The others contain at least one exotic channel. Notice, however, that for each of these three, a complete treatment of the spin and isospin states of the systems would multiply the number of functions by factors of four and three respectively—making 36 contributions in all! For obvious reasons therefore this was not attempted and a phenomenological kinematic factor κ was chosen to ensure the correct decay of the first resonances in each channel. This choice could only be done after some further simplifying assumptions were made regarding which were the dominant trajectories. For example, at the $p\bar{p}$ vertex the isospin $I = 1, \rho, A_2, \pi$, exchanges as well as the $I = 0, \omega, f^0$ exchanges should ideally be considered. (There is no evidence for a strong Pomeron coupling in these reactions). However, from prior knowledge that ω exchange dominates $Kp \to K^*p$ reactions. Chan *et al.* included only the exchange degenerate (EXD) ω, f^0 trajectories. Similarly only the N_α nucleon trajectory of the N^* trajectories was retained, and the $\Lambda_\alpha, \Lambda_\gamma$ EXD trajectories were preferred in the $K^- p$ channel over the $\Sigma_\beta, \Sigma_\delta$ trajectories.

With these assumptions the five point amplitudes is written as

$$A(s_{12}, s_{23}, s_{34}, s_{45}, s_{51}) = \beta\kappa(B_5(a) + B_5(b) + B_5(c)) \tag{3.2.27}$$

where the three B_5 graphs a, b, c are shown in Fig. 16. The relative weights and signs of the three terms were chosen to produce the correct signature in the N_α and Δ channels. The only "free parameter" then is the overall normalization β and κ is given by

$$\kappa = \sum_{\mu\nu\rho\sigma} p_\pi^\mu p_\kappa^\nu p_\kappa^\rho p_{\bar{p}}^\sigma \tag{3.2.28}$$

which ensures the first resonances on each trajectory decay into a p-wave as it should. This factor, however, is not sufficient to reproduce the correct asymptotic behaviour for baryon Regge exchanges. Chan *et al.* remedied this defect by simply shifting those arguments of the B_5 functions correspon-ding to *exchanged* baryons by half a unit.

The resulting amplitude (3.2.27) was found to provide a generally quali-tative fit to the experimental distributions, but not a quantitative fit. Some spurious daughter resonances were produced which are not manifest in the data, and the momentum–transfer distributions were not found to be sufficiently peripheral. Also some of the angular distributions did not agree with the empirical evidence. These are failings generally found in B_5 fits and are by no means special to the analysis of Chan *et al.* On the other hand, since most authors use very similar simplifying assumptions to those of Chan and collaborators it is not completely clear how much the basic B_5 model is at fault, if at all, and how much is due to the approximations employed. To its credit, it must be stressed that the simple model of (3.2.27) very nearly accounts for the relative normalization of the $K^+ p \to K^0 \pi^+ p$ and $K^- p \to \overline{K}^0 \pi^- p$ cross-sections and fails by only a factor of two in repro-ducing the very large difference between the size of the $K^+ p$ and $\pi^- p$ reac-tions, i.e.

$$\frac{\sigma(K^+ p \to K^0 \pi^+ p)}{\sigma(\pi^- p \to K^- K^0 p)} \approx 50.$$

We would conclude therefore that the model is by no means a ridiculous description of the production amplitude in its attempt to incorporate resonance production, Regge asymptotic behaviour and crossing symmetry. No doubt better quantitative fits could be achieved by improving some of the rather ad hoc approximations so far employed, and by the addition of further terms. However, it is likely that such a programme would not lead to any further insight into the dynamics of production processes. For instance, in the analysis of Hoyer *et al.* (1970) it was found that a detailed fit to the data favoured quark-duality diagrams which are forbidden by the Rosner–Harari rules (i.e. quark-diagrams with exotic internal states) over those diagrams allowed by these rules (Rosner, 1969; Harari, 1969). If substantiated this would clearly be a most interesting result; but it has been shown by Berger (1971) that all of the discrepancies in the fits with allowed diagrams are a direct consequence of the finite width approximation of the model. It is probably wiser, therefore, to restrict one's use of the B_5 model to a discussion of fairly general, but important, features of two-body to three-body scattering data.

As an example of this let us outline the discussion of the line reversal symmetry breaking effect observed in $Kp \to K^* p$ reactions (Berger 1971, Berger and Fox, 1971). Normally the concept of exact exchange degeneracy of Regge trajectories predicts that differential cross-sections should become asymptotically equal for a pair of processes related by line reversal, and these presumably include, for example, the reactions $K^+p \to K^*p$ and $K^-p \to \bar{K}^*p$. Experimentally however, even though the initial state has exotic quantum numbers the former reaction is found to be approximately 30% larger than the latter.

By considering the three diagrams of Fig. 16, Berger and Fox were able to explain this effect in the following way. Taking the asymptotic limits of these graphs for the process $K^+p \to K^0\pi^+p$ with $s_{K^0\pi^+}$ held fixed near the K^* (890) mass M^* we obtain

$$B_5(a) \to (s_{K^+p})^{\alpha(t)} f_a(t) g_a(s_{K^0\pi^+})$$

$$B_5(b) \to e^{-i\pi\alpha(t)}(s_{K^+p})^{\alpha(t)} f_b(t) g_b(s_{K^0\pi^+}) \qquad (3.2.29)$$

$$B_5(c) \to 0$$

as $s_{K^+p} \to \infty$; where $g_a(s_{K^0\pi^+})$ in the strict zero width B_5 limit is given as

$$g_a \propto \frac{-1}{1 - \alpha(s_{K^0\pi^+})} \qquad (3.2.30)$$

but g_b is some smooth featureless background. The form factors f_a, f_b at the $p\bar{p}$ vertex are both proportional to $\Gamma(1 - \alpha(t))$ where t is the momentum transfer at this vertex. Now if we move the K^+ pole off the real axis (3.2.30) will presumably go over to some simple Breit-Wigner form

$$\frac{-\Gamma M^*}{M^{*2} - s_{K^0\pi^+} - i\Gamma M^*} \qquad (3.2.31)$$

so that the asymptotic amplitude will look like

$$A(K^+p \to K^*p) \propto s_{Kp}^{\alpha(t)} \Gamma(1 - \alpha(t)) \left\{ \frac{-\Gamma M^*}{M^{*2} - s_{K\pi} - i\Gamma M^*} + g_b \exp\left[-i\pi\alpha(t)\right] \right\}$$

$$(3.2.32)$$

For the process $K^-p \to \bar{K}^*p$ on the other hand, a similar analysis of the production amplitude (3.2.27) suggests that

$$A(K^-p \to \bar{K}^*p) \propto \exp\left[-i\pi\alpha(t)\right] s_{Kp}^{\alpha(t)} \Gamma(1 - \alpha(t)) \left\{ \frac{-\Gamma M^*}{M^{*2} - s_{K\pi} - i\Gamma M^*} + g_b \right\}$$

$$(3.2.33)$$

Notice that near the point $s_{K\pi} = M^{*2}$ the Breit-Wigner form is approximately purely imaginary so that the interference of this term with the background will be mainly due to the imaginary part of the background. For the process $K^+p \to K^*p$, (3.2.32) indicates this interference will be large and constructive, while for $K^-p \to \bar{K}^*p$ (3.2.33) suggests the interference will be negligible. Although this analysis is not quite rigorous for reasons discussed by Berger, it is sufficient to show how the breaking of line-reversal symmetry may be achieved in terms of dual models. It should be stressed that this achievement is due to the prediction of the B_5 model that there exists a non-zero asymptotic background, i.e. $B_5(b)$, which interferes with the K^* resonance production term $B_5(a)$. As a corollary, therefore, dual models would suggest there is frequently no simple way to separate resonance production effects from non-resonance terms, i.e. the arbitrary separation of a non-interfering smooth background from a resonance peak must just simply be wrong. A study of the entire Dalitz plot is necessary.

One further point regarding the qualitative use of B_5 is relevant here. If the relative sign between graphs a and b in Fig. 16 had been negative instead of positive, i.e. if we had chosen different baryon trajectories with different signature, then there would have been maximum *destructive* interference between the resonance term and background in (3.2.32). Consequently the relative sizes of the two reactions would have been reversed (e.g. as in the case of $K^0 p \rightarrow K^* p$ and $\bar{K}^0 p \rightarrow \bar{K}^* p$). Thus even a qualitative discussion of the B_5 model such as this suggests there is a link between the dominant baryon exchanges in the five point amplitude and the sign of the line-reversal symmetry breaking effects. Furthermore, it is clear that such a conclusion would probably not be apparent from a straightforward unthinking numerical analysis.

3.2c. *Numerical computation of B_5*. Although we have tried to stress the advantages of the qualitative features of the Bardakci–Ruegg model over numerical fits, it is still often desirable to be able to compute the B_5 functions explicitly. The problem of numerical evaluation has been solved by Hopkinson (1969) who noted that although the integral representation (3.2.10) only converges provided Re $\alpha(s_{ij}) < 0$ for all i, j the series representation of B_5, i.e.

$$B(x_1, x_2, x_3, x_4, x_5)$$

$$= B(x_1, x_2)B(x_4, x_5) \sum_{k=0}^{\infty} \frac{\Gamma(x_2 + k)\Gamma(z_3 + k)\Gamma(x_4 + k)\Gamma(x_1 + x_2)\Gamma(x_4 + x_5)}{\Gamma(x_2)\Gamma(z_3)\Gamma(x_4)\Gamma(x_1 + x_2 + k)\Gamma(x_4 + x_5 + k)} \times \frac{1}{k!} \qquad (3.2.34)$$

converges provided

$$\mathrm{Re}\,(1 + (x_1 + x_2) + (x_4 + x_5) - x_2 - z_3 - x_4) > 1,$$

which is, to say Re$(x_3) > 0$. Here we have written $x_k = -\alpha(s_{ij})$ and $z_3 = x_1 - x_3 + x_5$. Similarly, z_i will be defined by cyclic permutation of the appropriate indices. Note that the series (3.2.34) is equivalent to the hyper-geometric form (3.2.14) after a trival cyclic re-arrangement of the arguments.

Because of this much larger region of convergence the series representation for B_5 therefore would seem a better starting point for a numerical calculation of the function than the integral representation. However, even this region is not large enough to accommodate all possible values of the arguments which may be required. To go beyond this region Hopkinson made repeated use of the recursion relations of the hypergeometric functions to relate the B_5 function of any arguments to the series representations (3.2.34) which do converge. Specifically his method of calculating B_5 for any general complex arguments x_1, x_2, x_3, x_4, x_5 uses the relation

$$B_5(x_1, x_2, x_3, x_4, x_5) = A_n^1 B_5(x_1 + n, x_2, x_3 + n, x_4, x_5)$$
$$+ A_n^2 B_5(x_1 + n, x_2, x_3 + n + 1, x_4, x_5), \qquad (3.2.35)$$

where the integer n is chosen such that $\mathrm{Re}(x_3) + n$ is not only positive but also provides the most effective convergence of the series representation (3.2.34). The coefficients A_n^1 and A_n^2 are given by the recursion relations

$$A_{k+1}^1 = A_k^1 \frac{(z_4 + k)(x_1 + k) + z_1(z_3 - 1)}{(x_1 + k)(x_3 + k)} + A_k^2 \frac{(z_3 - 1)}{(x_1 + k)}, \qquad A_0^1 = 1$$

$$\qquad (3.2.36)$$

$$A_{k+1}^2 = A_k^1 \frac{z_1(z_5 + k + 1)}{(x_1 + k)(x_3 + k)} + A_k^2 \frac{(z_5 + k + 1)}{(x_1 + k)}, \qquad A_0^2 = 0$$

and since the series does converge, it is possible also to give estimates of the errors involved in truncating the series after a given number of terms.

The fact that it is possible to evalue the B_5 function in this manner sets it apart from other dual models for higher order production processes, i.e. the B_n functions defined in the next chapter. For such functions there appears to be no *simple* way to enlarge the obvious region of convergence given by the integral representation, i.e. $\mathrm{Re}\,\alpha(s_{ij}) < 0$. Part of the problem is that the corresponding series expansion is in fact a multiple expansion which is difficult to handle analytically. Thus we know of no practical way at present to calculate numerically these higher order dual models. Consequently any phenomenological study of such models must be confined to a qualitative discussion in terms of the various asymptotic limits of the function.

3.2c. *Finite width dual resonance models*. Finally in this section on B_5 let us make a few remarks about some variations of the B_4 and B_5 models which allow the resonance poles to be moved off the real axis—i.e. given finite widths—without the introduction of ancestors. These models all have the same general integral form

$$\tilde{B}_4 = \int_0^1 du\, u^{-\alpha(s,\,u)}\,(1-u)^{-\alpha(t,\,1-u)}, \tag{3.2.37}$$

for the four point function; and

$$\tilde{B}_5 = \int_0^1 \cdots \int_0^1 du_1\, du_2\, du_3\, du_4\, du_5\, \delta(u_2 + u_1 u_3 + 1)\delta(u_3 + u_2 u_4 - 1)$$

$$\times\, \delta(u_5 + u_1 u_4 - 1) \sum_{i=1}^5 \{u_i^{-\alpha(s_i,\,i+1,\,u_i)}\} \tag{3.2.38}$$

for the five point function; where the generalized trajectory functions $\alpha(s, u)$ are continuous functions of u in $[0, 1]$ with the property

$$\alpha(s, 0) = \alpha(s) \equiv a + \alpha's + \Delta\alpha(s) \tag{3.2.39}$$

and

$$\Delta\alpha(s) = s/\pi \int_{s_0}^\infty \frac{ds'\, \text{Im}\,\alpha(s')}{s'(s'-s)}. \tag{3.2.40}$$

In the dual model proposed by Suzuki (1969), this property was obtained by choosing

$$\alpha(s, u) = \alpha(s) - f(u)\Delta(\alpha(s)) \tag{3.2.41}$$

where $f(u)$ is a van der Corput neutralizer defined by

$$f(u) = g(u)/g(1)$$

$$g(u) = \int_0^u dy\, (-\log y)^{\log y}\, [-\log(1-y)]^{\log(1-y)} \tag{3.2.42}$$

This neutralizer is defined so that

$$\left(\frac{d}{du}\right)^n f(u) = 0 \qquad \text{for } u = 0, 1\,; n = 1, 2, 3\dots$$

and has branch-points at $u = 0$ and $u = 1$ but is otherwise free of singularities. Hence $\alpha(s, u)$ also has these properties, and since

$$f(0) = 0, \qquad f(1) = 1$$

(3.2.39) is automatically satisfied, and $\alpha(s, 1)$ is just the linear part of trajectory, i.e.

$$\alpha(s, 1) = \alpha + \alpha's. \tag{3.2.43}$$

These properties ensure that, although the resonance poles are complex, their residues are still only fine polynomials in the crossed channel variables. Moreover, it has been shown (Suzuki 1969, Atkinson *et al.* 1972) that this model has the correct Regge asymptotic behaviour in the whole of the complex s-plane, except possibly along the positive real axis. Notice, there is still some freedom in the Suzuki parameterization. The imaginary parts of the trajectory functions are in general arbitrary as is the choice of neutralizer to some extent. Since the aim of the finite width modifications is to approximate the effects of unitarity on the scattering amplitudes this is probably not a bad thing.

A more ambitious dual model scheme has recently been proposed by Cohen-Tannondji *et al.* (1971). This scheme not only moves the poles off the real axis without the introduction of ancestors, but is also capable of giving the amplitudes a curved double-spectral-function boundary which can be made to coincide with the Mandelstam boundary. Specifically they choose generalized trajectory functions of the form

$$\alpha(s, u) = \alpha\left(s(1 - u)\right) - \log f\left(s(1 - u)\right)/\log u \qquad (3.2.44)$$

where $\alpha(y)$ is a trajectory satisfying (3.2.39), (3.2.40), and $f(y)$ is a suitable function which decreases faster than any inverse power as $|y| \to \infty$ in all directions. For example, we can take

$$f(y) = \exp\left[-\beta(s_0 - y)^{\frac{1}{4}}\right]$$

However (3.2.43) does not have the property that $\alpha(s, 1)$ is equal to $a + \alpha's$, nor does the four point function reduce to the B_4 function of (3.2.1) in the limit of $\mathrm{Im}\,\alpha(s) = 0$. On the other hand, the model of Cohen-Tannondji *et al.* has discontinuities arising from $\mathrm{Im}\,\alpha$ or $\mathrm{Im}\,f$ for $s > 4m^2$, $t < 4m^2$. This discontinuity itself will have a non-vanishing discontinuity in t when $t(1 - u) > 4m^2$ with $u > 4m^2/s$, or with the boundary curve

$$(1 - 4m^2/s)t \geqslant 4m^2$$

which is the conventional boundary of the double-spectral-function region for scalar meson scattering. Unfortunately, it has been shown by Atkinson *et al.* (1972) that this model does not have Regge asymptotic behaviour in the sector $|\arg s| < 30°$ unless the real part of the trajectory $\alpha(s)$ increases slower than s as $s \to \infty$. While this is quite possible it is not supported by the approximately linear form for the trajectories indicated by the known resonances. However, there is no reason to believe this must continue to be the case as we go to higher energies.

Several other finite width dual resonance models have also been proposed. See for instance Gaskell and Contogouris (1972), Bugrij *et al.* (1971) (1972), Schmidt (1971), Franzen (1971), Adjei *et al.* (1971b). For the most part they lie somewhere between the two models we have outlined above. None of them, however, are completely free from difficulties with regard to the Regge asymptotic behaviour in the right half of the complex s-plane. For instance, the models of Gaskell and Contogouris (1972) and Bugrij *et al.* (1971), which are particularly noteworthy because they also produce curved double-spectral-function boundaries, only have Regge asymptotic behaviour provided $|t| < s_0$ where s_0 is the physical threshold.

In conclusion therefore we can summarize the situation as follows. If we require that $\mathrm{Re}\,\alpha(s)/s = O(1)$ as $s \to \infty$ together with Mandalstam analyticity for the scattering amplitude there is, as yet, no finite width model which does not have some difficulty in producing Regge behaviour in the right half of the complex s-plane for all values of t. Nevertheless, it still seems possible that such a model can be constructed. In view of the important role these models have as theoretical laboratories this point seems well worthy of further study. As we have stressed throughout this section, this is particularly true for the five point function with its special implications for overlapping resonances and the problems of multiparticle production in general. Such a model may even be able to give us some insight into the analytic structure of the five point amplitude which, as we shall discuss next, is not a simple matter.

3.3. Analyticity and dispersion relations

It is worth reiterating that the purpose of any phenomenological study of particle scattering data is to try to deduce from it the structure of the scattering amplitudes. In this way the kinematic effects associated with phase-space can be factored out and the dynamical content of the reaction may be displayed. As we have already noted, however, most data on production phenomena represent partially integrated productions of scattering amplitudes. In this case, therefore, it will often be difficult to accurately deduce from them the individual scattering amplitudes. It may very well be possible that several rather different amplitudes can fit the same limited amount of data.

One way to overcome at least some of these difficulties is to ensure that the amplitudes have as many of the properties, such as unitarity, analyticity and crossing, of S-matrix theory as possible. In the last section we discussed dual resonance models which have the important property of crossing-symmetry; but only in some finite width versions has any attempt been made to include the known out-plane analytic structure of the amplitude. From two-body scattering, however, it has been found that by exploiting also this analyticity requirement, (i.e. that scattering amplitudes are the values of analytic functions on real boundaries), in the form of dispersion relations strong constraints can often be placed on the form of the amplitudes. Although these dispersion relations can be written in several ways they are generally used to relate integrals over the imaginary part of the amplitude either to the real part evaluated at some particular point, or to some low-energy constant such as a scattering length or a coupling constant. Alternatively, together with the assumption of duality they can be written in the form of finite energy sum-rules (FESR's) which relate the low energy amplitudes to some known functions of the high energy Regge parameters. It is noteworthy that most two-body phase-shift analyses these days also incorporate dispersion relation constraints in order to rule out some of the solutions which are otherwise compatible with the data.

In the same way it is likely that much more insight into the form of the production amplitudes can be achieved by a similar dispersion relation treatment. Before this can be attempted, however, it is obviously necessary to know the analytic structure of the amplitudes. For two-body scattering amplitudes these analytic properties have been extensively studied and are nowadays well understood and well founded on the general principles of local quantum field theory. Unfortunately, in the case of production amplitudes the situation is much more complicated and their analytic structure is not yet fully understood, although recently some significant work has been done by Alvarez Estrada (1971) on the five-point function, and by Bros *et al.* (1972) on the five- and n-point functions. We would refer the interested reader to these papers for a summary of the results so far obtained.

An alternative, though perhaps less rigorous, way of discussing the analyticity of amplitudes is to use perturbation theory based upon orders of Feynman graphs. This approach has been a useful tool in considering many of the analyticity problems in high energy physics in the past few years. See, for example, Eden *et al.* (1966). For our purposes which are based more on phenomenological considerations than on complete rigour these Feynman graph techniques are probably more than adequate. None of the results from quantum field theory so far invalidate those of perturbation theory and there is as yet no reason to suspect they might. With this assumption we can now outline some of the difficulties in deriving dispersion relations specifically for the 5-point amplitude.

The first consideration is to decide in terms of which variables one is going to consider the amplitude as a complex function. By analogy with two-body reactions let us express the production amplitudes as functions of the Lorentz scalars s, s_{12}, s_{23}, t_1, u_3 which are the generalization of the Mandelstam variables s and t. (Here we return to our original notation instead of the s_{ij}, $i = 1, \ldots, 5$ of the last section). Whatever other singularities the amplitudes possess they will have the normal threshold branch point singularities corresponding to the possible intermediate states in each channel, i.e. where $s > (m_a + m_b)^2$, $s_{12} > (m_1 + m_2)^2$, and also $s > (m_1 + m_2 + m_3)^2$

etc. However, if also in analogy with two-body dispersion relations we try to disperse in only one of these variables, e.g. s_{23}, keeping s, s_{12}, t_1 and u_3 fixed it is well known that when these are in the physical region the production amplitude has right and left hand cuts which overlap. The reason for this is that while s, s_{12}, t_1 and u_3 may be fixed there are five other non-independent Lorentz scalars s_{13}, t_3, u_1, t_2, u_2 of which s_{13}, u_1 and u_2 must also vary with s_{23}: cf (2.1.3). In particular the amplitude $T(s_{23}, s, t_1, u_3 s_{12}; s_{13}, t_3, u_1, t_2, u_2)$ will have one branch cut corresponding to physical values of s_{23}, i.e. from $(m_2 + m_3)^2$ to infinity, and a second cut corresponding to physical values of s_{13}. Thus assuming no other singularities we could write

$$T(s_{23}\ldots; s_{13}\ldots) = \frac{1}{2\pi i} \int_{(m_2+m_3)^2}^{\infty} ds'_{23} \frac{\Delta_1 T(s'_{23},\ldots; s'_{13}\ldots)}{s'_{23} - s_{23}}$$

$$+ \frac{1}{2\pi i} \int_{(m_1+m_3)^2}^{\infty} ds'_{13} \frac{\Delta_2 T(s'_{23},\ldots; s'_{13}\ldots)}{s'_{13} - s_{13}}. \tag{3.3.1}$$

where $\Delta_1 T$, $\Delta_2 T$ represent the discontinuities across the s_{23} and s_{13} cuts respectively. However, since

$$s_{13} = s - s_{23} - s_{12} + m_1^2 + m_2^2 + m_3^2$$

from energy-momentum conservation, the second integral in (3.3.1) can be rewritten in terms of a left-hand cut in s'_{23} as

$$\int_{-\infty}^{\bar{s}_{23}} ds'_{23} \frac{\Delta_2 T(s'_{23},\ldots; s'_{13}\ldots)}{s'_{23} - s_{23}}. \tag{3.3.2}$$

Here $\bar{s}_{23} = s - s_{12} + m_1^2 + m_2^2 + m_3^2 - (m_1 + m_3)^2$ which in the physical region with $s > (m_1 + m_2 + m_3)^2$ and $s_{12} > (m_1 + m_2)^2$ implies $\bar{s}_{23} > (m_2 + m_3)^2$. That is to say, the right and left hand cuts in s_{23} overlap. Of course, it may well be possible to distort one or both of these integral contours away from the real axis to avoid this overlapping but then it will be necessary to consider the scattering amplitude at complex values of its variables.

A second way in which complex singularities may arise can be seen by considering the triangle graph of Fig. 17. Using the reduced variable $z_{\alpha\beta}$ defined by

$$s_{\alpha\beta} = (p_\alpha + p_\beta)^2 = m_\alpha^2 + m_\beta^2 + 2m_\alpha m_\beta z_{\alpha\beta}, \tag{3.3.3}$$

where p_α, p_β are the four-momenta of particles α and β, this graph produces a line of singularities given by the equation

$$\det|z_{\alpha\beta}| = 0. \tag{3.3.4}$$

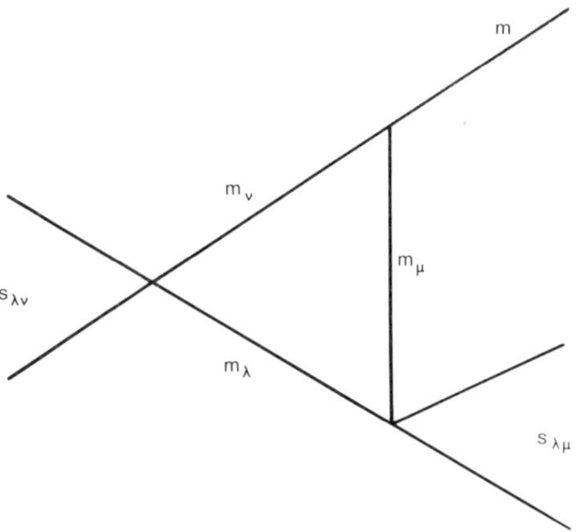

FIG. 17. Typical triangle diagram for the five-point amplitude.

where $z_{\alpha\beta} = z_{\beta\alpha}$, $z_{\alpha\alpha} = 1$. (Landshoff and Trieman, 1961). Equation (3.3.4) defines the ellipse

with

$$z_{\lambda\mu}^2 + z_{\lambda\nu}^2 + 2m_x z_{\lambda\mu} z_{\lambda\nu} + m_x^2 - 1 = 0 \qquad (3.3.4')$$

$$m_x^2 = (m^2 - m_\mu^2 - m_\nu^2)/(2m_\mu m_\nu)$$

which as shown in Fig. 18 is tangential to the normal thresholds $z_{\lambda\mu} = 1$, $z_{\lambda\nu} = 1$. When $s_{\lambda\mu}$ (and $z_{\lambda\mu}$) is kept fixed and below its normal threshold value the triangle diagram produces two real singularities given by the intersection of the line $z_{\lambda\mu} = $ constant with the ellipse (though physical singularities correspond only to that part of the curve AB which lies between the normal thresholds). However, as the value of $z_{\lambda\mu}$ is increased, the line of fixed $z_{\lambda\mu}$ moves to the right in Fig. 18 until it is tangential to the ellipse. At this point the two singularities coincide and above threshold ($z_{\lambda\mu} > 1$) they become a complex conjugate pair (one physical, one unphysical) the physical one giving rise to a complex branch cut in the physical $s_{\lambda\nu}$ plane. Also it will generally be possible to draw another triangle graph which depends on one of the dependent variables, e.g. t_γ such that $s_{\lambda\nu} = -t_\gamma + c$ where c is some fixed quantity of the other invariants. Thus complex singularities will occur in both halves of the $s_{\lambda\nu}$ plane and a dispersion relation in terms of one Lorentz scalar must always involve a complex contour. The situation is clearly going to become even more complicated for higher order diagrams. The $s_{\lambda\nu}$ plane is likely to become a veritable jungle of complex singularities if one were to consider all orders of Feynman graphs in the same manner.

In spite of these difficulties some attempts have been made to write down

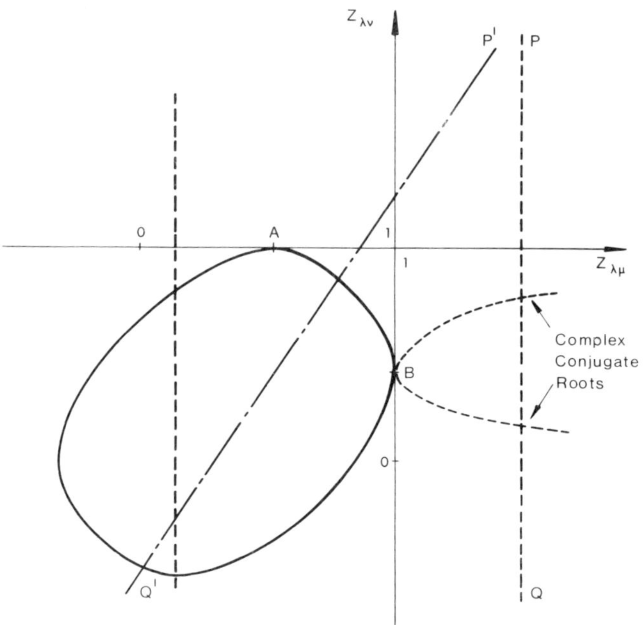

FIG. 18. Singularities of the triangle diagram in the $z_{\lambda\mu}$, $z_{\lambda\nu}$ plane.

dispersion relations in terms of one of the five independent Lorentz scalars which define the five point amplitude. However, in each case it is true to say that some simplifying assumptions have had to be made and many, if not all, of these complex branch points have been neglected. See, for example, Blankenbecler (1961), Ball *et al.* (1962).

In order to see how this problem of complex singularities might be overcome at least formally let us consider more closely the singularity diagram in Fig. 18. So long as $z_{\lambda\mu}$ is below the normal threshold $z_{\lambda\mu} < 1$ there is no difficulty; we have only real singularities given by the intersection of the line PQ with the ellipse. The difficulties arise when $z_{\lambda\mu} > 1$ and *PQ* lies completely outside the singularity contour so that its points of intersection must necessarily be complex. However, if the dispersion relation is to be of any practical use at least some of the $z_{\alpha\beta}$ must be above their threshold values at some point. From Fig. 18 we can see that this can be achieved without introducing complex singularities if we can tilt the line *PQ* onto the line *P'Q'* so that there is some point on it which lies inside the ellipse and hence *P'Q'* cuts the ellipse at two real points. In this case *both* $z_{\lambda\mu}$ and $z_{\lambda\nu}$ vary, rather keeping one or the other fixed; the variation being parameterized in terms of points on the line $P'Q'$. Of course, to eliminate all complex singularities we must consider all possible triangle graphs and thus generalize the line $P'Q'$ to a line through the whole five-dimensional space of Lorentz scalars. Now instead of having some point inside the $(z_{\lambda\mu}, z_{\lambda\nu})$ ellipse this line must also pass through the intersection of all ellipses arising from all these triangle diagrams. Provided this intersection is free from any singularities arising from higher order diagrams, we can write a parametric dispersion relation along this line which has only real singularities. Of course, the amplitude in principle will still have singularities corresponding to these triangle and higher order diagrams, but in this case they will all lie along the real axis.

Wu (1961) and Boyling (1963) have shown that such a singularity-free intersection does indeed exist *to all orders of perturbation theory* at least, for certain masses of the external particles, and is disjoint from the physical regions. Utilizing this information Branson, Landshoff and Taylor (1963) were able to write a dispersion relation for a production amplitude with five equal mass particles in the variable ξ where

$$s = 9m^2 + \xi$$
$$s_{ij} = 4m^2 + \xi/3$$
$$t_i = -m^2 - \xi/3 \qquad\qquad (3.3.5)$$
$$u_i = -m^2 - \xi/3$$

and m is the particle mass. The parametric line characterized by ξ was chosen so that it passed through the singularity free Wu region and also entered the

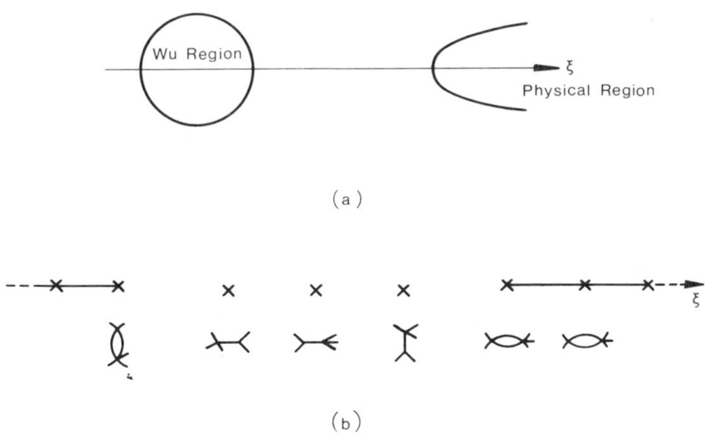

FIG. 19. (a) Diagrammatic representation of the ξ-line through the space of Lorentz scalars. (b) Typical singularities of the production amplitude as a function of ξ.

physical region at the point $\xi = 0$ as shown in Fig. 19a. However, the draw-back to this parameterization is that while it lies in the physical region for $\xi > 0$, the linear relations (3.3.5) necessarily mean that it must be in a very unphysical region for negative ξ, i.e. with some momentum transfers tending to $+\infty$. Therefore the left hand singularities, indicated in Fig. 19b, must involve values of the scattering amplitude at very unphysical points. Nevertheless, it is apparent that (3.3.5) is only one of many possible parametric forms for the invariants. In the following section we shall describe an alternative parameterization which we believe may be of some practical use in the study of the two-body to-three-body production amplitude.

3.3a. *Finite energy sum rules for five point amplitudes.* One of the successes of the dispersion relation approach to two-body reactions has been the derivation of FESR's and their definition of duality. However, because of the difficulties involved with dispersion relations for production processes the interpretation of duality for production amplitudes has rested solely on the formulation of mathematical models such as B_5. As we have seen, such models have to be modified somewhat to be useful in a practical comparison with the experimental data; and it is not clear whether their failure to account for the quantitative features of the data is due to these approximations or is a consequence of the dual models themselves. For this reason it would seem particularly important to be able to derive FESR's for five point amplitudes in which it is possible to examine the experimental information more closely. By this means we may be able to test whether the production data do or do not satisfy the resulting definition of duality. An interesting aspect of this approach, for instance, would be to investigate the dual nature of the Pomeron. According to the Harari–Freund conjecture the Pomeron is dual not to resonances but to background effects. For two-body reactions the definition of this background is at least intuitively clear even if in practice it is sometimes difficult to disentangle from the resonance part of the amplitude. For production processes, on the other hand, much of the background to the resonances in a final-state sub-energy channel comes from the effects (reflections) of *resonances* in other final state channels and the possibility of a duality between background and Pomeron becomes a rather more delicate question. This is the typical sort of problem which, if it can be resolved at all, can probably only be resolved by a detailed study of FESR's.

In order to develop a practical approach to dispersion relations and FESR's for five point amplitudes, let us adopt a middle course between the single invariant and parametric dispersion relations and write a semi-parametric form which leaves certain of the Lorentz scalars fixed while parameterizing others so that we do not encounter complex singularities or overlapping cuts. To be specific, let us consider the process $\pi_a N_b \to \pi_1 \pi_2 N_3$

and write

$$s = \bar{s} + z[(2\mu + m)^2 - \bar{s}]$$
$$s_{12} = \bar{s}_{12} + z[4\mu^2 - \bar{s}_{12}]$$
$$s_{23} = \bar{s}_{23} \tag{3.3.6}$$
$$t_1 = \bar{t}_1$$
$$u_3 = \bar{u}_3,$$

where m and μ are the mass of the nucleon and pion respectively. This form is chosen so that it has the physical region branch cuts in s and s_{12} for $z > 1$. Also with $\bar{s}$ and $\bar{s}_{12}$ given by

$$\bar{s} = \tfrac{1}{2}(m^2 + 2\mu^2 + \bar{s}_{23} - \bar{t}_1)$$
$$\bar{s}_{12} = \tfrac{1}{2}(3\mu^2 - \bar{t}_1 + \bar{u}_3), \tag{3.3.7}$$

it is a simple matter to show, from (3.3.6) or (2.1.3) that the crossed process $\bar{\pi}_1 N_b \to \bar{\pi}_a \pi_2 N_3$ gives rise to physical branch cuts in u_1 and t_2 at $z < -1$. However, it must be noted that the point $z = 0$, which we take to be the necessary point inside the Wu region corresponds to the point $(\bar{s}, \bar{s}_{23}, \bar{s}_{12}, \bar{t}_1, \bar{u}_3)$ in our space of invariants. Thus the value of $s_{23} = \bar{s}_{23}$ must always be slightly unphysical, i.e. below the $\pi_2 + N_3$ threshold. This is the price we apparently have to pay to obtain a left hand cut contribution which is not determined by *extremely* unphysical crossed channel processes, as in the dispersion relations of Branson *et al*. Although this is not very satisfactory let us point out that the extrapolation of the physical scattering amplitude to the unphysical point $s_{23} = \bar{s}_{23}$ can be very small, and indeed, is much less than the extrapolations often performed in two-body FESR's (e.g. where even positive momentum transfers are sometimes considered).

It is worth stressing that by the parameterization of (3.3.6), a dispersion relation in the variable z can be defined which is truly a dispersion relation for particle production amplitudes. One still has to consider final state effects between $\pi_1 N_3$ and $\pi_1 \pi_2$ *and thus it is not equivalent to a quasi two-body process*, even though one final state sub-energy is held fixed.

Figure 20a shows diagrammatically the z-line defined by (3.3.6) through the space of invariants such that it is inside the Wu region for the $z = 0$ and approaches asymptotically the physical regions for $\pi_a N_b \to \pi_1 \pi_2 N_3$ and the crossed process $\bar{\pi}_1 N_b \to \bar{\pi}_a \pi_2 N_3$. In Fig. 20b we indicate the corresponding cut structure of the production amplitudes $A(s, s_{jk}, t_i, u_i)$ considered as a function of z—i.e., $f(z)$. It should be noted that asymptotically the kinematics suggest that these amplitudes will be dominated by a few single-Regge exchange diagrams. This again differs from the dispersion relation of Branson

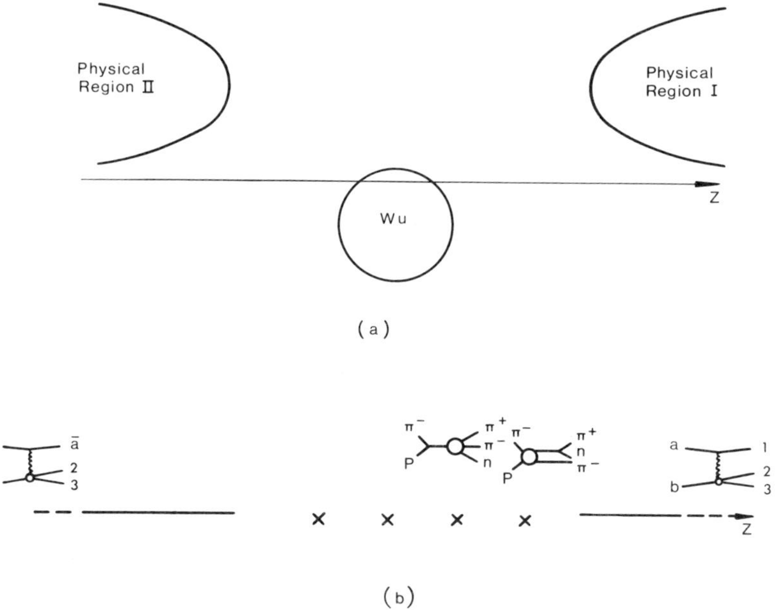

FIG. 20. (a) Diagrammatic representation of the z-line through the space of Lorentz scalars. (b) Pole terms and asymptotic limits for the production amplitude as a function of z.

et al. where asymptotically all the invariants tend to $\pm\infty$ so that no simple Regge parameterization is applicable. The semi-parametric dispersion relation or FESR follows by drawing a contour round these singularities, closed by two semi-circles at infinity and applying Cauchy's theorem. For instance, in terms of the modified amplitude $f(z) - f(z)_{\text{asymp}}$ where $f(z)_{\text{asymp}}$ is the asymptotic amplitude given in terms of the single Regge exchange graphs shown in Fig. 20b, i.e.,

$$f_{\text{asymp}} = \sum_i \gamma_i(t_1 u_3 s_{23}) \frac{1 + \tau e^{-i\pi\alpha_i}}{\sin \pi\alpha_i} (a + bz)^{\alpha_i(t_1)} \tag{3.3.8}$$

the production amplitude FESR can be written as

$$\int_{z_0}^{\Lambda} dz\, z^m \operatorname{Im} f(z) = \sum_i \int_{z_0}^{\Lambda} dz\, \gamma_i(t_1, u_3, s_{23})(a + bz)^{\alpha_i(t_1)} z^m \tag{3.3.9}$$

where it has been assumed that $f(z) \approx f_{\text{asymp}}(z)$ for $|z| \geqslant \Lambda$ and z_0 is the lower limit of the amplitude singularities in the variable z. Also $m = 2n$ or $2n + 1$ depending on whether f is an odd or even function of z. This, of course, depends on the spin and isospin of the system. However, provided the $\pi_a N_b \to \pi_1 \pi_2 N_3$ amplitude is symmetric in the isospin indices under $\pi_a \leftrightarrow \pi_1$ crossing its expansion into invariant amplitudes $\bar{A}, \bar{B}, \bar{C}, \bar{D}$, i.e.,

$$T = \bar{u}(q_3)\,\gamma_5[\bar{A} + \gamma \cdot k^- \bar{B} + \gamma \cdot k^+ \bar{C} + [\gamma \cdot r^+, \gamma \cdot r^-]_- \bar{D}]\,u(p_b) \quad (3.3.10)$$

(where γ_5, γ_μ and $\bar{u}$, u are the usual Dirac matrices and four-spinors respectively, and $k^- = p_a + q_1$, $k^+ = q_2$, $r^+ = k^- + k^+$, $r^- = k^- - k^+$) implies that $\bar{A}$ and $\bar{C}$ are also symmetric and $\bar{B}$ and $\bar{D}$ are antisymmetric. Notice, there is some freedom in the choice of crossing symmetric energy variable $(a + bz)$ used in (3.3.8). For two-body reactions the usual variable is taken to be $v = (s - u)/s_0$ where s_0 is typically 1 GeV^2. In the present case it is not obvious whether the corresponding variable should be $(s_{12} - t_2)/s_0$ or $(s - u_1)/s_0$. We have therefore covered both possibilities by the parameterization $a + bz$.

In order to utilize the sum-rules (3.3.9) to study the structure of the high energy amplitudes it is necessary to know in some detail the low-energy production amplitude. Thus initially it is necessary to analyse the low-energy data in terms of a partial wave expansion. This expansion can then be extrapolated to the line defined by the parameterization (3.3.6). In a recent paper (Humble 1973) a preliminary study of the production data by this means was reported. The results, although not conclusive, are quite consistent with the concept of duality. No very strong statements can be made, however, because of the necessarily approximate nature of the partial waves employed. Nevertheless we believe these dispersion relation techniques can be an extremely important development in the understanding of production dynamics. To this end let us stress the need for detailed, and accurate, partial wave analyses. Also, while there are still obvious drawbacks to the linear parameterization (3.3.6) it may well be possible, with some effort, to derive other parameterizations which define dispersion relations wholly in terms of *physical* amplitudes.

Here as throughout this chapter we have restricted ourselves to consideration of the five point amplitudes. For higher order production amplitudes the analytic structure is undoubtedly going to be even more complicated because of the increased number of variables. The most obvious complication is the quadratic relation, from energy-momentum conservation, between the Lorentz scalars of the system. Thus kinematic singularities are introduced in the complex space of invariants which mean that no singularity-free Wu region exists. We know of no similar method at present, therefore, for writing dispersion relations for general n-body production amplitudes.

4

High Energy Exclusive Reactions

4.1. Longitudinal phase-space plot

We now come to the question of particle production at high energy. Here, in many ways, the problems of presentation and interpretation of the data are even more acute than at low energies. We can no longer assume dominance of the production amplitude by a few partial waves, but must consider it instead truly as a function of all of its independent variables. Moreover, while resonance production gives a sizeable contribution to the cross sections in the appropriate regions of phase-space, there are still, apparently, a large number of multi-particle events that do not explicitly involve final state resonances. In order to analyse these genuine multi-particle production reactions we must be able to present the data in such a way as to provide a critical test of any dynamical assumption we may make. With the conventional single or double differential distributions in eqns (2.3.8) to (2.3.10) for 3-body processes this is not always possible. The two or three integrations that are required must in general smear out some of the effects of the dynamics. For n-particle final states the situation will become progressively worse as the number of integrations increases.

Of course, if we know beforehand what effect we are looking for we may be able to choose a distribution which is sensitive to this effect. Unfortunately the integrations cover such a wide range of values of the invariants that such techniques are not often conclusive. However, the idea of using what knowledge we have of particle dynamics to simplify the analysis is very appealing. In the past few years Van Hove (1969) has utilized this approach to suggest a method of presenting production data on a longitudinal phase space (LPS) plot which does not involve the combination of events with widely different 3-momenta. That is to say, while we must still integrate over some of the variables (in order to reduce the dimensionality of the problem) these variables are physically only important over a small range.

The usefulness of the LPS plot rests on the empirical observation that in any multi-body production reaction at high energies the transverse momentum of each of the outgoing particles is small, i.e. its momentum component at right angles to the incident beam direction usually averages around 0·35 GeV/c.

This situation is shown diagrammatically for a 3-body reaction in Fig. 21.

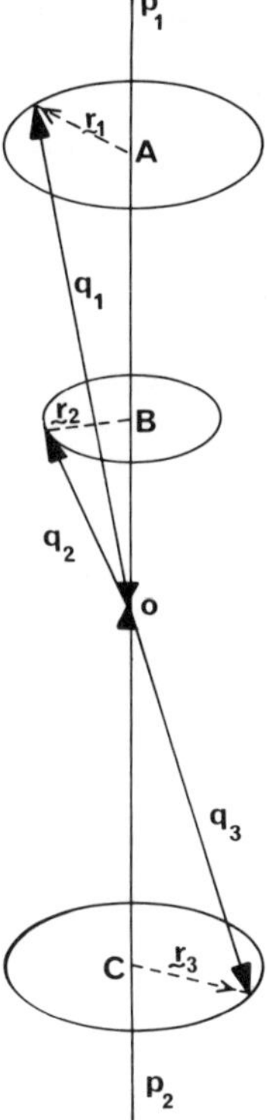

FIG. 21. Single particle production in the c.m. system.

The whole of phase-space is represented by the three spheres described by $\mathbf{q}_1$, $\mathbf{q}_2$, $\mathbf{q}_3$ restricted only by the requirement of energy momentum conservation. However, as indicated, all the particles are produced with only a small transverse momentum, so that the mapping of these events at high energy onto the line OP_1P_2 will introduce only a small distortion of the dynamics while greatly simplifying the dimensionality problems of phase-space.

Let us consider the reaction $a + b \rightarrow c_1 \ldots + c_n$ with a total c.m. energy W. We denote the longitudinal momentum of c_i (i.e. its momentum parallel to the incident particle a) by q_i and its transverse momentum by $\mathbf{r}_i$ such that $|\mathbf{r}_i| = r_i$. Then the c.m. energy of each particle c_i of mass m_i is given by

$$w_i = (m_i^2 + r_i^2 + q_i^2)^{\frac{1}{2}} = (m_i'^2 + q_i^2), \tag{4.1.1}$$

where $m_i'^2 = m_i^2 + r_i^2$ can be thought of as its effective mass for longitudinal scattering. Imposing energy–momentum conservation we have

$$\sum_{i=1}^{n} q_i = 0 \tag{4.1.2}$$

$$\sum_{i=1}^{n} \mathbf{r}_i = 0 \tag{4.1.3}$$

$$\sum_{i=1}^{n} w_i = W. \tag{4.1.4}$$

Thus if we are going to represent each n-body event by the longitudinal components of its produced particles, eqn (4.1.2) indicates that such events lie on an $(n-1)$ hypersurface L_{n-1} which is called the longitudinal phase-space. The dimension is $n-1$ because of this condition connecting the n variables q_i. However, for any given values of r_i, and hence of the effective masses m_i', eqn (4.1.4) provides a second condition on the q_i. Therefore for fixed r_i the longitudinal phase space reduces to an $n-2$ hypersurface K_{n-2} defined in L_{n-1} by the equation

$$\sum_{i=1}^{n} (m_i'^2 + q_i^2)^{\frac{1}{2}} = W, \tag{4.1.5}$$

which for all $q_i^2 \gg m_i'^2$ becomes

$$\sum_{1}^{n} |q_i| = W. \tag{4.1.6}$$

Equation (4.1.6) defines a polyhedron H_{n-2} in L_{n-1}, its "faces" corresponding to all $q_i \neq 0$. Its "sides" and "vertices" correspond to some number of $q_i = 0$, so that it is on these sides and vertices that the LPS for fixed r_i deviates most from the polyhedron H_{n-2}; i.e. where eqn (4.1.6) is no longer a good approximation to eqn (4.1.5).

All this sounds a little technical. As usual, however, things should become clearer if we consider a specific example. For instance, for a 3-body reaction, $n = 3$, the LPS is just the plane defined by q_1 and q_2. ($q_3 = -q_1 - q_2$ by eqn (4.1.2).) Imposing also energy conservation for the limiting case of all $r_i = 0$, i.e.

$$\sum_i (m_i^2 + q_i^2)^{\frac{1}{2}} = W \tag{4.1.7}$$

we obtain the rounded-off hexagon H_I shown in Fig. 22. Each physical

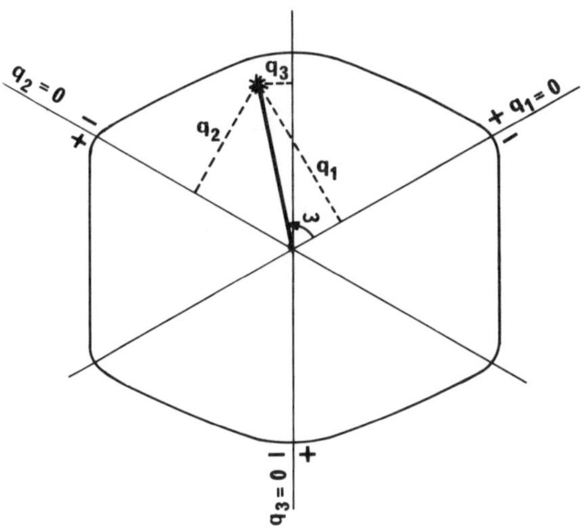

FIG. 22. LPS hexagon for three-body process.

event will have $r_i \geqslant 0$ and so lie inside this hexagon. However, since the r_i are physically known to be small the great majority of events will lie near to this boundary. Thus it should be very reasonable to consider the 3-body cross-section as a distribution only in the angle ω (which is defined in Fig. 22), i.e. as a one-dimensional distribution.

The longitudinal momenta q_i are given in terms of this angle by

$$q_1 = \sqrt{\tfrac{2}{3}}q \sin \omega$$
$$q_2 = \sqrt{\tfrac{2}{3}}q[-\tfrac{1}{2}\sin \omega - \tfrac{1}{2}\sqrt{3}\cos \omega] \tag{4.1.8}$$
$$q_3 = \sqrt{\tfrac{2}{3}}q[-\tfrac{1}{2}\sin \omega + \tfrac{1}{2}\sqrt{3}\cos \omega]$$

The implied summation or integration of the cross-section is over the values of r_i for fixed q_i and thus is not expected to affect the gross dynamical features too much.

One of the rather nice features of the LPS plots is that they explicitly display the distributions of events corresponding to different configurations of the final state particles. For example, in Fig. 22 for 3-body reactions as the angle ω varies from 0 to 2π the six sectors $2(n-1)\pi/6 \leqslant \omega \leqslant 2n\pi/6$, $n = 1$ to 6, cover all $2^3 - 2$ possibilities of $c_1 c_2 c_3$ being produced forward or backward, i.e. with q_i positive or negative. [The -2 possibilities correspond to all three particles forward or backward which are forbidden by momentum conservation]. Similarly, for 4-body reactions the 2^4-2 configurations of the four final state momenta correspond to the 14 sides of the polyhedron shown in Fig. 23. The square faces correspond to those cases where two of the longitudinal momenta are positive and two negative, while the triangular faces

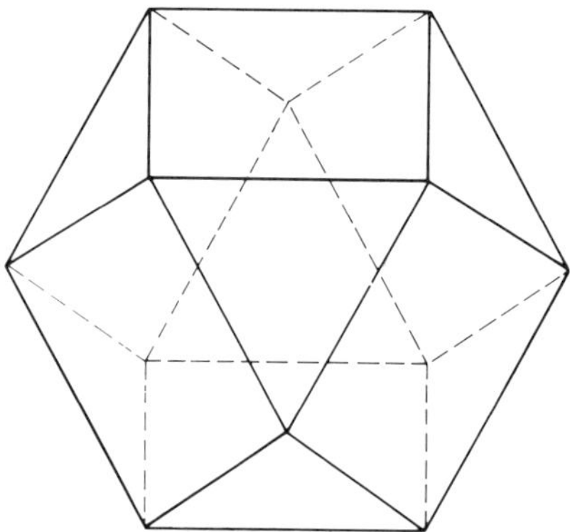

FIG. 23. LPS cuboctahedron for four-body process.

come from configurations with three particles in one direction (either forward or backward) and the fourth produced in the opposite direction. This cuboctahedron is the polyhedron H_2 given by eqns (4.1.2) and (4.1.4) which surrounds the 4-body LPS plot.

In the same way we can consider (at least mathematically), the bounding polyhedra of any n-body reaction. But in our three dimensional world we are obviously going to run into trouble trying to draw them. However, it may be possible to again use some physical insight to simplify some of these difficulties. As an example, let us return to the 3-body reaction and identify $a + b \rightarrow c_1 + c_2 + c_3$ with the reaction $\pi N \rightarrow \pi_1 \pi_2 N_3$. From Fig. 22 we see that for $4\pi/3 \leqslant \omega \leqslant 7\pi/3 q_3$ is positive, i.e. N_3 is produced in the opposite direction, in the c.m. frame, to the initial state nucleon. This backward scattering situation, of course, is expected to be quite strongly suppressed and often that part of the LPS can be ignored. A typical LPS distribution showing this sparsity of backward events is given in Fig. 24 (Bartsch *et al.*, 1970a). Also, we may note that for values of the angle ω near $2\pi/3$, q_1 is

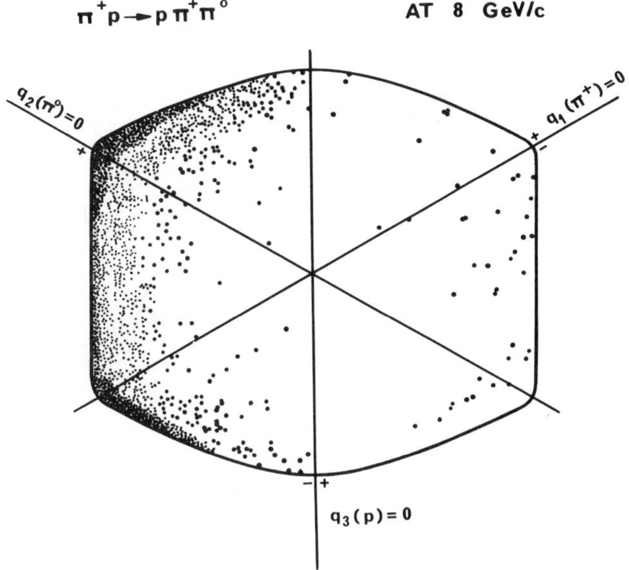

FIG. 24. Typical scatter plot of events on the LPS plot (Bartsch *et al.* 1970a).

maximal and approximately equal to the incident pion momentum. Thus (quantum numbers permitting) this region should correspond predominately to diffraction dissociation phenomena.

Similarly, in the case of 4-body processes, e.g. $\pi^- p \to \pi^+ \pi^- \pi^- p$ of the fourteen faces of the cuboctahedron, seven correspond to baryon exchange and again are found to provide a negligible contribution to the cross-section. In fact, by ordering the two π^-'s according to their longitudinal momenta, i.e. $q_f^- > q_s^-$, where f and s refer to the faster and slower of the two negative pions, it is observed that only four faces give a sizeable contribution to the cross-section (Kittel *et al.*, 1971). These are shown in Fig. 25a. Here it is more conventional to use the reduced longitudinal momenta defined in the c.m. system as

$$x_i = q_i \left/ \left(\tfrac{1}{2} \sum_{j=1}^{n} |q_j| \right) \right. , \qquad (4.1.9)$$

which from (4.1.2) are seen to satisfy the constraints

$$\sum_{i=1}^{n} x_i = 0, \quad \text{with} \quad \sum_{i=1}^{n} |x_i| = 2. \qquad (4.1.10)$$

Many higher order processes can also be presented in the same systematic

D

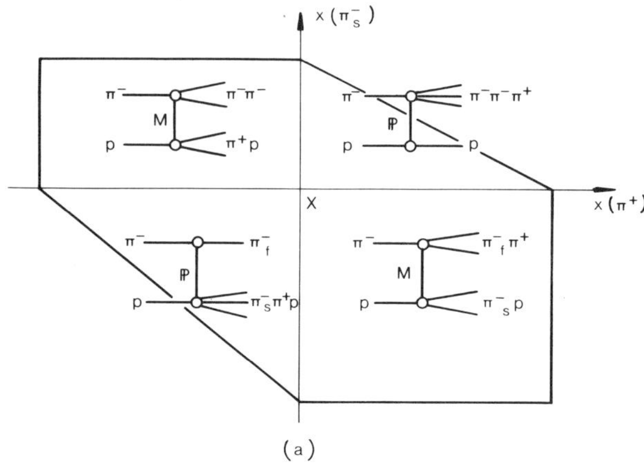

(a)

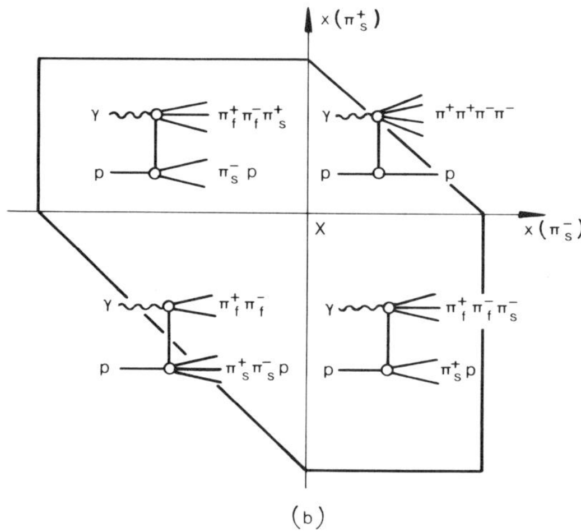

(b)

FIG. 25. The dominant faces of the LPS distributions for the processes (a) $\pi^- p \to \pi^- \pi^+ \pi^- p$, (b) $\gamma p \to \pi^+ \pi^- \pi^+ \pi^- p$.

way as 3- or 4-body reactions. In Fig. 25b we show the dominant regions for the process $\gamma p \to \pi_f^+ \pi_s^+ \pi_f^- \pi_s^- p$ which have recently been studied by Liu *et al.* (1972). Notice in Fig. 25 we have indicated the Regge exchange graph which is likely to be operative in that particular part of the LPS plot. A comparison of these exchange graphs with the experimental distributions shows that a sizeable part of the cross-section comes from those regions involving vacuum

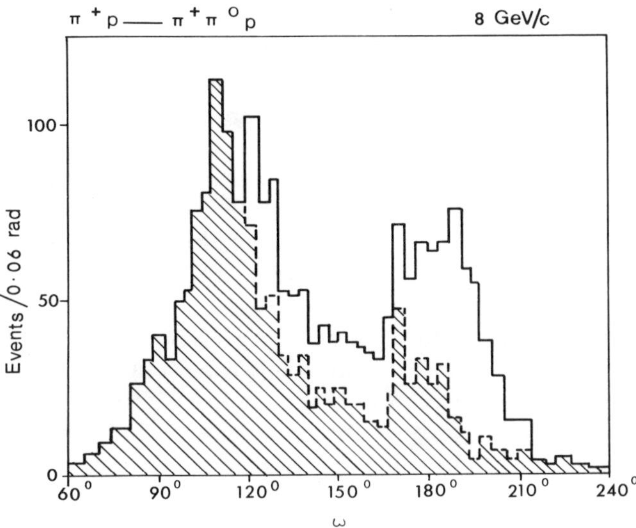

FIG. 26. LPS w-angle distribution for the process $\pi^+ p \rightarrow \pi^+ \pi^0 p$ at 8 GeV/c. Shaded area shows events remaining after the ρ and Δ resonance bands have been subtracted (Bartsch *et al.*, 1970a).

exchange. By examining the distributions in these LPS regions at different energies it is found that this part of the cross-section has very little energy dependence. For instance, parameterizing this dependence in the conventional way by p_{lab}^{-n} where p_{lab} is the incident particle momentum in the laboratory frame, the exponent n is found to be about 0·2 for the vacuum exchange regions compared with a typical value of 1·0 for charge exchange processes. Such fairly constant vacuum exchange mechanisms are usually given the title of diffraction dissociation reactions although in terms of Regge language they presumably correspond to some sort of Pomeron exchange graph.

Even with higher multiplicity production processes, therefore, it may be possible through the LPS analysis to isolate regions of phase-space which are of particular dynamical interest. This is one of the major achievements of the LPS approach to multiparticle phenomenology since it allows the classification of events in a fairly model independent way. It should be noticed also that this way of presenting the data provides a more reliable analysis of possible correlations among produced particles, particularly resonance formation. For example, a selection of events contributing to the bottom right-hand sector of Fig. 25a show strong resonance peaks in both the $\pi_f^- \pi^+$ and $\pi_s^- p$ effective mass distributions. On the other hand, without this prior selection, or "cut" of the data such resonance peaks are considerably obscured by background effects arising from kinematic reflections of other dynamical features. It is therefore likely that the LPS type of analysis can also be extremely valuable in identifying the production of some final state resonances which might otherwise be swamped by the background signal. With this in mind it must be pointed out that although the LPS plot is a very elegant and informative picture of multiparticle reactions, especially with regard to exchange mechanisms, it must be complemented in a phenomenological study by other distributions, particularly invariant mass plots. To illustrate this we show in Fig. 26 a typical LPS histogram of events for the process $\pi^+ p \to \pi^+ \pi^0 p$ at 8 GeV/c (Bartsch *et al.*, 1970a). The shaded area represents those events which remain after cuts have been made on the data to exclude possible Δ and ρ formation. It can be seen that these resonance effects are rather spread out over a range of values of the Van Hove angle ω and in any case do not give rise to any clearly identifiable peaks in the distributions. Recently Pless and his collaborators have tried to analyse 3-body data in terms of ω and the energies of three particles by means of a "prism plot" (Brau *et al.*, 1971). The base of the prism is a triangle describing the constraints on these three energies, and the height of the prism represents the range of the angle ω. The resulting three-dimensional distribution, which can be displayed for instance on a TV screen, provides a very clear identification in the data of the various exchange *and* resonance formation regions.

While we have tried to stress the utility of the LPS approach to the phenomenology of production processes we must add one cautionary note. Although the LPS plots do tend to separate the data into well-defined regions there is bound to be some overlap between these regions, certainly at finite energies. To simplify the dimensionality of the process we integrate over the transverse components of the final state momenta and though empirically these are generally observed to be small, they are still finite. Moreover, the exchange mechanisms and resonance effects which we are trying to investigate also extend over somewhat ill-defined regions of phase-space. We should therefore not be too surprised to find that certain features in one region of the LPS plot are sometimes manifested in other regions of the plot. See for instance Beaupré *et al.* (1972).

From a theorist's or phenomenologist's point of view this overlapping may well be fortuitous, since it allows a comparison between his model for one type of dynamical situation to his model for another, thus providing further constraints on both. Indeed, in some model fits it has been suggested that the data may be well described only by strong interference between effects from two regions of phase-space (Humble, 1971). However, since the LPS analysis does still possess this residual ambiguity, it could be argued that there is nothing in the LPS technique which could not be obtained by careful cuts of the data presented in the more orthodox manner of invariant mass plots and Chew-Low momentum transfer distributions. This is generally true, of course, though there are at least three arguments we might put forward in defence of the LPS approach. The first is the relative ease with which a meaningful separation of the data can be achieved. As we have tried to show the various faces of the LPS plot correspond quite naturally to the various configurations of the final state particles. The second point is that the LPS analysis provides a universal language in which to discuss multiparticle reactions. Unlike most other techniques there is little need to specify the cuts made on the data in an attempt to isolate the underlying dynamical mechanisms. The LPS analysis implicitly performs these cuts in a well-defined and easily understood way. Thirdly, since the data are displayed on the surface of an n-2 polyhedron in a continuous fashion, the LPS approach displays the transitional region between two different dynamical situations. Therefore, if each face of the polyhedron corresponds to some type of single Regge exchange it is obviously advantageous to be able to examine the interesting transitional region. For example, is it compatible with *double* Regge exchange? This is particularly true for the LPS plots in Figs. 25a, b where two faces correspond to single Pomeron exchange. At the point of contact of these two faces, i.e. at the point X, we might expect a transitional mechanism of double Pomeron exchange to be applicable. As we shall discuss later, there is considerable theoretical interest in the existence or non-existence of this double Pomeron graph. It is desirable therefore not only to be able to identify accurately the region of phase-space corresponding to this possible mechanism, but also to know how this region is influenced by effects in adjoining areas of phase-space.

In order to establish the existence or otherwise of double Pomeron exchange it is necessary to study the energy dependence of the central distribution (Yamdagni and Gavrilas, 1971) or possibly the corresponding $\pi^+\pi^-$ mass distribution (Humble, 1972b). However, as always, before we can establish the structure of the scattering amplitude we must first know the behaviour which is implied purely by phase-space. Let us consider therefore the element of phase-space given way back in (1.2.9). For the general n-particle final state this can be written as

$$dR_n = \delta^2\left(\sum_{i=1}^{n} \mathbf{r}_i\right) \prod_{i=1}^{n} d^2\mathbf{r}_i \delta\left(\sum_{i=1}^{n} q_i\right) \delta\left(\sum_i w_i - \sqrt{s}\right) \prod_{i=1}^{n} \frac{dq_i}{w_i}. \qquad (4.1.11)$$

By introducing the quantities

$$Q = \frac{1}{\sqrt{n}} \sum_{i=1}^{n} q_i$$

$$q = \left(\sum_{i=1}^{n} q_i^2 - Q^2\right)^{\frac{1}{2}}, \qquad (4.1.12)$$

which are the distance of the event point $(\mathbf{q}_1 \ldots \mathbf{q}_n)$ from the hyperplane L_{n-1} and the radial length in L_{n-1} respectively, R_n can be transformed into the generalized polar co-ordinate system as

$$dR_n = \left\{ \delta^2 \left(\sum_i \mathbf{r}_i \right) \prod_i \frac{d^2 \mathbf{r}_i}{w_i} \right\} \frac{1}{\sqrt{n}} \, \delta(Q) \, dQ \, \delta(q - q_0) \, dq \, D_n \, d^{n-2}\omega \qquad (4.1.13)$$

Here the functions q_0 and D_n are defined by

$$\sum_{i=1}^{n} (m_i^2 + r_i^2 + q_0^2 \gamma_i)^{\frac{1}{2}} = W, \qquad q_0 \geqslant 0 \qquad (4.1.14)$$

$$D_n = \frac{q^{n-2}}{\dfrac{\partial}{\partial q} \left\{ \sum_i (m_i^2 + r_i^2 + q^2 \gamma_i)^{\frac{1}{2}} \right\}_{\text{at } q = q_0}} \qquad (4.1.15)$$

where γ_i are the direction cosines of the q_i such that

$$\gamma_i = \frac{q_i}{q^2 + Q^2} \quad \text{and} \quad \sum_i \gamma_i^2 = 1 \qquad (4.1.16)$$

In (4.1.13) $d^{n-2}\omega$ is the element of solid angle $d\omega_1 \ldots d\omega_{n-2}$ subtended on the longitudinal phase-space L_{n-1}. Of course, for $n = 3$, ω_1 is just the angle ω defined in Fig. 22.

Unlike, say, the Dalitz plot of (2.3.16) the phase-space element (4.1.13) with the weight factor of $D_n \prod_i w_i$ does not produce an isotropic LPS distribution for a constant T-matrix element. For instance, in the case of $\pi N \to \pi\pi N$ with $W = 4 \text{ GeV}$ and $r_i = 0 \cdot 4, 0 \cdot 4, 0 \cdot 5 \text{ GeV}/c$ $i = 1, 2, 3$ the function $D_3 / \Pi w_i$ is found to vary by as much as 50% as a function of ω. Its detailed structure is shown in Fig. 27. Such variations in phase-space, however, are not a serious handicap and can be simply eliminated by giving each data point an appropriate weighting factor. See, for example, Kittel, Ratti and Van Hove (1970), Kittel (1972). To obtain this factor it is convenient, particularly for high order multiplicities (i.e. $n > 3$), to re-write the phase-space volume as

$$dR_n = \tilde{\omega}_n^{-1} \, dV_0$$

where

$$dV_0 = \left\{ \delta^2 \left(\sum_1^n \mathbf{r}_i \right) \prod_1^n d^2 \mathbf{r}_i \right\} \delta \left(\sum_1^n x_i \right) \delta \left(1 - \tfrac{1}{2} \sum_1^n |x_i| \right) \prod_1^n dx_i \qquad (4.1.17)$$

and

$$\tilde{\omega}_n = \left[\sum_1^n (x_i^2 / w_i) \right] \left[\sum_1^n |q_i|/2 \right]^{3-n} \left(\prod_1^n w_i \right) \qquad (4.1.18)$$

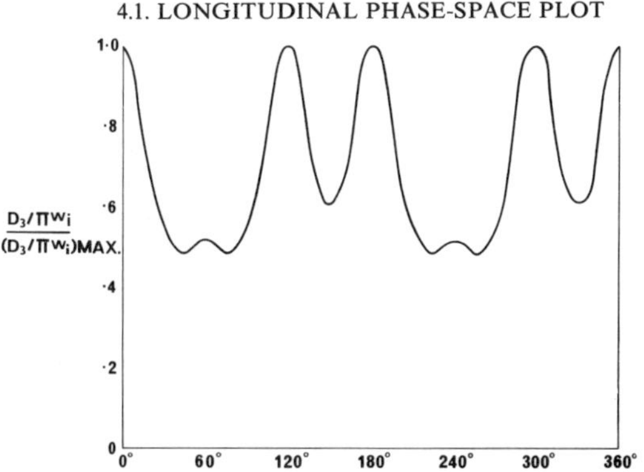

FIG. 27. Typical phase-space distribution for the process $\pi N \to \pi\pi N$ at $w = 4$ GeV and $r_i = 0.4$, 0·4, 0·5 GeV/c, $i = 1, 2, 3$.

dV_0 is independent of the total energy of the system $\sqrt{s}$, but $\tilde{\omega}_n$ has quite a fairly pronounced s-dependence, particularly when one or more of the x_i is very small. We can display this dependence explicitly by writing out (4.1.18) more fully as

$$\tilde{\omega}_n = \frac{s}{4}\left[\sum_1^n x_i^2/(x_i^2 + 4m_i'^2/s)^{\frac{1}{2}}\right]\left[\sum_1^n (x_i)/2\right]^{3-n}\prod_1^n (x_i^2 + 4m_i'^2/s)^{\frac{1}{2}} \quad (4.1.18')$$

where m_1' is the longitudinal mass $(m_i^2 + r_i^2)^{\frac{1}{2}}$, and $x_i \approx 2q_i/\sqrt{s}$. See (4.1.9).

Now within each region δV of phase-space, i.e. a region in the space of variables $x_1, \ldots, x_n, \ldots, \mathbf{r}_1, \ldots, \mathbf{r}_n$ subject to their constraints, the weighted LPS distribution is defined as

$$\Delta_\omega = \sigma_0 N_t^{-1} s^{-1} \sum_k \tilde{\omega}_n^k \quad (4.1.19)$$

compared to the unweighted distribution

$$\Delta = \sigma_0 N_t^{-1} \sum_k 1. \quad (4.1.20)$$

In these expressions σ_0 denotes the channel cross-section; that is to say, the n-particle cross-section integrated over all allowed values of phase-space. Similarly N_t represents the total number of events in an experiment, where by the term "event" we mean the occurrence of the particular n-particle reaction we are considering. (N_t is presumed to be corrected for any experimental inefficiencies so that it counts *all* events). The summations run over all events k falling in the region δV. In Δ each such event contributes 1 and the sum is simply the number of events δN_t in δV. In Δ_ω, on the other hand, each event k contributes a weight $\tilde{\omega}_n$ appropriate to that event corresponding to the measured momenta of the particles and is given by (4.1.18). In this way the anisotropy of the LPS phase-space can be suppressed, and the dependence of the LPS distribution on the dynamics of the process can be more easily displayed.

In an analogous manner weighted averages over δV can also be defined, e.g.

$$\langle r_i \rangle_\omega = \sum_k (r_i^k \tilde{\omega}_n^k) \bigg/ \sum_k \tilde{\omega}_n^k \tag{4.1.21}$$

compared with the unweighted average

$$\langle r_i \rangle = \sum_k r_i^k \bigg/ \sum_k 1 \tag{4.1.22}$$

where r_i^k refers to the transferse momentum of particle i in the kth event in δV. From the point of view of phenomenology of production processes it will be useful to relate such observable quantities to the matrix element T. It is a simple matter to show that

$$\overline{\Delta} = p_c^{-1} s^{-\frac{1}{2}} \int_{\delta V} |T|^2 dR_n = \delta\sigma$$

$$\overline{\Delta}_\omega = p_c^{-1} s^{-\frac{3}{2}} \int_{\delta V} |T|^2 \, \tilde{\omega}_n dR_n - p_c^{-1} s^{-\frac{3}{2}} \int_{\delta V} |T|^2 \, dV_0$$

$$\overline{\langle r_i \rangle} = \int_{\delta V} |T|^2 r_i \, dR_n \bigg/ \int_{\delta V} |T|^2 \, dR_n \tag{4.1.23}$$

$$\overline{\langle r_i \rangle}_\omega = \int_{\delta V} |T|^2 r_i \, dV_0 \bigg/ \int_{\delta V} |T|^2 \, dV_0$$

Hence while $\overline{\Delta}$ is just the partial cross-section σ for the region δv, the weighted quantity $\overline{\Delta}_\omega$ gives the integral of $|T|^2$ over the *energy independent* variables x_i and r_i with the volume element dV_0. Similarly for $\langle r_i \rangle_\omega$. Thus we can see explicitly that the weighting procedure has removed the energy dependence from the phase-space integration, and any s-dependence in $\langle r_i \rangle_\omega$ and $p_c s^{\frac{3}{2}}\overline{\Delta}_\omega$ must be due to the structure of the production amplitude T.

4.1a. *Monte Carlo techniques.* At this point it would seem appropriate to say a few words about the numerical computational techniques for phase-space integrals of n-body processes. As was pointed out in Chapter 1, we know of no way of performing these integrals analytically when there are more than three particles in the final state. This is the case even without the additional complication of possible structure in the production amplitudes which weight the integrals. Thus for an arbitrary n-body production reaction it will generally be necessary to resort to numerical techniques to estimate the phase-space integrals taken over any domain of allowed variables. That is to say, we must be able to compute the integral of the quantity

$$|T|^2 \bigg/ \prod_{i=1}^{n} w_i \qquad (4.1.24)$$

over any desired range of the variables $\mathbf{q}_1, \ldots, \mathbf{q}_n$ which are restricted by the energy-momentum conserving delta functions, i.e. $\delta^4(\sqrt{s} - q_1 - \ldots - q_n)$ in the c.m. system.

If we take the energy of the system as fixed and average over polarization effects, so that there is rotational symmetry about the initial particle momentum direction, this leaves us with $3n - 5$ independent integration variables. We could, of course, consider a set of N_k points in each of these k variables and employ some type of numerical integration routine, e.g. the trapezium rule, or Runge–Kutta method, etc. In general, however, there may be no reason to suppose that the number of integration points for each of these variables should be widely different. Hence it is probable that we should take each $N_k \approx N$, to stand a chance of obtaining a reasonable numerical estimate. The implication of this approach is therefore that we must consider N^n points in a lattice spanning the phase-space domain, in order to compute the integral. Clearly even when n is still relatively small, i.e. $\gtrsim 4$, N does not have to be too large before the number of points becomes prohibitively large.

An alternative approach is to perform (on a computer) an idealized scattering experiment—a Monte Carlo experiment—in which one generates randomly events in phase-space, and weights each event (i.e. set of vectors $\mathbf{q}_1 \ldots \mathbf{q}_n$) by the factor (4.1.24), (James, 1970). The numerical computation of the integral is then given by the average measure of these weighted events, and while the accuracy of this calculation no doubt increases as the number of events increases we are no longer restricted to taking the huge increments from N^n to $(N + 1)^n$ in order to improve the estimate. Moreover, the idea of randomly generating events in phase-space is quite appealing when dealing, as we are, with quantum mechanical systems.

Although these Monte Carlo techniques are extremely useful at low energies, as the energy of the system increases the numerical evaluation of such integrals over the whole of phase-space demands excessive periods of computer time. This is simply because the volume of phase-space increases with energy and more events must be generated to cover it adequately. On the other hand, any reasonable model for the higher energy amplitude T which might be employed will have the physically observed feature of damping events with large transverse momenta. Thus most of the events generated to cover the phase-space volume, when weighted by the square of the scattering amplitude will give only a negligible contribution to the integral. Although this situation is clearly energy dependent we might illustrate the point by suggesting that of any 20 events generated perhaps only one or two will have the transverse momentum of all produced particles sufficiently small to give a non-negligible contribution. Therefore, if we require a few thousand significant events, twenty times as many events must be generated to obtain them—a somewhat wasteful procedure.

In view of this several authors have suggested modifications to the Monte Carlo approach which incorporate the idea of importance sampling into the generation of phase-space points in order to minimize this wasteful use of computer time. See, for example, Byckling and Kajantie (1969), Pene and Krzywicki (1969) and Kopylov (1962). Here we shall describe one of the most useful of these methods due to Van Hove (1969), and Kittel *et al.* (1970). The basis of this modified Monte Carlo calculation is the introduction of the gaussian distribution

$$\Phi \equiv \Phi\,(\mathbf{r}_1,\ldots,\mathbf{r}_n) = \exp\left(-\sum_1^n |\mathbf{r}_i|^2/2R^2\right), \qquad (4.1.25)$$

where the parameter R is chosen to reflect as far as possible the transverse-momentum dependence of the squared modulus of the production amplitude. In practical calculations therefore R will probably be about 0·4 or 0·5 GeV/c. The integral to be evaluated can then be rewritten as

$$J = \int \delta^2\left(\sum_1^n \mathbf{r}_i\right)\Phi \prod_1^n \mathrm{d}^2\mathbf{r}_i \int F\delta\left(\sum_1^n q_i\right)\delta\left(\sqrt{s} - \sum_1^n w_i\right)\prod_1^n \mathrm{d}q_i, \quad (4.1.26)$$

where

$$F = |T|^2\Phi^{-1}\prod_1^n w_i^{-1}.$$

The momentum conserving delta functions are eliminated by substituting

$$q_i = \sum_{j=1}^n O_{ij}k_j$$

$$\mathbf{r}_i = \sum_{j=1}^n O_{ij}\mathbf{u}_j, \qquad (4.1.27)$$

where k_j and $\mathbf{u}_j$ are longitudinal and transverse vectors, and O_{ij} is a real orthogonal $n \times n$ matrix with the last column given by

$$O_{in} = 1/\sqrt{n} \qquad (4.1.28)$$

This condition together with the orthogonality property of O_{ij}, i.e.

$$\sum_{i=1}^n O_{ij}O_{ik} = \delta_{jk}, \qquad j,k = 1,\ldots,n$$

imply

$$\sum_1^n q_i = \sqrt{n}\,k_n$$

$$\sum_1^n \mathbf{r}_i = \sqrt{n}\,\mathbf{u}_n, \tag{4.1.29}$$

so that

$$J = n^{\frac{3}{2}} \int \Phi \prod_1^{n-1} \mathrm{d}^2\mathbf{u}_i \int F\delta\left(\sqrt{s} - \sum_1^n w_i\right) \prod_1^{n-1} \mathrm{d}k_i, \tag{4.1.30}$$

where k_n and $\mathbf{u}_n$ are put everywhere equal to zero (including of course the relations (4.1.27)). The orthogonality of O_{ij} also implies that Φ becomes in terms of $\mathbf{u}_i$,

$$\Phi = \exp\left(-\sum_1^{n-1} |\mathbf{u}_i|^2/2R^2\right) \tag{4.1.31}$$

Notice that, if desired, (4.1.30) can be written in terms of the polar angles $\omega_1 \dots \omega_{n-2}$ of (4.1.13) as

$$J = \int \tilde{F}\delta(Q)\,\mathrm{d}Q\left(\Phi \prod_1^{n-1} \mathrm{d}^2\mathbf{u}_i\right)\mathrm{d}^{n-2}\omega$$

where

$$\tilde{F} = D_n n^{-\frac{3}{2}}\Phi^{-1}\,|T|^2 \prod_1^n w_i^{-1} \tag{4.1.32}$$

and D_n is given by (4.1.15).

The modified Monte Carlo procedure now consists of generating random vectors $u_1,\dots,u_{n-1}$ and random polar angles $\omega_1,\dots,\omega_{n-2}$ according to the probability distribution

$$\mathrm{d}N = \left(\Phi \prod_1^{n-1} \mathrm{d}^2\mathbf{u}_i\right)\mathrm{d}^{n-2}\omega$$

where Φ is given by (4.1.31). From these $\mathbf{u}_i$ and ω_i through (4.1.16) (considered for $q = q_0$ and $Q = 0$), and (4.1.27) the corresponding longitudinal and transverse momentum components can be calculated and hence the Monte Carlo estimate for $J = \int \tilde{F}\,\mathrm{d}N$ is obtained.

Finally let us mention that random values for the $\mathbf{u}_i$ can be generated by writing

$$u_i^{(x)} = u\xi_{2i-1}, \qquad u_i^{(y)} = u\xi_{2i} \qquad (i = 1,\dots,n-1)$$

with

$$u = \left(\sum_{1}^{n-1} \mathbf{u}_i \right)^{\frac{1}{2}}, \qquad \sum_{1}^{2n-2} \xi_j^2 = 1$$

Thus

$$\Phi \prod_{1}^{n-1} d^2 \mathbf{u}_i = u^{2n-3} \exp\left(-u^2/2R\right) du\, d^{2n-3}\xi$$

where $d^{2n-3}\xi$ is the solid angle element in $(2n - 2)$ dimensional space. Therefore the required random distributions of u and ξ_j are given by taking a uniform distribution for the variable

$$\bar{u} = \int_0^u y^{2n-3} \exp\left(-y^2/2R^2\right) dy$$

as well as a uniform distribution on the unit ξ-sphere of $(2n - 2)$ dimensional space.

In this way, by suitably choosing the parameter R it is possible to define a numerical integration method which concentrates mainly on the region of small transverse momenta though otherwise is free of unwarranted bias. It is designed to be efficient at high energies where the cut-off of the transverse momenta is observed, but of course, it is applicable also at low energies where none of the momenta can become very large.

4.2. Multiperipheral and multi-Regge models

Having outlined some of the difficulties in presenting high energy production data let us now return to the problem of constructing suitable models for these n-body reactions. Here, as in the last section we shall find it convenient to consider intitially a process involving only three particles in the final state. Also, although we are not advocating it as the sole means of analysing high energy data we shall illustrate our discussion by means of the Dalitz Plot shown in Fig. 28. The closed contour on this plot is the boundary for the two dimensional distribution $d^2\sigma/ds_{12}\, ds_{23}$ given in eqn (2.3.17). Points outside this boundary correspond to unphysical values for one or more of the variables s_{12}, s_{23} and s_{13} which for fixed s are related (see eqn (2.1.3)) by the equation

$$s_{12} + s_{23} + s_{13} = m_1^2 + m_2^2 + m_3^2 + s.$$

The shaded regions I, II and III in Fig. 28a are the regions of small s_{12}, s_{23}, s_{13} respectively. In such areas of the Dalitz plot we may expect the process to be dominated by the production of resonances, i.e. to be quasi two particle reactions. Also for the important area of phase-space where the momentum transfer is small these quasi two particle amplitudes are likely to be dominated

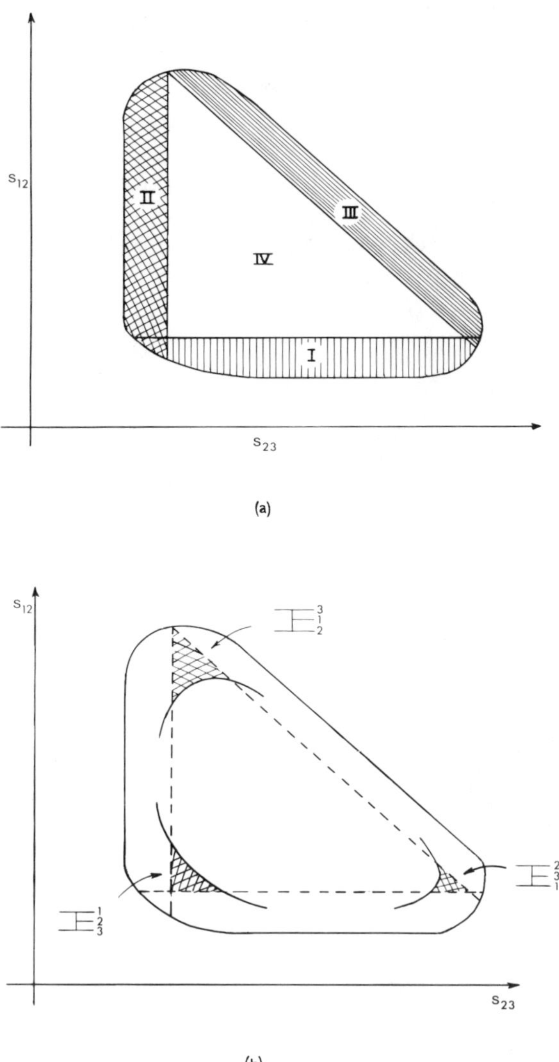

FIG. 28. Dalitz plot indicating (a) resonance production regions. (b) areas corresponding to double Regge pole exchanges.

by the nearest singularity in the crossed channel, i.e. the longest range force. Quantum numbers permitting, this singularity will be the pion pole, and the peripheral model of Section 2.3a should be the operative model in these regions of phase-space.

Let us recall that the peripheral model for a typical process $a + b \rightarrow c_1 +$

$c_2 + c_3$ consists of a pion propagator, a function which describes the bc_3 coupling to the virtual exchanged pion "π", and a scattering amplitude for the process $a + $"$\pi$"$ \to c_1 + c_2$. Now, however, if we let s_{12}, the invariant mass squared of the $c_1 + c_2$ system, also become large we might expect that this latter amplitude will also be dominated by crossed channel exhanges. In region IV of the Dalitz plot in Fig. 28 therefore, where all the sub-energies are simultaneously large, it seems plausible that diagrams like that in Fig. 29 should hold, i.e. with two particles exchanged, provided the corresponding momentum transfers are small.

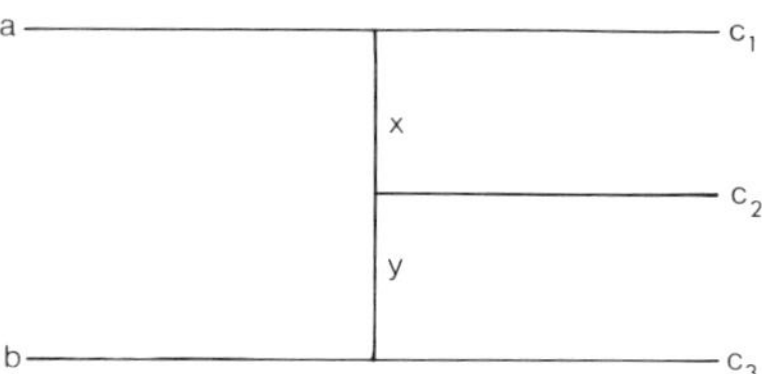

FIG. 29. Double pole exchange diagram.

As we have indicated, these exchanges may be pions, (quantum numbers of the particles permitting), in which case the model is traditionally given the title of the double peripheral model. On the other hand, if only more massive particles are allowed to be exchanged, e.g. the ρ or f_0 mesons with non-zero spins, it is more useful to consider a whole tower of resonances in the form of a Regge pole exchange. Thus, by analogy with two-body reactions, we might guess that the amplitude for the process $a + b \to c_1 + c_2 + c_3$ in region IV has the form, (Chan *et al.*, 1967)†

$$\beta_x(t_1)\, \xi_x(t_1) s_{12}^{\alpha_x(t_1)}\, \gamma(t_1, u_3, \phi)\, \beta_y(u_3)\, \xi_y(u_3)\, s_{23}^{\alpha_y(u_3)} \qquad (4.2.1)$$

where β_x, β_y are the couplings of the Regge poles x and y to particles ac_1 and bc_3 respectively and in general will be functions of the helicities of these particles. In two body reactions these couplings are usually found to be rapidly decreasing functions of the momentum transfers. ξ_x, ξ_y are the corresponding signature factors such that

$$\xi_x = e^{-i\pi\alpha_x} + \tau_x \qquad (4.2.2)$$

where τ_x is the signature. $\gamma(t_1, u_3, \phi)$ is the coupling of the two Regge poles x and y to particle c_2, and ϕ is the azimuthal angle between planes defined by

† The Regge propagators are actually of the form $(s_{ij}/s_0)^{\alpha(t)}$ where s_0 is some scale factor. However, in analogy with two-body reactions where s_0 is typically $\sim 1\,\text{GeV}^2$ we shall assume s_{ij} is given in units of GeV^2 and hence neglect this scale factor.

$(\mathbf{p}_a, \mathbf{q}_1)$ and $(\mathbf{p}_b, \mathbf{q}_3)$ in the rest frame of particle c_2, i.e.

$$\cos\phi = (\mathbf{p}_a \times \mathbf{q}_1)(\mathbf{p}_a \times \mathbf{q}_3)/(|\mathbf{p}_a \times q_1||p_a \times q_3|) \quad \text{with} \quad \mathbf{q}_2 = 0. \quad (4.2.3)$$

It is expected that γ should be a function of the Regge pole "masses" t_1 and u_3, but its dependence on their "relative polarization" described by ϕ is not so transparent. This inuitive approach. however. can be confirmed by considering the crossed reaction $b + \bar{c}_3 \to \bar{a} + c_1 + c_2$ and performing a partial wave analysis as in Section 2.4 where the three particle final state is initially decomposed into angular momentum states of the $\bar{a}c_1$ system. A double Sommerfeld–Watson transformation is then performed to analytically continue the partial wave expansion to the physical region of the original channel. (Ter-Martirosyan, 1963; Kibble, 1963; Roberts and Frazer, 1967). The high energy limit of this manipulation leads to the form (4.2.1) and ϕ is seen to be just the analytic continuation of the azimuthal angle ϕ' between planes $(\mathbf{p}_{\bar{a}}, \mathbf{q}_1)$ and $\mathbf{p}_{\bar{b}}, \mathbf{q}_3)$ in the $(\mathbf{p}_a, \mathbf{q}_1)$ c.m. system: (see the definition of σ_k after eqn (2.4.7)).

In phenomenological calculations the explicit dependence of γ on ϕ is not known and is frequently taken to be constant. This can be partially justified by noting that ϕ has the physical significance of bounding the region of phase-space corresponding to this double pole exchange. That is to say, at high energy we can write, (Chan *et al.*, 1967)

$$\cos\phi = \frac{1}{2(t_1 u_3)^{\frac{1}{2}}}\left((t_1 + u_3 - m_2^2) + \frac{s}{s_{12}s_{23}}(t_1 - u_3)^2 - 2m_2^2(t_1 + u_3) + m_2^4\right)$$

$$(4.2.4)$$

so that for physical values of ϕ such that $\cos\phi \geqslant -1$ this leads to the inequality

$$\frac{s_{12}s_{23}}{s} < m_2^2 - t_1 - u_3 + 2(t_1 u_3)^{\frac{1}{2}} = m_2^2 + g(t_1, u_3) \quad (4.2.5)$$

which is just the result (2.2.5). (This indicates what we already know, that there are many equivalent ways to derive the physical region. It is purely a question of taste whether one uses this Toller variable ϕ or ϕ_2^c as in (2.2.5)). It might now be argued that since the double pole exchange (4.2.1) strongly emphasises the region of phase-space corresponding to small momentum transfers t_1 and u_3 we are always close to the physical region boundary, and hence $\cos\phi$ is normally close to -1. Any dependence of γ on ϕ, on the other hand, will occur only as we go away from the boundary, i.e. in depopulated regions of phase-space, and will therefore only give rise to a small, probably negligible effect.

This is clearly only an approximate argument which may or may not be adequate for phenomenological purposes. However, there are good reasons to expect that the Reggeon–Reggeon-particle coupling function in general will indeed possess some dependence on the Toller variable. We have already seen in the previous chapter that the B_5 function gives rise to such a such a dependence, (or at least a dependence on the ratio

$$\eta - s_{12}s_{23}/s, \qquad (4.2.6)$$

which from (4.2.4) we see is linearly related to cos ϕ). Similarly, Drummond, Landshoff and Zakrzewski (1969) have shown that a similar dependence is also found in other models—specifically Gribov's hybrid perturbation theory model and a model in which the Reggeons are generated by ladders. In each case it can be shown that the coupling function has a dependence on η of the form

$$\gamma_{\alpha_x\alpha_y} = n^{-\alpha_x}A(\eta) + \eta^{-\alpha_y}B(\eta), \qquad (4.2.7)$$

where A and B are entire functions of η and depend on α_x and α_y. (See for example the double Regge limit of the B_5 function given in (3.2.24)). These functions are such that when α_x, say, takes a value corresponding to a physical particle of spin j, $\gamma_{\alpha_x\alpha_y}$ reduces to a polynomial of degree j in η^{-1}. This is necessary in order that γ may then be identified with an ordinary Regge residue function.

Similar model arguments also show that the effects of signature on the phase of the amplitude are not quite as simple as indicated by (4.2.1). This is because besides the signature factors ξ_x, ξ_y the coupling $\gamma_{\alpha_x\alpha_y}$ also generally has a phase associated with it. By considering the normal threshold cuts of the production amplitude in the s, s_{12} and s_{23} variables along the positive real axis it is found that the inclusion of signature leads to a vertex function of the form

$$\gamma_{\alpha_x\alpha_y} = \hat{\gamma}(\eta + i\varepsilon) + \tau_x\tau_y R^{(\eta)}_{\alpha_x\alpha_y}, \qquad (4.2.8)$$

with

$$R_{\alpha_x\alpha_y}(\eta) = \xi_{\alpha_x}^{-1}\xi_{\alpha_y}^{-1}\left|\hat{\gamma}(\eta - i\varepsilon) - \hat{\gamma}(\eta + i\varepsilon)\right|,$$

and γ has singularities in η of the type shown in (4.2.7). (Besides Drummond, Landshoff and Zakrzewski (1969) see also Goddard and White (1971) and Landshoff and Polkinghorne (1972) for a discussion of this point). Thus there is an additional phase introduced by the second term in (4.2.8).

When α_x is a positive integer the discontinuity of $\hat{\gamma}$ vanishes. This means that at right signature points of α_x or α_y the remainder term $R(\eta)$ vanishes and nonsense decoupling occurs. At wrong signature points, on the other hand, the corresponding signature factors ξ_x or ξ_y also vanish so $R(\eta)$ remains finite and there is no nonsense decoupling at wrong signature integers. A phenomenological consequence of this is that there is no reason to expect nonsense wrong signature dips to be observed in momentum transfer distributions of certain differential cross-sections which would otherwise be anticipated by considering only the first term in (4.2.8).

Returning now to the bound given by (4.2.5) we seen that the double pole expression which we believe is the dominant mechanism when certain of the momentum transfers are small is thereby only applicable in rather restricted areas of the Dalitz plot, even in region IV. It is important to note that these regions, shown in Fig. 28b corresponding to the three cases

$$s_{12}s_{23}/s < m_2^2 + g(t_1, u_3)$$
$$s_{12}s_{13}/s < m_1^2 + g(t_3, u_2) \qquad (4.2.9)$$
$$s_{23}s_{13}/s < m_3^2 + g(t_2, u_1)$$

will in general be disjoint and correspond to a different permutation of the final state particles. Thus, if we assume dominance in each of these regions by a single double Regge graph, the corresponding cross-section may be well approximated by the incoherent sum of such terms, the interference effects being negligible. It is therefore perhaps possible to ignore spin effects for these multi-Regge models and to replace the Regge residues β_x, β_y by their spin averaged values. We might also note at this point that assuming these residues are strongly damped functions for large momentum transfers—e.g. exponentials or gaussian functions, the models can reproduce the observed cut-off in the transverse momentum r_i of any produced particle. To see this we have only to express r_i as $q_i \cos \theta_i$ and remember from (2.1.6c) that $\cos \theta_i$ depends linearly on the momentum transfer u_i (or equivalently t_i).

Numerous comparisons with experimental data have been made with the double Regge exchange model in the past few years. It would be impossible and probably not very instructive to give a detailed list of references to this work, (though a partial list is given in the review by Jackson (1970)). Instead let us make a few remarks of a fairly general nature regarding these calculations. Firstly we must again note that the region of applicability of the multi exchange model represents only a very small part of phase-space. Just how small we shall try to emphasize later. However, it has been suggested by Chew and Pignotti (1968) that the use of the asymptotic Regge form even down to threshold in one or other sub-energies may be justified by duality arguments. If this is allowed then it can be shown that the form of the multi-Regge amplitude and kinematics are such that if $\alpha_x(0) < \alpha_y(0)$, say, the mass distribution of particles c_1 and c_2 peaks near threshold while the mass distribution of particles c_2 and c_3 is much broader and tends to peak at higher masses (Ranft, 1969). This is, of course, just a manifestation of the Deck effect (Deck, 1964) which was first calculated for the process $\pi p \to \rho \pi p$. Indeed, even now it is still not clear whether the observed peaking in the $\rho \pi$

mass distribution around $1\cdot1$ GeV is due to this Deck mechanism, which incidentally is predominantly in a 1^+ s-wave state (Froggatt and Ranft 1969), or whether it is in fact the A_1 resonance. (It could be argued from duality that one is just a first approximation to the other. However, it is possible to think of situations where low mass Deck enhancements are produced in certain two-body final state channels which have exotic quantum numbers, and hence are presumably not equivalent to resonances. (See Frampton and Törnqvist (1972).)

While discussing the particular process $\pi p \to \rho \pi p$ we should mention that in the Reggeized version of Deck's original calculation by Berger (1968) the $\rho \pi$ mass enhancement is achieved by the exchange of a pion trajectory which has a low intercept ($\alpha_\pi(0) \approx 0\cdot0$) compared to the other trajectory coupling to the $\bar{p}p$ vertex, i.e. either the Pomeron with $\alpha_p(0) = 1$ or meson trajectory $\alpha_M(0) \approx 0\cdot5$. This enhancement is found to be much more pronounced than that obtained by Deck using elementary pion exchange. Also, since the pion propagator is now $s_{12}^{\alpha_\pi(t_1)}$ instead of the simple $1/(t - m_\pi^2)$, the double Regge model amplitude gives rise to a strong dependence on the azimuthal (Treiman–Yang) angular dependence. This is to be contrasted with the null dependence given by the peripheral (elementary pion) exchange model. See our remarks in Section 2.3a. The fact that a strong $\pi\rho$ mass enhancement and dependence on the azimuthal angle are both observed in the data tends to support the belief that there does, indeed, exist a pion trajectory.

It is also worth noting that the Regge propagator $s_{12}^{\alpha(t_1)}$ in the $a + b \to c_1 + (c_2 c_3)$ amplitude in (4.2.1) is actually the large s_{12} limit of $(\cos \xi)^{\alpha(t_1)}$ where $\cos \xi$ is the c.m. scattering angle in the crossed channel reaction $b + \bar{c}_1 \to \bar{a} + (c_2 c_3)$ such that

$$\cos \xi = - \frac{2t_1}{h_1 h_2} [s_{12} - u_3 - m_2^2 - \tfrac{1}{2}t_1^{-1}(m_1^2 - m_2^2 - t_1)(t_1 + u_3 - m_2^2)],$$

$$(4.2.10)$$

where

$$h_1^2 = t_1^2 + u_3^2 + m_2^4 - 2t_1 u_3 - 2m_2^2(t_1 + u_3)$$

$$h_2^2 = m_1^4 + m_2^4 + t_1^2 - 2m_1^2 m_2^2 - 2t_1(m_1^2 + m_2^2).$$

It has been suggested (Berger, 1968) that at the energies normally considered in phenomenological calculations, (4.2.10) is perhaps a better function to use rather than just s_{12}. Although the evidence is limited at present comparative studies such as that by Coleman (1971) tend to support this view.

Let us now expand our three-body discussion to a typical n-body reaction with $n > 3$. The generalization of (4.2.1) has been performed by Bali et $al.$, (1967) using a method developed by Toller (see Toller 1968). The result, which is not unexpected, can be written in the high energy limit in which all sub-energies $s_{i,j+1}$ are large and the momentum transfers t_i are small as

$$\beta_a(t_a)\, s_{12}^{\alpha_a(t_a)}\, \gamma_{ab}(t_a t_b, \phi_{12})\, s_{23}^{\alpha_a(t_b)}\, \gamma_{bc}(t_b, t_c, \phi_{23}) \ldots s_{n-1,n}^{\alpha_z(t_z-1)}\, \beta^{((t_z-1))} \quad (4.2.11)$$

where the notation is explained in Fig. 30. Although in (4.2.11) we have written the same two Reggeon coupling functions as in (4.2.1) there has been some

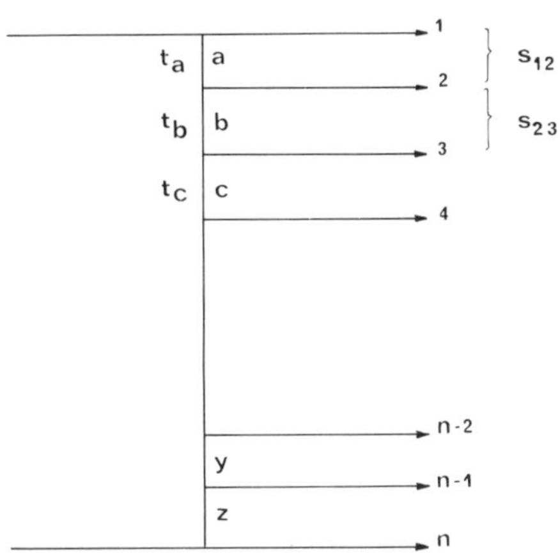

FIG. 30. Multi-Regge pole exchange diagram.

discussion in the past whether this should be valid. This question arises because of the quadratic relations which relate the Lorentz scalars when there are more than three particles in the final state. However, it has recently been shown by Weis (1971) and Halliday (1971) that this is not really a problem if tackled in the right way and that (4.2.11) does indeed follow. Specifically Weis and Halliday ignore these quadratic relations, by pretending there are more than four dimensions in the system, until they have taken the desired high energy limit.

The importance of such multi-exchange models has not been in their detailed fits to particular n-body reactions, but rather as a theoretical tool for investigating production dynamics in general. For example, they can be used to study the way in which inelastic processes are related via unitarity to the elastic amplitude. Before the advent of the multi-Regge scheme (4.2.11), much work had already been done in this direction by Amati *et al.* (1962) using the simple multi-pion exchange diagram shown in Fig. 31 for the n-pion production process. In their original work the dipion amplitudes which are produced along the chain were approximated by low energy resonances. Some years later this work was taken up by Chew and Snider

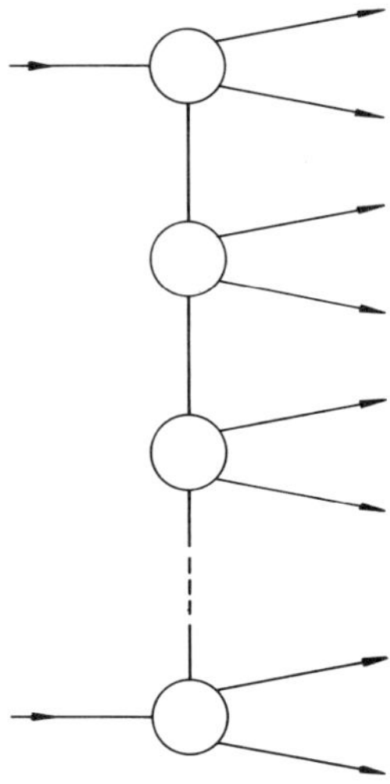

FIG. 31. The AFS multipion exchange diagram.

(1970) who added a high-energy tail to the $\pi\pi$ amplitude represented by Pomeron exchange. By relating these n-body amplitudes to the elastic two-body amplitude through the unitarity condition they found that the model was capable of producing a doublet of vacuum Regge exchange in the elastic amplitude which they identified with the Pomeron and f^0 trajectories. The interesting feature of this doublet is that for $t < 0$ the Pomeron slope is small, but for $t > 0$ the slope is large, $d\alpha/dt \approx 1.0\,\mathrm{GeV}^{-2}$ and particles usually associated with the f^0 trajectory actually lie on this trajectory instead. The f^0 trajectory on the other hand, is nearly flat for positive t and only has the normal large slope for negative t.

Ball and Marchesini (1969) carried out a similar calculation, but they replaced the low energy resonance part of the amplitude by the exchange of the ρ and f^0 meson trajectories. In this way they hoped, via duality, to be able to include the effects not only of one or two low mass resonances but of all of them. However, it turns out that this modified resonance part of the $\pi\pi$ amplitude is too weak to reproduce the doublet behaviour of Chew and Snider's so-called "schitzophrenic Pomeron". In view of this we should point out that Chew and Snider obtained their result by arbitrarily using a low energy resonance term which is much larger than the experimentally observed on-mass-shell $\pi\pi$ amplitude.

Similar results are obtained if one uses the multi-Regge model of (4.2.11) instead of the multiperipheral model. Indeed, both models have the same general features of rapidly damping the momentrum transfer dependence between successive links in the chain; and factorization, i.e. the ability to write the full amplitude as a product of factors which describe the dynamics of local regions along the chain. The main difference therefore is only in the form of the propagator describing the exchange, i.e. whether it is $1/(t_i - m_\pi^2)$ or $(s_{ij})^{\alpha(t_i)}$.

In order to illustrate some of the theoretical features of such models in more detail, let us for simplicity consider a multi-Regge model and calculate the resulting n-body cross-sections. By again noting that these models strongly limit the transverse momentum of each produced particle it should not be too extreme an approximation to simply ignore the transverse momentum degrees of freedom and following De Tar (1971) formulate the model in terms of the longitudinal momenta only. In this case, in the laboratory system we can write the four momenta of the initial and final state particles for the process $a + b \rightarrow c_1 + \ldots + c_n$ in terms of the rapidity variables y_i as

$$p_{a_\mu} = (m_a, 0, 0, 0)$$

$$p_{b_\mu} = (m_b \cosh y_b, 0, 0, m_b \sinh y_b) \tag{4.2.12}$$

$$q_{i_\mu} = (m_i' \cosh y_i, \mathbf{r}_i, m_i' \sinh y_i), \qquad i = 1, \ldots, n,$$

where the rapidity y_i is defined in terms of the longitudinal momentum q_i as

$$y_i = \sinh^{-1}(q_i/m_i') \tag{4.2.13}$$

and m_i' is the longitudinal mass

$$m_i' = (m_i^2 + r_i^2)$$

At high energies the total energy squared of the system is related to y_b by

$$s = m_a^2 + m_b^2 + 2m_a m_b \cosh y_b \simeq m_a m_b \, e^{y_b} \tag{4.2.14}$$

and in the so-called "strong ordering" limit (Zachariasen and Zweig, 1967) in which

$$w_i \ll w_{i+1}, w_i \gg m_i' \quad \text{and hence} \quad y_i \ll y_{i+1} \tag{4.2.15}$$

it follows that

$$s_{i,i+1} = 2m_i' m_{i+1}' \cosh(y_i - y_{i+1}) + m_i^2 + m_{i+1}^2 - 2\mathbf{r}_i \cdot \mathbf{r}_{i+1}$$

$$\simeq m_i' m_{i+1}' \exp(y_{i+1} - y_i) \tag{4.2.16}$$

A similar argument leads to the result for the momentum transfer in the ith link in the chain

$$t_i \approx m_i' m_{i+1}' \exp(y_i - y_{i+1}) - \sum_{j=1}^{i} \mathbf{r}_i^2 \tag{4.2.17}$$

which is therefore negligibly small in this limit. Hence the Regge propagators $(s_{i, i+1})^{\alpha(t_i)}$ can be approximated by

$$(m'_i m'_{i+1})^{\alpha} \exp\left[\alpha(y_{i+1} - y_i)\right] \tag{4.2.18}$$

where α is now some constant effective trajectory.

The n-particle production cross-section is given by squaring (4.2.18) and integrating the product of these terms over all regions of phase-space so that

$$\sigma_n \propto \exp\left(-y_b\right) \int g^{2n-2} \prod_i \exp\left[2(y_{i+1} - y_i)\right] dR_n \tag{4.2.19}$$

where g is the coupling constant at the Reggeon–Reggeon vertex and we have omitted various other unimportant constant factors. Here the factor $\exp(-y_b)$ has been included to take account of the usual flux factor $(m_a p_b^{\text{lab}})^{-1} \approx s^{-1}$. In order to evaluate the integral in (4.2.19) it is convenient to write the differential element of phase-space for n-particle production, i.e. dR_n of (1.2.9) in terms of the longitudinal rapidities as

$$dR_n = \tfrac{1}{2} \prod_{i=1}^{n} d^2 \mathbf{r}_i \, dy_i \delta^2 \left(\sum_1^n \mathbf{r}_i\right) \delta\left(\sum_1^n m'_i \exp\left(-y_c\right) - m_a - m_b \exp\left(-y_b\right)\right)$$

$$\times \delta\left(\sum_1^n m'_i \exp\left(y_i\right) - m_a - m_b \exp\left(y_b\right)\right) \tag{4.2.20}$$

Now using the strong ordering limit of (4.2.15) the last two delta functions in (4.2.20), i.e. (those expressing energy and longitudinal momentum conservation) can be approximated by

$$\frac{\exp\left(-y_b\right)}{m_a m_b} \delta(y_1 - \log(m'_1/m_a)) \, \delta(y_b - y_n - \log(m'_n/m_b)). \tag{4.2.21}$$

This shows what a very special limit the strong ordering limit is, since (4.2.21) implies that the first and last particles in the chain, i.e. c_1 and c_n, carry away practically all of the energy provided m_1 and m_n are not too different from m_a and m_b. Even so we see by considering the approximations for s and $s_{i, i+1}$ given by (4.2.14) and (4.2.16) that the energy–momentum conservation conditions of (4.2.21) lead to the result

$$s \simeq (s_{12} s_{23} \cdots s_{n-1, n})/(m'^2_1 m'^2_2 \cdots m'^2_n) \tag{4.2.22}$$

which is just the generalization of the double Regge exchange condition obtained in (2.2.4), (2.2.5) and also (4.2.5). The resulting n-body cross-section given by the multi-Regge model therefore (ignoring the transverse momenta) is proportional to

$$\sigma_n \propto g^{2n-2} \exp\left[(2\alpha - 2) y_b\right] \int \prod_2^n dy_i \delta(y_b - y_n)$$

Because of the limit (4.2.15) which we have assumed, this can be written as

$$\sigma_n \propto g^{2n-2} \exp\left[(2\alpha - 2)\,y_b\right] \int^{y_b} dy_{n-1} \int^{y_{n-1}} dy_{n-2} \cdots \int^{y_2} dy_1$$

which is to say

$$\sigma_n \propto g^{2n-2} \exp\left[(2\alpha - 2)\,y_b\right]\frac{(y_b)^{n-2}}{(n-2)!}$$

$$\alpha \propto s^{2\alpha-2}\frac{(g^2 \log s)^{n-2}}{(n-2)!}.$$

$$(4.2.23)$$

Having obtained the cross-section for each n-particle production process in this way we can also calculate the total cross-section simply by adding up all these n-body cross-sections, i.e.†

$$\sigma_{\text{tot}} = \sum_{n=2}^{\infty} \sigma_n \propto s^{2\alpha-2+g^2}. \tag{4.2.24}$$

If we believe that the dominant contribution to the cross-section comes from the multiperipheral part of phase-space, we would like the resulting total cross-section to have the experimentally observed constant (or nearly constant) behaviour. This can be achieved provided

$$2\alpha - 2 + g^2 = 0. \tag{4.2.25}$$

Inserting this relation into (4.2.23) we find

$$\sigma_{n+2} = \text{const } (g^2 \log s)^n \exp\left(-g^2 \log s\right)/n! \tag{4.2.26}$$

which is just a Poisson distribution with a mean multiplicity growing logarithmically with s, i.e.

$$\langle n \rangle = g^2 \log s \tag{4.2.27}$$

Although this De Tar formulation of the multi-Regge model is clearly a rather approximate treatment it nevertheless displays most of the general features of the model which can be verified by more realistic calculations (Chew and Pignotti, 1968; Chew, Rogers, and Snider, 1970). The resulting form of the total cross-section is particularly significant since it implies that a model for the n-particle production amplitudes defined solely in terms of Regge exchanges induces Regge behaviour in the total cross-section and hence in the forward elastic amplitude. This raises the more general possibility of constructing multi-Regge bootstrap schemes, i.e. where the repeated

† Notice that at any finite value of s, there will only be a finite number of channels open. Hence by taking an infinite sum in (4.2.24) we are calculating the asymptotic value of the total cross-section.

exchange of some Regge trajectory $\alpha_{in}(t)$ (or set of trajectories) in the multi-Regge chain produces, via the unitarity condition, a similar trajectory $\alpha_{out}(t)$ (or set of trajectories) in the imaginary part of the elastic amplitude. This is illustrated diagrammatically in Fig. 32.

In the above simple calculation the bootstrap leads to the relation

$$\alpha_{out}(0) = 2\alpha_{in} - 1 + g^2 \tag{4.2.28}$$

between the output trajectory at $t = 0$ and the effective input trajectory. Notice, however, if the input trajectory is the Pomeron with unit intercept (4.2.28) gives $\alpha_{out} = 1 + g^2$ which violates the Froissart bound. This is just the famous Finkelstein–Kajantie effect (Finkelstein and Kajantie, 1968). Hence we must either set the Pomeron–Pomeron-particle coupling g equal to zero (Gribov, 1967) or else take the Pomeron intercept less than unity

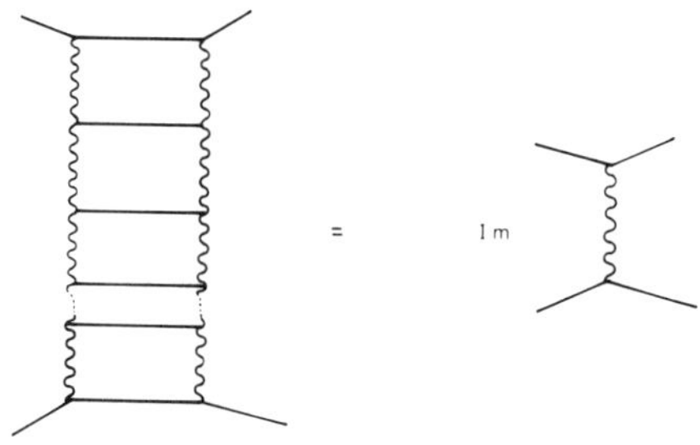

FIG. 32. Illustration of the multi-Regge bootstrap scheme.

(i.e. $\alpha(0) = 1 - \delta$) in which case the total cross-section would decrease like $s^{-\delta}$. However, the coupling g determines the multiplicity, (see (4.2.27)), and a fit to the experimentally observed multiplicities suggests $g^2 \approx 1\cdot3$. For a self-consistent solution to (4.2.28) therefore in which $\alpha_{in} \approx \alpha_{out} \approx 1 - \delta$ this value for g^2 implies $\delta \approx 1\cdot3$ which is certainly not consistent with the energy dependence of hadronic total cross-sections. On the other hand, if we assume the input trajectory is some effective *meson* trajectory and exclude repeated Pomeron exchange in the multi-Regge chain, this value of g^2 (which is now the Reggeon–Reggeon-particle coupling) gives a constant total cross-section for $\alpha_{in} \approx 0\cdot35$—a very reasonable value.

In order to study these bootstrap conditions more generally and obtain the t-dependence of the Regge singularities, Chew, Goldberger and Low (1969) wrote these conditions in the form of an integral equation. In particular they used the factorization property of the multi-Regge model to relate the amplitude for $(n + 1)$-particle production to the n-particle amplitude plus an additional link in the chain—another Regge exchange. From these equations it is possible to make a generalized partial wave projection which allows an investigation of the J-plane singularities of the scattering amplitude. By this means it can be seen, for instance, that the Pomeron and f^0 in the schizophrenic Pomeron solution of Chew and Snider (1970) are essentially the same trajectory but considered on different sheets of the complex J-plane.

In fact it turns out that the J-plane structure is really quite complicated. Amati *et al.* (1962), for instance, found that the multi-peripheral model generates, via the unitarity relations, a Regge cut contribution—the so-called AFS cut—as well as a Regge pole. Although Mandelstam (1963) showed that this cut is, in fact, cancelled on the physical sheet there are other cuts, generated by diagrams in which some of the rungs of the multi-peripheral chain are crossed, which do occur on the physical sheet. Moreover, it is found that for certain values of the moment transfer (i.e. squared mass of the Regge trajectory) the J-plane poles collide with the cuts to produce complex conjugate pairs of trajectories. (See for example Zachariasen, 1971.) Indeed, the schizophrenic Pomeron of Chew and Snider (1970, 1971) is produced because a single pole in the spectrum of the unperturbed kernel of the Chew–Goldberger–Low equation occurs near a weak AFS branch point of the perturbing kernel. The result of this is to generate a pole in the perturbed spectrum on each sheet of the Riemann surface associated with the branch point. This is clearly rather technical and we would refer the interested reader to papers by Chew and Snider (1970), (1971), Frazer and Mehta (1970) for further details. In conclusion, however, let us just mention some interesting recent work by Auerbach *et al.* (1972). They show that by using a model based on the multi-peripheral model but which guarantees the full multi-particle S-matrix is unitary at high energies (i.e. including absorptive effects) yet another J-plane cut is obtained in the elastic amplitude. This "unitarity" cut in the angular momentum plane has the attractive feature of enforcing the Froissart bound since any pole with $\alpha(t) > 1$ is always on an unphysical sheet because it has passed through the unitarity cut. Although some of the details of this work may be in error, the general approach of using the unitarity condition to enforce the Froissart bound is certainly very appealing.

While these results from a theoretical standpoint are most interesting, it is not clear how useful the multi-Regge or multi-peripheral description of production amplitudes may be in practice. It is presumably only valid to assume dominant Regge or peripheral exchanges when the sub-energies are large and corresponding momentum transfers small. However, from (4.2.22) we see that this last criterion leads to a strong bound on the size of the sub-energies and that on average

$$\frac{s_{i,\,i+1}}{\overline{m}} \simeq (s/\overline{m})^{1/(n-1)}$$

where $\overline{m}$ is some typical mean mass of the system. If we believe that Regge behaviour applies only for $s_{i, i+1} > 4\,\text{GeV}^2$ (say) this means that s must be of the order of 4^{n-1} or larger for the multi-Regge regime to hold. On the other hand, if we also take into account that the average multiplicity of produced particles increases with s such that

$$\langle n \rangle \propto g^2 \log s$$

then for the relevant production amplitudes

$$\log s_{i, i+1} \lesssim \frac{\log s}{\langle n - 1 \rangle} \approx \text{constant}$$

and we may never attain a dominant multi-Regge limit if g^2 is too large. Indeed, experimentally the *average* $s_{i, i+1}$ is found to be quite small $<1\,\text{GeV}^2$ and correspondingly only a portion of the events $(<20\%)$ are in the multi-Regge region. In this sense the multi-Regge model is self-destructive in that it restricts itself to a region of phase-space where it predicts there will be very few events. A corollary to this is that while the model gives a contribution to each n-body cross-section and hence the total cross-section of the form (4.2.23), (4.2.24) it could well be that this contribution is not the dominant part of the cross-sections, and hence should not be equated directly to the data. Of course, the model may have a large region of validity in some average sense because of duality, though many of the steps in (4.2.12) to (4.2.24) must then be modified because the strong ordering limit will no longer be applicable. Moreover, although duality may work well for the imaginary part of the sub-amplitude, it is not clear how good it should be for the real part.

In view of these remarks, in order to describe the majority of production events, therefore, we would like to be able to supplement the multi-Regge model with a model which is applicable in those regions of phase-space where particles are produced in clusters, i.e. in groups where the relative momenta between the particles is small. These particles will tend to stay close to each other for a relatively long time and hence give rise to strong final state interactions. Clearly therefore there is a need to devise models which incorporate both Regge behaviour and these final state effects. This is the problem we tackle in the next section.†

4.3. Reggeized cluster models

The division of the Dalitz plot in Fig. 28 into four regions I, II, III and IV may lead us to anticipate that we can simply approximate the three-body production reaction as a multi-Regge amplitude in region IV and resonance (or cluster) production amplitudes in regions I, II and III. This assumes that

---

† In Chapter 6 we shall discuss a much simpler version of a cluster model which is applicable for the study of inclusive distributions and correlations; i.e. where one is interested in fairly general features of the data rather than any specific n-particle process.

the boundaries between the different regions are rigid and distinct and that there is no "shading" between them. To see that this is not the case and why it is essential to consider the regions simultaneously can be most elegantly displayed by the three-body LPS plot. We remind ourselves that most production events have only small transverse momenta r_i. Then taking the example of $\pi N \to \pi_1 \pi_2 N_3$ to fix our particle masses and assuming typically small values of $r_i = 0\cdot4,\ 0\cdot4,\ 0\cdot5$ GeV/c for $i = 1, 2, 3$ respectively, we can calculate the invariants $s_{ij},\ t_i,\ u_i$ of (2.1.1) as values of the LPS angle ω (that is as functions of the longitudinal momenta q_i). The values are shown graphically in Fig. 33 for $\sqrt{s} = 16$ GeV. From this we can see that the region of the LPS

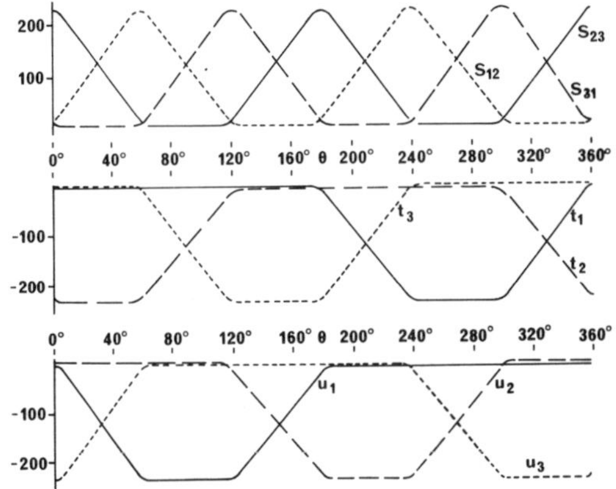

FIG. 33. Typical values of the Lorentz invariants on the LPS plot.

plot which corresponds to the double Regge region with s_{12} and s_{23} "large" and t_1 and u_3 small is restricted to the neighbourhood of the point $\omega = 120°$, though even here s_{12} and s_{23} appear to be quite small ($\sim s^{\frac{1}{2}}$). Either side of this point one of the sub-energies decreases and the Regge limit is no longer satisfied. However as we have already said, for low-sub-energies we must be able to take into account some effect of final state interactions. Therefore we would like a model for the production amplitude which can be continued smoothly from these low sub-energy regimes to the multi-Regge limit.

4.3a. *B_n dual model.* One model of this type of course is the many body dual amplitude of Chan (1969) and others (e.g. Chan and Tsou, 1969, Koba and Nielson 1969) which in analogy with the Veneziano and Bardakci–Ruegg models has been given the title of B_n.

As with these other dual models the amplitude is written in the form

$$T = \sum B_n(1, 2, 3, \ldots, n), \tag{4.3.1}$$

where the sum runs over all non-equivalent orderings of the external lines. Since we wish to impose complete crossing symmetry on the production amplitude the B_n functions must be constructed so that they are invariant under either cyclic permutations of the indices $1, 2, \ldots, n$ or a complete reversal in ordering of the external lines. This means that there will in general be $(n - 1)!/2$ terms contributing to (4.3.1). For each of these contributions, e.g. $B_n (1, 2, 3, \ldots, n)$, we must construct a function which contains poles corresponding to some trajectory α in all possible Mandelstam channels which can be formed without interchanging the order of the external lines. For the four and five point functions these Mandelstam channels simply correspond to the Lorentz scalars formed by adding and squaring the four-momenta of two adjacent particles, i.e. $s_{i, i+1} = (p_i + p_{i+1})^2$, $i = 1, \ldots, n - 1$ in the notation of Section 3.2. However for $n > 5$ other Lorentz scalars can be

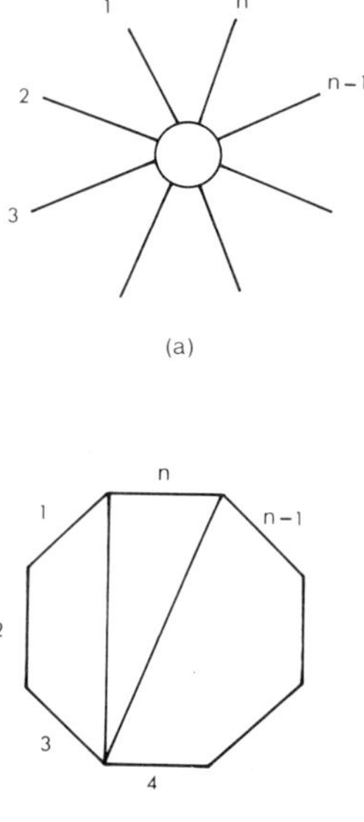

FIG. 34. (a) An n-point amplitude and (b) its dual diagram.

formed, by considering the momenta of three, four, five etc. adjacent particles, which also correspond to possible Mandelstam channels. In order to obtain *all* $3n - 4$ possible independent Lorentz scalars it is convenient to associate with any n-body scattering configuration an n-sided polygon (called a "dual diagram") whose vertices are labelled by the particle indices $i = 1, 2, \ldots, n$ as shown in Fig. 34. Then the diagonals of this polygon connecting vertices

j, k (say) define the Mandelstam channels for this particular n-body scattering configuration, and we denote the corresponding Lorentz invariants by the symbols s_{jk}.

Having obtained the relevant channels in this way we now associate with each one a trajectory α_{jk} defined as

$$\alpha_{jk} = \alpha_0 + \alpha' s_{jk}, \tag{4.3.2}$$

where

$$s_{jk} = (p_j + p_{j+1} \cdots + p_k)^2, \tag{4.3.3}$$

and by analogy with previous dual models construct an integral representation for $B_n(1, 2, \ldots, n)$ as

$$B_n = \int_0^1 \prod_{jk \in P} \left[\mathrm{d}u_{jk} u_{jk}^{-1-\alpha_{jk}} \right] f(u_{jk}) \tag{4.3.4}$$

Here P runs over j, k for all Mandelstam channels for the ordering $1, 2, \ldots, n$, and by construction we have ensured that in general B_n will have poles whenever α_{jk} passes through a non-negative integer. However while channels which are *not* dual to each other can have coincident poles, this is not allowed for dual channels. Therefore the function $f(u_{jk})$ in (4.3.4) must be constructed so as to kill concident poles which may otherwise occur in these dual variables. (A few moments thought will show that dual channels correspond to those diagonals in the "dual"-diagram which intersect each other). Therefore by ensuring that $f(u_{jk})$ gives rise to the constraints

$$u_{jk} = 1 - \prod_P u_{il} \quad \text{for all} \quad jk \in P, \tag{4.3.5}$$

where $\overline{P}$ runs over all channels dual to jk we see that no two integration variables corresponding to dual channels can vanish simultaneously. (This is just the generalization of the constraints (3.2.9) for the five-point function).

The equations (4.3.5) give rise to $n - 3$ non-dual independent variables which for definiteness we can take as u_{1m} ($m = 2, 3, \ldots, n - 2$) corresponding to the poles in the multiperipheral (or equivalently the resonance decay) chain shown in Fig. 30. In terms of these variables (4.3.5) can be solved to give

$$u_{jk} = \frac{(1 - V_{j,k-1})(1 - V_{j-1,k})}{(1 - V_{j-1,k-1})(1 - V_{j,k})} \tag{4.3.6}$$

where

$$V_{il} = u_{1,i} u_{1,i+1} \cdots u_{1,l}$$

with

$$u_{1,1} = u_{1,m-1} \equiv 0.$$

Hence writing $f(u_{jk})$ as a product of delta functions expressing the constraints (4.3.5), the dual amplitude is given by

$$B_n(1, 2, \ldots, n) = \int_0^1 \prod_{jk \in P} [du_{jk} u_{jk}^{-1-\alpha_{jk}}] \prod_{jk \neq 1m} \delta\left(u_{jk} - 1 + \prod_{\overline{P}} u_{il}\right) \qquad (4.3.7)$$

which can be partially evaluated as

$$B_n(1, 2, \ldots, n) = \int_0^1 \prod_{m=2}^{n-2} du_{1m} \left(\frac{1}{J}\right) \prod_{jk \in P} u_{jk}^{-1-\alpha_{jk}}, \qquad (4.3.8)$$

where

$$J = \prod_{i<j} (u_{ij})^{j-i-1} \qquad [i = 2, \ldots, n-2; j = 3, \ldots, n-1].$$

The function (4.3.8) has been constructed so that it is invariant under cyclic permutations or reversal of the indices and is analytic apart from poles at non-negative values of the trajectories α_{jk}. One further property we require of B_n is that at the $(p+1)$th pole in α_{jk}, the residue is a polynomial of total degree p in the trajectories α_{il}, where (i, l) runs over those channels which are dual to (j, k). Since these channels represent momentum transfers for the jk system this means that the pole at $\alpha_{jk} = p$ has maximum (finite) spin p and hence B_n will not give rise to ancestors. It is a simple matter to see that this requirement is satisfied since the residue of the $(p+1)$th pole in the α_{1j} channel (say) is given by

$$\operatorname*{Res}_{\alpha_{1j} = P} B_n(1, 2, \ldots, n) = \frac{1}{p!} \frac{\partial^p}{\partial u_{1j}^p} R \bigg|_{u_{1j} = 0} \qquad (4.3.9)$$

where

$$R = \int_0^1 \prod_{\substack{m=2 \\ m \neq j}}^{n-2} du_{1m} \left(\frac{1}{J}\right) \prod_{\substack{ik \neq 1 \\ j \in P}} (u_{ik})^{-1-\alpha_{ik}}.$$

Now since $u_{il} = 1$ whenever $u_{1j} = 0$ for all (i, l) dual to $(1, j)$ the differentiation can give a polynomial of at most degree p in the corresponding trajectories α_{il}. Although not completely obvious it can be shown that B_n also gives rise to the various possible Regge high energy limits. For example in the full multi-Regge limit where 1 and n are the incident particles and $\alpha_{i, i+1} \to -\infty$ for fixed values of $\alpha_{1, i}$ $(i = 2, 3, \ldots, n-2)$ and fixed values of the ratios $\eta_j = (\alpha_{j-1, j} \alpha_{j, j+1})/(\alpha_{j-1, j+1})$ (which are of course related to the Toller angles)

$$B_n \to \prod_{i=2}^{n-2} (-\alpha_{i, i+1})^{\alpha_{1 \cdot i}} V_n(\alpha_{1, 2}, \ldots, \alpha_{1, n-1}; \eta_3, \ldots, \eta_{n-2}) \qquad (4.3.10)$$

where V_n has the factorizing form

$$V_n = \Gamma(-\alpha_{1,2}) \, V(\alpha_{12}, \alpha_{13}; \eta_3) \, \Gamma(-\alpha_{1,3}) \, V(\alpha_{13}, \alpha_{14}; \eta_4) \ldots$$

$$\ldots \Gamma(-\alpha_{1,n-3}) \, V(\alpha_{1,n-3}, \alpha_{1,n-2}; \eta_{n-2}) \, \Gamma(-\alpha_{1,n-2}) \qquad (4.3.11)$$

and the vertices

$$V(x, y; z) = \frac{1}{\Gamma(-x)\Gamma(-y)} \int_0^\infty d\sigma_1 \, d\sigma_2 \, \sigma_1^{-1-x} \sigma_2^{-1-y}$$

$$\times \exp\left[-\left(\sigma_1 + \sigma_2 + \frac{\sigma_1 \sigma_2}{z} \right) \right] \qquad (4.3.12)$$

In view of these very nice features B_n, like the Bardakci–Ruegg B_5 function, would seem to be an excellent mathematical framework for examining many of the problems of multiparticle production dynamics. Besides the important properties of crossing symmetry, resonance pole saturation ("analyticity"), and Regge asymptotic behaviour the dual model scheme also incorporates the bootstrap hypothesis, i.e. that all particles are bound states of each other. For example, the residue at $\alpha_{1j} = p$ in (4.3.9) can be interpreted as an amplitude for a $p + 1$ particle process where one of the particles has spin p which subsequently decays in $n - p$ spinless ones. In this way it is clear that the amplitude for an n-point process involving any number of particles with integer spin can be derived. However the same residue in (4.3.9) besides describing the amplitude with a particle of spin p also contains daughter states with spin less than p. It turns out that there is a considerable degeneracy of these states, and although any pole in the function such as that at $\alpha_{1j} = p$ can be interpreted as a superposition of a finite number of factorized levels (i.e. degeneracies) the number of levels is extremely large. In fact the number of factors for a given p is given by the number of ways a set of non-negative integers q_k can be chosen to satisfy the partition equation

$$q_1 + 2q_2 + 3q_3 + \ldots + kq_k = p$$

which is $\exp(2\pi\sqrt{p/6})$ for p large. (This incidentally is similar to the number of levels required by certain statistical models of high energy collisions, e.g. the Hagedorn thermodynamical model which we shall consider in Chapter 6.) Unfortunately on fairly general grounds it follows that many of these states will have negative residues and hence correspond to ghost states "decaying backwards in time". (For further discussion of the B_n model and its level structure see for example Chan (1969), Fubini and Veneziano (1969), Bardakci and Mandelstam (1969).)

In view of the difficulty with ghost states together with the other problems we discussed for the B_5 function, which of course also apply to these B_n functions, it would appear we must be less ambitious and more pragmatic in our approach to multiparticle phenomenology. This conclusion is reinforced when one considers that while the integral representation (4.3.8) is useful for discussing the various mathematical limits of the B_n function it can only be calculated in a very restricted region in which each $\alpha_{ij} < 0$. In the special case of B_5 this region was enlarged by considering the series representation of Hopkinson and Plahte (1969). However for $n > 5$ the region of convergence of this series is still unknown and there is as yet no practical method of numerically evaluating the B_n functions. Since this is the first criterion for any phenomenological model of many particle production dynamics it is clear that we must look for other descriptions of the n-body amplitude which while hopefully incorporating many of the features of dual models also allow a numerical comparison with data.

4.3b. *CLA model and others*. The first tentative step in this direction was suggested by Chan, Loskiewicz and Allison (1968) who wrote the modulus squared of the production amplitude as a sum of terms, i.e.

$$|T|^2 = \sum c_N^2 (1, 2, \ldots, N) \tag{4.3.13}$$

where the sum runs over all permutations of the N-outgoing particles (cf. 4.3.1). That is to say they considered an incoherent sum of terms for the cross section. Now however for each of the functions c_N corresponding to a particular ordering of these outgoing particles they used the factorization property of the multi-Regge model to write

$$c_N = \prod_{i=1}^{N-1} c_{i,i+1} \tag{4.3.14}$$

where $c_{i,i+1}$ corresponds to the various two-body final states in the multi-Regge chain of Fig. 30. Instead of the simple Regge propagators, *CLA* proposed that each of these $c_{i,i+1}$ factors be approximated by the form

$$c_{i,i+1} = (\beta_x v_{i,i+1} + c)(v_{i,i+1} + 1)^{\alpha_x - 1} (e^{\lambda_x} v_{i,i+1} + 1)^{\alpha' t_x}, \tag{4.3.15}$$

where v_{ij} is the reduced sub-energy squared

$$v_{ij} = (s_{ij} - (m_i + m_j)^2)/s_0$$

and s_0 is some scale parameter. For large $s_{i,i+1}$ (4.3.15) becomes

$$c_{i,i+1} \approx \beta_x (s/s_0)^{(\alpha_x + \alpha' t_x)} e^{\lambda_x t_x} \tag{4.3.16}$$

which is the x Regge pole contribution to the multi-Regge chain with an exponential residue function $\beta_x e^{\lambda_x t_x}$ and linear approximation for the trajectory $(\alpha_x(t_x) = \alpha_x + \alpha' t_x)$. On the other hand when $s_{i,i+1}$ is near threshold, $v_{i,i+1} \approx 0$ and (4.3.15) becomes

$$c_{i,i+1} \approx c \tag{4.3.17}$$

i.e. the ith term in the chain is approximated by a constant. In effect the CLA model assumes that c represents the effect of *each* $(i, i + 1)$ two-body final state interaction.

It is clear that the sum of terms c_N^2 for all permutation of the final state particles together with form of $c_{i, i+1}$ given by (4.3.15) provides a model for high energy production processes over *all* regions of phase-space whether dominated in certain regions by Regge exchange or cluster formation or both. Initially several quite satisfactory fits to the data were made by Chan *et al.* with the model though, in later applications, some of its shortcomings became apparent. The simple statistical assumption that each low energy two body interaction can be approximated by some constant c is probably adequate when one considers a fairly large ensemble of particles. It is then less likely that the effect of any one particular final state interaction will grossly affect the dynamics of the process. However when one considers a few body system it is very plausible that the distributions depend quite critically on the assumed form of the final state interactions. This was indeed found to be the case in an analysis of some three body reactions presented in the form of LPS distributions by Bartsch *et al.* (1970a). In order to fit the experimental distribu-tions they observed that large amounts of baryon exchange had to be allowed in certain of the reactions. On the other hand, such contributions, as far as can be discerned, where required not in the appropriate Regge region but instead at low-sub-energies to better describe the low-energy two-body final state effects. That is to say, roughly speaking, they found that backward scattering in certain of the two-body amplitudes in the chain dominated over forward scattering. This strongly suggests that the low energy approximation to $c_{i, i+1}$, i.e. (4.3.17) is too crude for detailed analyses and must be modified. Moreover since, as Bartsch *et al.* found, different terms in the sum (4.3.13) can contribute to the same region of phase-space at least as displayed on the LPS plot, it is possible that interference occurs. Thus it would be desirable to have a model for the amplitude rather than the cross section which in some natural way predicts the phase and hence the interference between the various contributions.

An extension of the CLA model which overcomes this problem was devised by Plahte and Roberts (1969). It has the advantage over the CLA model in that
  (i) it allows for possible resonance formation at low sub-energies, and
  (ii) it introduces a phase into the amplitude, which at high sub-energies is just the phase of the corresponding Regge pole.
Briefly the PR model amplitude is a sum of terms

$$T = \sum_{\text{all perms}} P_N(1, 2, \ldots, N) \tag{4.3.18}$$

where P_n like (4.3.14) is given by a product of terms $P_{i, i+1}$. In this case

E

however each $P_{i,\,i+1}$ is given by the expression

$$P_{i,\,i+1} = \beta(v + 1)^{\alpha}\,(e^{\lambda}v + 1)^{\alpha't}$$

$$\times \left\{ \frac{1 + \cos \pi\,(\alpha(s) - \sigma_s)}{\sin \pi(\alpha(s) - \sigma_s)} \sin \pi(\alpha(t) - \sigma_t) + 1 + \cos \pi(\alpha(t) - \sigma_t) \right\} \quad (4.3.19)$$

where we have dropped the $i, i + 1$ and x subscripts on s, v and t respectively and $\alpha(t) = \alpha + \alpha't$ as before. $\alpha(s)$ is one complex trajectory carrying the s-channel quantum numbers of the $(i, i + 1)$ two-body amplitude, i.e. the amplitude for $R_w + R_y \to c_i + c_{i+1}$. Here R_w, R_y are the assumed Reggeons in the adjacent links to the $(i, i + 1)$ system. The parameter σ_s is determined by the spin of the lowest resonance on this trajectory so that $P_{i,\,i+1}$ will have poles whenever $\alpha(s) = \sigma_s + 2k$ $(k = 1, 2, 3, \ldots)$ arising from the factor

$$\frac{1 + \cos \pi(\alpha(s) - \sigma_s)}{\sin \pi(\alpha(s) - \sigma_s)} \quad (4.3.20)$$

Also if $\alpha(s)$ has an imaginary part above the $i, i + 1$ threshold, e.g.

$$\alpha(s) = \alpha + \alpha's + i\,d \log(1 + v)\,\theta(v) \quad (4.3.21)$$

then these poles will become second sheet resonance poles, whose widths are related to the parameter d in (4.3.21). Notice, if $\mathrm{Im}\ \alpha(s) \to \infty$ as $s \to \infty$ as indicated by this parametrization then the factor (4.3.20) at high sub-energies will tend to the constant $-i$ and the term in curly brackets in (4.3.19) reduces to

$$1 + \exp\left[-i\pi(\alpha(t) - \sigma_t)\right]. \quad (4.3.22)$$

Thus by suitably defining the parameter σ_t this expression is just the signature factor corresponding to the $\alpha(t)$ trajectory.

It should be mentioned that (4.3.19) was initially derived by considering a typical four point function as a sum of three Veneziano terms, i.e.

$$P_{i,\,i+1} = B_4(s, t) + B_4(t, u) + B_4(u, s)$$

although by the time it appears in the form (4.3.19) with $\alpha(s)$ given by (4.3.21) it has only a passing resemblance to a dual model. Furthermore the product of these factors which describe the production amplitude for a certain configuration of the final state particles is also somewhat different to the dual B_n amplitude of Section 4.3.a. For instance the complicated off-mass dependence in each of the two-body factors of the B_n amplitude (see (4.3.11) and (4.3.12)) is approximated in the PR model by the simple exponential factor $e^{\lambda t}$; and any dependence on the Toller angles has been ignored. Also as discussed in the last section the phase of the multi-Regge graph is not expected to be just a product of simple Regge phases.

In spite of these failings the model of Plahte and Roberts represents an honest attempt at incorporating many of the features of a dual model into a calculable phenomenological scheme for high energy exclusive reactions. We would refer the reader to the paper by Humble (1971) for a discussion of a detailed fit to some three body reactions. However when one considers the model for processes in which a large number of particles are produced two further drawbacks become apparent. Firstly we see that there is nothing in the Plahte–Roberts model to prevent resonances being produced in adjacent sub-energy channels. It will be remembered that we went to considerable trouble to avoid this unphysical situation in constructing the B_n amplitude. In the particular case of three particle final states at high energies, it is unlikely that adjacent sub-energies can simultaneously be in the resonance region so that this failing is not obvious for such reactions. However for higher multiplicities adjacent sub-energies can indeed both be small and one must somehow ensure that simultaneous resonance production is prohibited. Satz and Thomas (1971) constructed such a modification to the PR model which applies to the first resonance in each channel. The generalization of this to all resonances is straightforward. For example writing each P_N of (4.3.18) in the form

$$P_N(1, 2, \ldots, N) = \prod_{k=1}^{N-1} g_k \exp\left[d(t_{k-1} + t_{k+1})\right]$$

$$\times \left[(s_k + a)^{\alpha(t)} + \Gamma(\sigma_k^t - \alpha(t_k)) \frac{\Gamma(\sigma_k^s - \alpha(s_k))}{\Gamma(\sigma_k^s - \alpha(s) - \sigma(t_k))} \right]$$

$$\times \prod_{j=1}^{N-2} 2i \left[\frac{\exp(ix_j)\sin\pi(\sigma_j^s - \alpha(s_j)) + \exp(ix_{j+1})\sin\pi(\sigma_{j+1}^s - \alpha(s_{j+1}))}{\exp(y_j) + \exp(y_{j+1})} \right]$$

$$\tag{4.3.23}$$

where

$$x_j = \operatorname{Re} \pi(\sigma_j^s - \alpha_j(s_j))$$

$$y_j = \operatorname{Im} \pi(\sigma_j^s - \alpha_j(s_j))$$

$$\alpha_j(s_j) = a + bs_j + i \operatorname{Im} \alpha_j(s_j)$$

with $\operatorname{Im} \alpha_j(s_j)$ defines so that it tends to infinity with s_j, (the sub-energy in the $j, j + 1$ channel), we have a model incorporating the features of the Plahte–Roberts model, but which avoids simultaneous resonances in adjacent channels. To see how this happens let us symbolically put (4.2.23) in the form

$$P_n = \Pi A_k \, \Pi B_j \tag{4.3.23'}$$

where the factors A_k, as before, have poles at $\alpha(s_k) - \sigma_k^s = m$. In this case however the additional terms B_j have *zeroes* when $\alpha(s_j) - \sigma_j^s$ *and* $\alpha(s_{j+1}) - \sigma_{j+1}^s$ are integers. Thus these additional terms ensure that P_n cannot have poles occurring simultaneously in adjacent A_k and A_{k+1} functions. Also at high sub-energies each B_j just tends to unity while A_k has the Regge limit

$$g_k \exp\left[d(t_{k-1} + t_{k+1})\right] \left(1 + \exp\left[-i\pi\alpha(t_k) - \sigma_k^t\right]\right) (s_k)^{\alpha(t_k)} \qquad (4.3.24)$$

where the initial exponential factor describes the off-mass-shell behaviour of the adjoining Reggeon legs. Notice that the unit term in the Regge phase factor in (4.3.24) arises from $(s_k + a)^{\alpha(t)}$ in (4.3.23). This is a typical background contribution arising from something like a $B_4(t, u)$ Veneziano four point function. In the spirit of the CLA model we have introduced the parameter a in order to provide some flexibility in phenomenological calculations as to the size of this background effect.

One interesting point about the Satz–Thomas modification, i.e. (4.3.23') is that the amplitudes $P_N(1, 2, \ldots, N)$ unlike the multiperipheral amplitudes, no longer factorize. That is say while the multiperipheral, CLA and PR models only allow correlations between the momenta of every *pair* of adjacent produced particles, the Satz–Thomas model also introduces correlations between every *three* adjacent particles. This brings us to a further problem encountered in the phenomenological application of this set of models to high multiplicity reactions. So far we have been concerned in describing possible resonance intermediate states in the production amplitude which decay into *two* of the final state particles. However it is well known that many resonances decay predominantly into more than two particles, e.g. the ω, A_1, A_2 and most of the nucleon resonances. Therefore in order to incorporate such resonance decays into the cluster models it is necessary to consider larger groupings than the pair-wise interactions considered so far. Thus further correlations between adjacent sets of particles must somehow be introduced into the models. A moment's thought however will show how formidable a problem this is if one is to avoid double counting errors in the number of ways in which the n-body final state is achieved.

The final problem concerning the phenomenological analysis of high energy production data we consider in this section is the difficulty of including the effects of the spin of the external particles. In Section 4.2 we mentioned that the residue functions in the multi-Regge model where, in fact, functions of these external particle helicities. However since the multi-Regge model is applicable in only very limited regions of phase space this dependence on the external spins can often be neglected† provided there is no competing mechanism in these regions of phase-space to produce interference effects. Clearly a discussion of such interference terms would require a careful consideration of the possible helicity states of the system. The purpose of constructing Reggeized cluster models on the other hand was to be able to describe multiparticle phenomena in *all* regions of phase-space; i.e. in both the multi-Regge and resonance production regions. It is clear that a final state resonance produced say in the s_{23} channel (region II of the Dalitz plot in Fig. 28a) can be obtained by extrapolating from either of the adjacent multi-Regge diagrams in Fig. 28b. Thus it will be necessary to consider interference effects between the two extrapolations—i.e. the two cluster model amplitudes corresponding to the two different orderings of the final state particles. Moreover even supposing we can isolate one particular ordering of the particles (for example on the LPS plot), if we believe that more than one Regge pole is exchanged in any sub-energy channel we must again take into account possible interference effects. Indeed we have already stressed the advantage of having the phase factor in the Plahte–Roberts model in this connection.

In spite of these compelling arguments for considering explicitly the spin of the external particles, most models for high energy production amplitudes are still constructed on the assumption that all particles have zero spin. The most obvious reason for this is expediency. Without this assumption there would often just be too many amplitudes with too many parameters trying to analyse a very limited amount of experimental data. So long as we are concerned with rather gross features—such as energy dependences rather than more subtle interference effects this assumption is probably adequate. However it is probably true to say that until we can properly take account of the spin of the external particles all models must be treated with some scepticism. For instance the whole trend in two-body phenomenology in the past decade has been the realisation that spin is *not* an "inessential complication", and that many features in the data depend critically on a correct treatment of the spin amplitudes. It is very likely that this will also be the case for production processes. In fact if one considers a production process as a production of two high-spin quasi-particles (or system of particles) it is possible that spin effects should be even *more* important than in two-body reactions.

4.3c. *Pokorski–Satz model.* In view of this it seems worthwhile mentioning here one or two very preliminary attempts at incorporating at least some effect of spin into production amplitudes. The first of these was by Pokorski and Satz (1970) who suggested trying to isolate the Pomeron exchange contribution to the process $pp \to p\pi^+ n$ (see Fig. 35a). By assuming that the exchanged Pomeron behaves like a $J^P = 0^+$ particle (i.e. by making specific

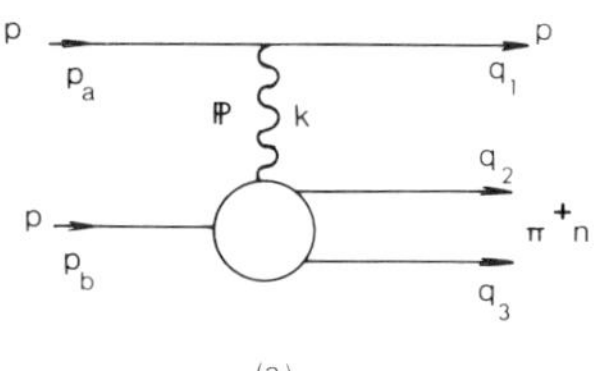

(a)

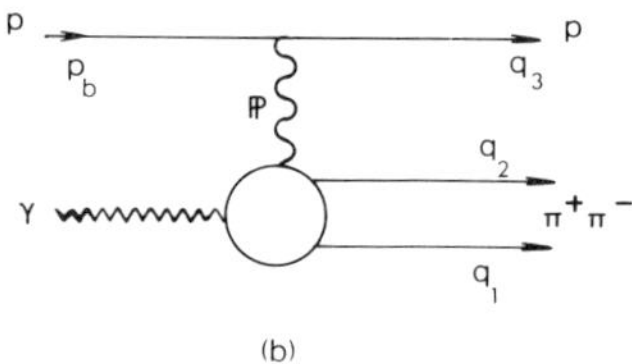

(b)

FIG. 35. (a) Illustration of the Pokorski–Satz model for the process $pp \to p\pi^+ n$. (b) Illustration of the Satz–Schilling model for the process $\gamma p \to \pi^+ \pi^- p$.

assumptions about which helicity states it couples to) they can write the production amplitude in the form†

$$T = \exp\left(\tfrac{1}{2}at_{pp}\right)s^{\alpha(t_{pp})}\bar{u}(q_3)\gamma_5\left(A + \tfrac{1}{2}B\gamma_\mu(q_2 + k)^\mu\right)u(p_b), \qquad (4.3.25)$$

where

$$s = (p_a + p_b)^2$$

$$t_{pp} = (p_a - q_1)^2$$

$$k_\mu = (p_a - q_1)_\mu$$

$\alpha(t_{pp})$ is the Pomeron trajectory, taken to be unity; and $u, \bar{u}$; $\gamma_5\gamma_\mu$ are the usual Dirac spinors and matrices. For the Pomeron $+ p \to \pi^+ + n$ invariant amplitude A they consider a Veneziano B_4 form written in terms of the nucleon and pion trajectories; while by assuming that A_1 exchange is negligible compared with pion exchange they can set $B = 0$.

One of the drawbacks of this dual model parametrization is that the baryon resonances are "parity doubled". That is to say the model gives rise to reso-- nances of both parities at each spin value—a feature which apparently is not supported by the data. A second application of this approach by Satz and Schilling (1970) to the process $\gamma p \to \pi^+\pi^- p$ (Fig. 35b) does not have this drawback, since the corresponding four-point amplitude now describes

---

† The spin averaging over the proton states has already been assumed in (4.3.25).

meson–meson scattering. Specifically they write the production matrix element (again averaged over nucleon spins) as

$$T = \exp\left(\tfrac{1}{2}at_{pp}\right)s^{\alpha(t_{pp})}\beta\{(k^+ \cdot p_b)k_\mu^- - (k^- \cdot p_b)k_\mu^+\}\varepsilon_\gamma^\mu\{A/(k^+ \cdot p_b)\} \quad (4.3.26)$$

where $k^\pm = q_1 \pm q_2$ and ε_γ^μ is the photon polarization vector, and A is a kinematic-singularity free invariant amplitude which they also parametrize in terms of B_4 functions, this time containing the ρ and π trajectories.

These model analyses together with a third by Bartsch *et al.* (1970) for the process $K^-p \to K^-\pi^+\pi^-p$ (which uses a B_5 parametrization!) are found to give fairly acceptable fits to the experimental distributions. In general they describe at least qualitatively the momentum transfer distributions and the invariant mass distributions of the produced systems. However they do have some fairly obvious defects. For example, while the Pomeron trajectory has been taken to have the constant value of 1, it has at the same time been treated as a virtual spin *zero* particle. Moreover it can be seen, for example in (4.3.25) that if we extrapolate the dual amplitude A to the nucleon pole $s_{12} = (q_1 + q_2)^2 = m_p^2$ we do not recover the correct elastic pp amplitude, because a form factor $(\exp\left(\tfrac{1}{2}at_{pp}\right))$ is missing. Also on a more quantiative level it is likely that the assignment of the Pomeron as a $J^P = 0^+$ particle will be shown to be in error by a careful examination of the Treiman–Yang angular distribution which of course is a sensitive test of this assumption.

In conclusion it should be mentioned that the models of Pokorski and Satz and also of Satz and Schilling were propsed not so much for their treatment of spin, but rather to investigate the phenomenon of diffraction dissociation. We have presented them here therefore in a somewhat unfamiliar context. However this will be remedied in the next section when we shall again return to these models as examples of schemes to incorporate diffraction (Pomeron exchange) into the duality hypothesis. For the moment let us merely point out that the Pokorski–Satz model is applicable (if at all) in only a very limited region of phase-space. How one extends their ideas to construct a global model for production amplitudes which takes account of spin effects in all regions of phase-space is as yet an open question. We believe this represents one of the most important current problems of production dynamics, since from our previous remarks it is likely that detailed phenomenological analyses must await the advent of such models.

4.4. Diffraction dissociation

In any discussion of high energy hadronic reactions one is led, sooner or later, to a consideration of the diffraction dissociation processes. However since this term has at times been used to describe more than one feature of particle dynamics let us first try to define what we mean by diffraction processes.

To do this let us repeat that one of the most obvious trends to be found in reactions in which only a few particles are produced is peripherality, i.e. the tendency of produced particles to gather into two sets, each set moving predominantly in the direction of the incident particles. If in a given data sample this is not always found to be the case then let us suppose that we have selected those events for which this peripherality condition holds. This class of reactions can then be depicted by the diagram in Fig. 36 where we denote the two groups of produced particles by A and B. The distribution of the momentum transfer

$$t_{aA} = t_{bB} = (p_a - q_A)^2 = (p_b - q_B)^2$$

is peaked around its minimal value, which will be close to zero for low mass A, B states. It is now possible to define diffractive processes as those reactions $a + b \to A + B$ in which the internal quantum numbers of the a, A, and b, B systems are conserved. That is to say, in Fig. 36 there is no isospin,

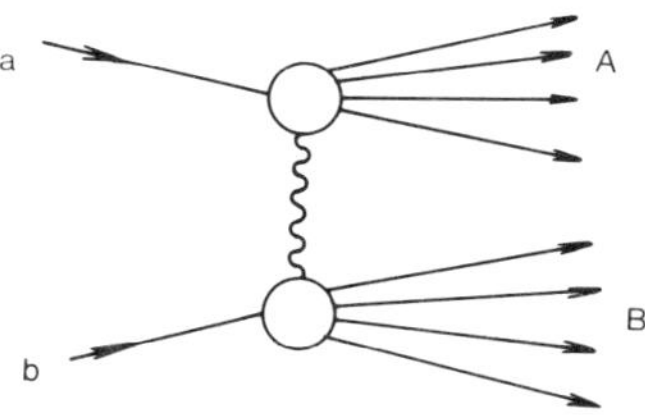

FIG. 36. A typical diffraction dissociation process $a + b \to A + B$.

strangeness, baryon number, charge or G-parity exchanged between the aA and bB vertices, and hence the "diffractively produced systems" A and B differ only from the initial particles a and b in their mass and spin content. All other reactions in which quantum numbers are exchanged are not unnaturally called exchange reactions.

This classification of few-body reactions may seem somewhat academic until one observes that there is a marked difference between the observed distributions of the two types of reactions. For example, it is found that the cross-sections for exchange processes decrease with increasing momentum approximately according to the power behaviour

$$\sigma_{\text{exchange}} \sim p_{\text{lab}}^{-n}$$

where $n > 1$. The exact size of this exponent tends to depend on the exchanged quantum numbers. For those reactions which are classed as diffractive on the other hand, the corresponding decrease is much slower, and it is often believed the diffractive cross-sections will eventually level off and approach non-zero constant values. (Constant that is except for possible logarithmic factors.) The most obvious examples of diffractive processes are elastic πN and NN scattering for which $a = A$, $b = B$; and ρ^0 photoproduction. All of these do indeed show very little energy dependence as one goes to high energies. In fact the energy independence property has often been used as a test of the diffractive nature of an interaction; so much so that nowadays the two terms *energy independence* and *diffraction* are frequently considered to be synonymous with the same dynamical mechanism. One reason for this, as we shall see, is that many models of diffractive processes incorporate Pomeron exchange, and while it is not at all clear exactly what the Pomeron looks like as a J-plane singularity it is defined to have the quantum numbers of the vacuum and to produce little energy dependence in the cross-sections.

Besides the elastic two-body reactions several diffractive phenomena have been observed in multi-particle processes. The most widely studied of these are

$$\pi p \to (\pi\pi\pi)p$$

where the 3 pions from the A_1 (1100)-bump

$$Kp \to (K\pi\pi)p$$

where the $K\pi\pi$ system produces the $Q(1300)$-bump, and

$$pp \to (p\pi\pi)p$$

which forms the $B(1450)$-bump. The cross-section for A_1-production shows very little energy dependence above 10 GeV but the cross-sections for Q-production and B-production continue to decrease over the whole energy range considered to date, i.e. up to 20 GeV. However, even for these reactions the energy dependence is much smaller than for typical exchange processes. Moreover, by examining not only the $Kp \to Qp$ and $pp \to Bp$ reactions but also the line reversed $\bar{K}p \to \bar{Q}p$ and $\bar{p}p \to \bar{B}p$ reactions it is found that there is a cross-over effect in the momentum transfer distributions $d\sigma/dt_{aA}$. For example, the forward differential cross-sections for $\bar{K}^0p \to \bar{Q}^0p$ and $K^-p \to Q^-p$ are observed to be larger than their corresponding anti-particle reactions $K^0p \to Q^0p$ and $K^+p \to Q^+p$. The former reactions have more rapid fall-off with momentum transfer than the latter, however, so that their distributions cross over at some point†, typically around $t \approx 0.15\,\text{GeV}^2$. This cross-over phenomenon is strong evidence for other non-diffractive components

---

† The cross-over point can only be determined from an accurate knowledge of the normalizations. For Q^0, $\bar{Q}^0$ production this is achieved because they come from the same K^0_L bea., but for $Q^\pm$ production the relative normalizations are only approximate.

in these reactions, such as ω or ρ Regge pole exchange. Thus it is very possible that the energy dependence of these cross-sections is due to the Regge (non-diffractive) term which should eventually die away at higher energies, leaving only the diffractive cross-section. It is this presumed dominance of the diffractive mechanism at high energies which has been one of the chief reasons for the keen interest in these reactions. Simply by going to higher and higher energies it should be possible to unambiguously separate this mechanism from other effects. It represents, therefore, one of the very few dynamical effects in hadron scattering which can be isolated in a fairly clean manner.

Interest in diffraction dissociation processes is further heightened when one considers some of the strange features of the experimental distributions. The first of these, oddly enough, is the approximate energy independence not only of the elastic, but also of the inelastic diffractive processes which we have already mentioned. For elastic scattering this might be considered fairly plausible because elastic unitarity relates the imaginary part of the amplitude at $t = 0$ to a sum of positive terms, i.e. symbolically

$$\operatorname{Im} T_{ab \to ab} = \sum \left| T_{ab \to n} \right|^2. \tag{4.4.1}$$

Since the major contribution to the elastic cross-section is found to come from this imaginary amplitude which is peaked near $t = 0$, constancy of the elastic cross-section would appear to be due to the fact that as the energy increases more and more channels contribute via (4.4.1) to

$$\operatorname{Im} T_{ab \to ab}.$$

Thus even though most separate contributions, $\left| T_{ab \to n} \right|^2$, may tend to fall off eventually with energy, the net effect is to produce a fairly energy-independent elastic cross-section. This explanation, however, needs much more thought, since it does not apparently suggest the existence of constant inelastic diffractive cross-sections. For such reactions the unitary condition can be written as

$$\operatorname{Im} T_{ab \to AB} = \sum_n T_{ab \to n} T^*_{AB \to n}. \tag{4.4.2}$$

where the sum is now over a sum of terms which are not necessarily positive. Therefore (4.4.2) allows a partial cancellation between the n-body contributions and there is consequently little reason, *a priori*, to expect that the sum will give the desired constant cross-section for the process $ab \to AB$.

An alternative approach, of course, is to consider t-channel exchanges rather than s-channel effects, in which case the presence of a factorizing J-plane Regge pole with the quantum numbers of the vacuum and intercept close to unity would lead to fairly energy-independent diffractive cross-sections. The Pomeranchuk pole with unit intercept was invented precisely for this purpose—to explain the apparent constancy of elastic cross-sections. It is worth pointing out, therefore, that this is only one possibility which may or may not be fruitful. If this leading J-plane singularity, the "Pomeron", does not factorize there is nothing in its s-channel description in terms of the unitarity conditions (4.4.1) (4.4.2) to ensure that it is even the same singularity (or set of colliding singularities) for each diffraction process. Thus while the description of diffractive processes in terms of a "Pomeron" t-channel exchange is an attractive simplifying assumption we must understand the nature of Pomeron much better before we know how useful this description is in practice.

Another interesting feature of inelastic diffraction processes is the correlation found between the mass of the dissociated system and the slope of the momentum transfer distributions. For example, in the process $pp \rightarrow p(\pi^+ n)$, $d\sigma/dt_{pp}$ at small t has an exponential form e^{bt}, where the slope b varies from approximately 17.5 ± 2.5 for a $\pi^+ n$ effective mass of 1.2 GeV to 5.0 ± 1.0 for a mass of 1.6 GeV. This correlation is not restricted to this one reaction, but it is found to be a common effect, more or less, in all diffractive processes. As we shall see shortly this feature is rather difficult to describe quantitatively in most of the models for diffraction dissociation which have so far been proposed. It remains therefore an intriguing example of how little we understand of the dynamics of production processes.

Although in our introductory remarks in this section we mentioned the particular reactions $\pi p \rightarrow A_1 p$, $Kp \rightarrow Qp$ and $pp \rightarrow Bp$ we did not specify precisely what we meant by the A_1 Q and B "bumps". In fact these bumps are enhancements in the 3π, $2\pi k$ and $2\pi N$ invariant mass spectra centred around the values 1.1 GeV, 1.3 GeV and 1.45 GeV respectively. Also, as might have been guessed from our definition of diffractive processes, and the observation that the corresponding cross-sections have little or no energy dependence, the A_1 Q and B systems are found to have the same internal quantum numbers as the beam particles, π, K and p respectively. Thus the only way these diffractively produced systems may differ from the beam particles, besides in mass, is in their spin and parity content. It turns out, however, that in each case these bumps are dominated by a *single* well-defined spin-parity state, i.e.

$$A_1(J^P) = A_1(1^+), \qquad Q(J^P) = Q(1^+), \qquad B(J^P) = B(\tfrac{1}{2}^+).$$

This in itself is a rather peculiar feature since there is little evidence to suggest that these bumps are actually single resonances which might give rise to one dominant state. See for instance Diebold (1972). On the other hand, if these systems are merely kinematical effects, we might expect *several* spin-parity states to contribute.

Even more striking is the observation that the change in J^P from the initial particle to the diffractively produced system apparently follows the natural spin-parity series, i.e.

$$\frac{P_{\text{initial}}}{P_{\text{final}}} = (-1)^{J_{\text{final}} - J_{\text{initial}}} \tag{4.4.3}$$

This is known as the Gribov–Morrison rule, though as yet there is little theoretical justification for it being called a "rule". It is merely an empirical observation which, if it continues to hold true as further diffractively produced systems are analysed in detail, will be yet another tantilizing piece of the puzzle of production dynamics.

For diffraction production of baryon resonances the Gribov-Morrison rule (4.4.3) would mean that only those resonances in the spin-parity sequence

$$\tfrac{1}{2}^+, \tfrac{3}{2}^-, \tfrac{5}{2}^+, \tfrac{7}{2}^-, \dots$$

could be produced. To date this would seem to be the case. For the three-pion system, besides the A_1 with $J^P = 1^+$, (4.4.3) implies that only the sequence

$$1^+, 2^-, 3^+, 4^-, \dots$$

would be diffractively produced. However, in the three-pion system a second bump is observed, at a little higher mass than the A_1, which is called the $A_2(1310)$ system. The spin and parity of this A_2 resonance is found to be 2^+ and therefore from (4.4.3) would not be expected to be produced diffractively. In fact, as discussed by Leith (1972), the energy dependence of the $\pi p \to A_2 p$ cross-section is found to be very similar to the cross-section for the process $\pi p \to A_1 p$, i.e. $\sigma \sim p_{\text{lab}}^{-0.6}$. Hence, using the energy-dependence as a test of diffraction dissociation we might conclude that A_2 production represents a violation of the Gribov–Morrison rule.† In view of the "cross-over" phenomena, however, which indicates the presence of "ordinary" Regge pole exchange contributions, (at least in some reactions), it is probably wiser to defer any definite conclusions until the processes have been analysed at much higher energies.

One further feature of diffraction dissociation which has frequently been commented on is the apparent property of factorization. That is to say, if the amplitude for a diffractive reaction can be written as

$$T_{\text{Diff}} = V_A(a, A)P . V_B(b, B), \tag{4.4.4}$$

---

† There are other possible examples in the literature. For instance, one often quoted violation of the Gribov–Morrison rule is the photoproduction of the B-meson. See Leith (1972) for a further discussion.

where V_A, V_B are vertex functions and P is a universal function, i.e. the same for all processes, then it follows that

$$\frac{\sigma(\pi p \to \pi p)}{\sigma(Kp \to Kp)} = \frac{\sigma(\pi p \to \pi N^*)}{\sigma(Kp \to kN^*)}$$

and $\hspace{10cm}$ (4.4.5)

$$\frac{\sigma(\pi p \to \pi p)}{\sigma(pp \to pp)} = \frac{\sigma(\pi p \to \pi N^*)}{\sigma(pp \to pN^*)},$$

where N^* is any diffractively produced baryon system not necessarily resonant. Similar predictions, of course, can be constructed for pion or kaon dissociated systems.

This factorization hypothesis for diffraction dissociated systems is, of course, a statement regarding the nature of the vacuum exchange mechanism. In terms of the Regge J-plane language this mechanism would again presumably be identified with the Pomeron, and the factorization property would result if the Pomeron were a simple J-plane pole. The fact that relations like those in (4.4.5) are indeed found to hold at least to within the experimental accuracy of $\sim 10\%$ is therefore quite surprising. As we have already said, the nature of the Pomeron is not known, but it seems very likely that it cannot just be a simple pole. The multiperipheral bootstrap ideas considered in Section 4.2 for example tend to suggest that there may well be several (colliding) J-plane singularities involved in determining the asymptotic behaviour of elastic amplitudes. In this case we would not expect the factorization property to hold in general. Alternatively, if the Pomeron is not a simple pole this apprent factorization of diffraction processes seems difficult to explain. Notice, however, that since such processes peak near $t' = 0$, it may be sufficient (to the present experimental accuracy) for the amplitudes to factorize in the forward direction, which is clearly a much less stringent requirement.

Although the factorization hypothesis is a simple enough statement at infinite energies, at normal laboratory energies there is some difficulty in how one should interpret factorization. In (4.4.4) we wrote the diffractive amplitude as a product of two vertices and a propagator P. However, although we believe P is a function of the momentum transfer and, to some extent, an energy variable it is not altogether clear which energy variable this should be. For example, if $B = b$ and $A = c_1 + c_2$ where the invariant mass squared of the $c_1 + c_2$ system, s_{12}, is finite, the energy variable could be either $s(=(p_a + p_b)^2)$ or s_{1B} or s_{2B} depending on the exact meaning we give to the hypothesis of factorization. If A and B are resonant states then s would seem to be the appropriate variable to choose, but in terms of a multiperipheral model s_{1B} or s_{2B} might be a more appropriate choice depending on the relative sizes of the longitudinal momenta of particles c_1 and c_2. Asymptotically, of course, this will make no difference, since for finite s_{12}, both $s/s_{1B} \to 1$, and $s/s_{2B} \to 1$, but at finite energies quite considerable differences could occur in relating the elastic to the inelastic cross-sections in (4.4.5).

It should also be noticed that the factorization hypothesis implies that double inelastic diffraction processes should be strongly suppressed at current lab. energies. Here we define a double diffraction process to be one in which both produced systems A and B contain more than one particle. To see how this comes about let us consider the two processes

$$(1) \quad a + b \to a + B$$

$$(2) \quad a + b \to A + b.$$

By assuming factorization for each reaction it is a trivial matter to deduce that

$$\sigma(a + b \to A + B) = \left(\frac{\sigma_1}{\sigma_{\mathrm{el}}}\right)\left(\frac{\sigma_2}{\sigma_{\mathrm{el}}}\right)\sigma_{\mathrm{el}}$$

where σ_{el} represents the elastic cross-section $\sigma(a + b \to a + b)$. Now empirically it is observed that $\sigma_i/\sigma_{\mathrm{el}}$ tends to be very small—of the order of a few per cent—for any inelastic system A or B. Thus the double diffraction reaction will be *two* orders of magnitude down on the elastic cross-section. Experimentally such double diffraction cross-sections are indeed found to be very small, though they have not yet been determined accurately enough to provide a further test of factorization.

Finally in our catalogue of the phenomenological features of diffraction dissociation we must mention the question of the possible spin structure of diffractive amplitudes which has received so much attention recently. The chief reason for this is the observation by a SLAC–Berkeley–Tufts collaboration in 1970 that in a diffractive ρ-meson photoproduction experiment using polarized photons the t-channel parity was almost 100% natural and that s-channel helicity was conserved in the $\gamma \to \rho$ transition. This discovery, together with the finite-energy sum-rule results by Barger and Phillips (1968) and Höhler and Strauss (1970) that the helicity flip amplitude in πN scattering vanishes, prompted the hypothesis that s-channel helicity conservation (SCHC) holds in *all* diffractive reactions whether they are elastic, resonance production or many-particle production processes. (Gilman *et al.* 1970). Here by SCHC we mean that the helicity amplitude of a reaction $a + b \to c + d$ in its s-channel centre of mass system has the structure

$$T^{(s)}_{\lambda_c, \lambda_d; \lambda_a, \lambda_b} = \delta_{\lambda_a, \lambda_c}\, \delta_{\lambda_b, \lambda_d}\, t^{(s)}_{\lambda_a, \lambda_b} \tag{4.4.6}$$

i.e. that the helicities on both vertices $(a \to c)$, $(b \to d)$ are *separately* conserved, though of course the spins of a, c; b and d need not be equal.

Clearly the proposition of SCHC from such limited evidence is extremely tentative. Indeed (4.4.6) need not even be true for the process $\gamma p \to \rho p$ since only one vertex has so far been examined. It is considerably more difficult experimentally to examine the spin structure at the nucleon vertex. Thus one arrives at (4.4.6) from our present knowledge of $\gamma p \to \rho p$ and $\pi N \to \pi N$ only after the further assumption of factorization (4.4.4). Nevertheless, once the hypothesis of SCHC had been made its obvious simplifying effect on diffractive amplitudes made it a very appealing idea, and many experiments were set up to test the hypothesis further on other diffraction processes.

It is worth noting that multi-particle production processes have some considerable advantages over elastic processes in determining the spin structure of the amplitudes. For elastic scattering, most polarization experiments use either a polarized target or measure only the recoil polarization, but not both. Thus one only measures $\langle \vec{\sigma} \cdot \vec{n} \rangle$ where $\vec{n}$ is the *normal* to the production plane and $\vec{\sigma}$ is the spin operator on the initial or final nucleon. It is therefore impossible without spin correlation measurements to discriminate between the spin structure *within* the reaction plane; e.g. is s-channel helicity conserved or is t-channel helicity conserved (TCHC)? By t-channel helicity conservation we mean the scattering amplitude (written in terms of the helicities of the particles defined in the t-channel) can be expressed as

$$T^{(t)}_{\lambda_a,\,\lambda_c;\,\lambda_b,\,\lambda_d} = \delta_{\lambda_b,\lambda_d}\,\delta_{\lambda_a\,\lambda_c}\,t^{(t)}_{\lambda_b\lambda_d} \tag{4.4.7}$$

For the diffractive production of unstable particles, or unstable states, the decay angular distributions already carry information on the polarization of the decaying system, that is to say, they "analyse themselves" without the recourse to double polarization measurements. It is this fact which enabled SCHC to be postulated from the process $\gamma p \to \rho p \to \pi^+ \pi^- p$, while for elastic πN scattering one required the more theoretical input of FESR's.

The results of the spin-structure analyses of many-particle diffractive systems are somewhat unclear. For the two most widely studied processes, i.e. A_1 and Q production SCHC is certainly found to be strongly violated. However, for some time it was thought that the alternative proposal of TCHC could be satisfied. Various cuts have to be made on the data to try to ensure that only pure $J^P = 1^+$ diffractively produced states are analysed. Such cuts, of course, mean that any conclusions cannot be 100% rigorous. Now, however, the concensus of opinion would seem to be that while TCHC is better satisfied for inelastic reactions than SCHC, neither can, in general, be claimed to be valid for most diffractively produced systems.† See for example Rushbrooke (1972).

4.4a. *Models for diffraction dissociation.* Having outlined some of the distinctive phenomenological features of diffraction processes let us now turn to some of the theoretical models which have been suggested to explain them.

The first of these, at least for historical reasons if no others, must be the model of Good and Walker (1960) who based their arguments on an analogy with the scattering of light by a target. It is a well-known optical phenomenon that the absorption of the incident light by a target can lead to an elastic scattering of the wave. However, as Good and Walker pointed out, this absorption or *diffraction* of the incident wave can lead not only to elastic but also to *inelastic* scattering of the incident wave. Moreover, since diffractive elastic scattering is characteristic to the interaction of *any* wave with an absorbing medium they proposed that it might be natural also to expect a phenomenon similar to the inelastic scattering of the incident wave to be found, for example, in high-energy particle scattering. This phenomenon was therefore termed diffraction dissociation. Moreover, since the elastic scattering of high energy particles is seen to be fairly energy independent, it would seem plausible that this energy independence should also be a feature of inelastic diffraction dissociation.

To be more specific Good and Walker proposed that the scattering of a high energy projectile off a nucleon target could proceed through the elastic (diffractive) scattering of a pion in the nucleon's pion cloud, where the pion is excited from a virtual to a real state. This is shown diagrammatically in Fig. 37a. The behaviour in the momentum transfer variable t_1 is taken as the t-dependence of the elastic scattering amplitude and must be peripheral. The t_2 dependence of the model is not so clearly defined, though some fall-off is required for a finite cross-section.

Although Fig. 37a does not represent the full story, and more recent work by Bialas, Czyz and Kotanski (1971) has stressed the importance of other contributions which are indicated in Fig. 37b, the close link between this picture of diffraction dissociation and the Deck model, and also with the multi-Regge model with at least one Pomeron exchanged should be clear. In fact these models with various modifications and amendments represent most of the theoretical suggestions for the diffraction mechanism. By assuming some particlar properties for the Pomeron couplings, for example, it has

---

† Since SCHC and TCHC do not appear to hold for most diffractive reactions, at least at present accelerator energies, we have omitted the details of how one goes about analysing the experimental distributions to test these hypotheses. Instead we would refer the interested reader to Cohen-Tannoudji *et al.* (1970) and Bialas, Dabkowski and Van Hove (1971) as well as to any of the many excellent reviews of diffraction dissociation, e.g. Satz and Schilling (1971).

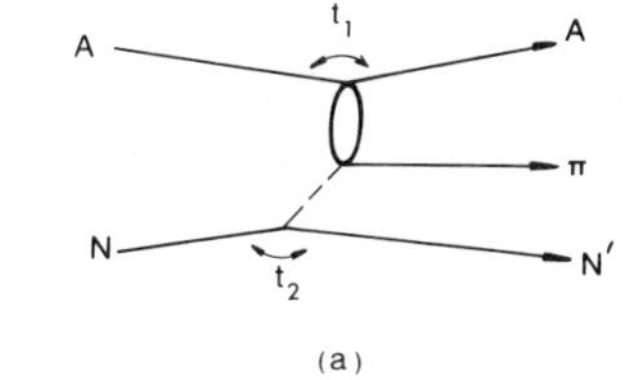

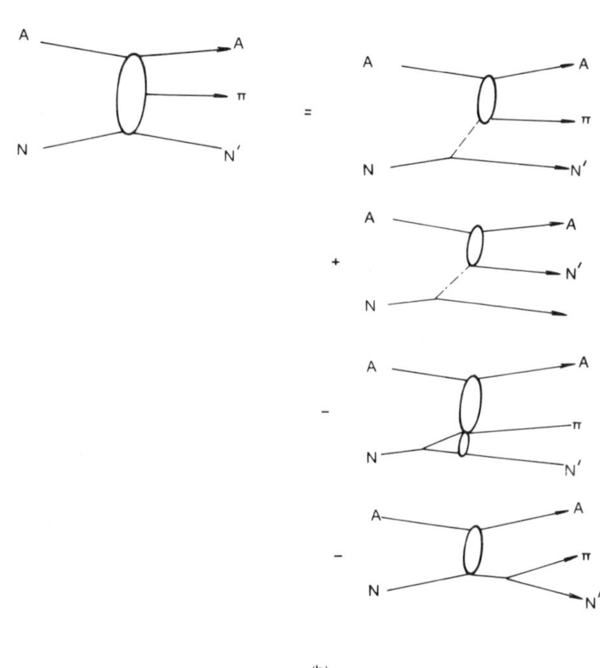

FIG. 37. (a) Pion production in inelastic diffractive $A + N$ collisions (b) Diagrams suggested by Bialas *et al.*, (1971) for inelastic diffractive $A + N$ scattering.

been possible to obtain many of the features which are thought to characterise diffraction dissociation. The TCHC hypothesis is readily explained, for instance, by assuming the t-channel Pomeron exchange couples like a O^+ particle. (Pokorski and Satz, 1970). Such an assignment, incidentally, also automatically ensures the Gribov–Morrison rule for boson dissociated systems though not for baryon systems. SCHC on the other hand, imposes less trivial constraints on the Pomeron residues, which nevertheless can still be obeyed for quasi-two body final states and lead to predictions of the behaviour at nonsense wrong signature points (Hite, 1971). A more recent proposal by Gault and Walters (1972) suggests that the Pomeron couples with a γ_μ coupling at the baryon vertices. This leads to SCHC for elastic scattering, but gives different predictions for higher helicity N^* states which are consistent with the existing data for N^* (1520) and N^* (1690).

In view of the considerable interest in these diffraction models, let us now consider if such a description can account for the very dramatic correlation between the mass of the dissociated system and the slope of the momentum transfer distributions. It had been thought that the Deck model or multi-Regge model could achieve this correlation simply through the kinematics of the situation. That is to say, by assuming, for example, that

$$|T|^2 = \exp(At_1)(s_{23})^{at_2} \exp(bt_2),$$

or more generally that

$$|T|^2 = \exp(At_1)g(s_{12}, s_{23}, t_2), \tag{4.4.8}$$

where the function g is strongly damped, (peripheral) for large values of $|t_2|$, it was believed that phase-space could ensure a large variation of the experimentally observed momentum transfer slope with the mass of the produced system. Indeed, it can be shown that peripherality at one vertex can be reflected as an apparently peripheral effect at the other vertex, because of kinematical simple requirements.

To illustrate this let us present the argument proposed by Pokorski and Satz for the apparent success of their diffractive dual model in describing this slope-mass correlation, although the argument is more general than their model. Remembering the form of the Pokorski–Satz model given in Section 4.3, the double differential distribution in the relevant invariant mass and momentum transfer variables is given by

$$\frac{\partial^2 \sigma}{\partial s_{23}\, \partial t_1} \approx \exp(At_1) \int_{-1}^{1} \mathrm{d}\,(\cos\theta_3^s)\, t_2 \,|V(s_{23}, t_2)|^2 \tag{4.4.9}$$

where V depends on t_1 only as an external mass; θ_3^s denotes the angle between the incoming proton and outgoing neutron in the $n\pi^+$ c.m. frame. For very large s_{23}, V becomes independent of all external masses and hence of t_1. Hence for large s_{23} the t_1 dependence comes solely from the first term $\exp(At_1)$. Near the s_{23} threshold, on the other hand,

$$t_2 \to m_N(t_1 - m_\pi^2)/(m_N + m_\pi)$$

and therefore becomes independent of θ_3^s. The amplitude $V(s_{23}, t_2)$ is peripheral in t_2, and so for low values s_{23}, will contribute an extra term $\exp(bt_2) \approx c\exp(bt_1)$ to the t_1 distribution. Thus the correlation between

s_{23} and the t_1 slope is obtained because of the peripherality at the other vertex and the kinematics of the reaction.

However, a recent detailed *four-dimensional* $(t_1, t_2, s_{12}, s_{23})$ analysis by Meittinen and Pirilä (1972) of the process $pp \rightarrow p\pi^+ n$ has shown that this effect is much too weak to reproduce the observed effect, and in any case, it is only really present at all when they take extremely peripheral functions of t_2. (The preceding argument, of course, was only approximate). They conclude that the traditional explanation of the slope-mass correlation does not arise from a slope A in (4.4.9) which is independent of s_{23}.

The Meittinen–Pirilä analysis carries with it a very significant message. (Unless one can look very closely at the fully differential distributions it is possible that models may be thought to be satisfactory simply because they can describe the averaged, partially integrated distributions, even though they do not reflect the full dynamical features of the data). Even before the Meittinen and Pirilä results there was clearly something wrong with the Deck or multi-Regge model descriptions when one considered the cross-over effects. For

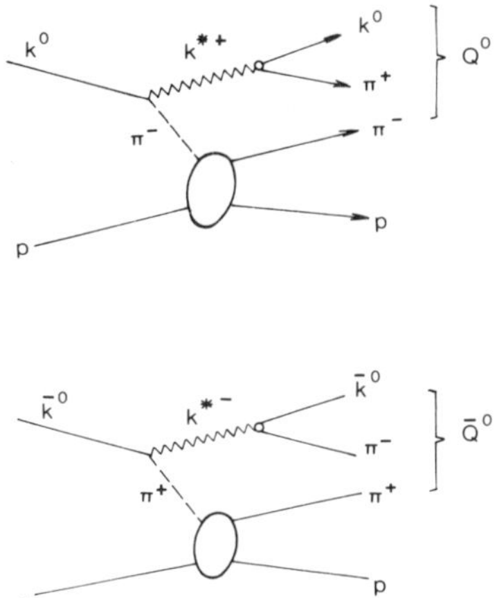

FIG. 38. Q^0 and $\bar{Q}^0$ production in the Deck model.

instance, in Fig. 38 we show the relevant diagrams for Q^0 and $\bar{Q}_0$ production from $K^0_L p$ interactions. The lower parts of these diagrams contain the diffraction mechanism together with whatever additions are necessary to produce the cross-over effects. From these diagrams we can see that since $\pi^- p$ elastic scattering has a steeper slope than $\pi^+ p$ elastic scattering we would expect that $K^0 \rightarrow Q^0$ would have a steeper differential cross-section than $\bar{K}^0 \rightarrow \bar{Q}^0$ which is in contradiction to the data.

In view of these failings of the simple picture of diffraction dissociation it is worth mentioning a rather different approach suggested recently by Kane (1972) which is somewhat more flexible. Basically his approach is to consider the diffractive amplitudes in terms of s-channel helicity amplitudes. Thus for the process $a + b \to A + B$ he considers the amplitudes $T_{\lambda_A, \lambda_B, \lambda_a, \lambda_b}$, where λ_A, λ_B, λ_a, λ_b are the s-channel helicities of the final state systems and initial state particles respectively. Notice, it is not necessary to consider a dissociated system A, or B with only one value of angular momentum, several values can be combined together to give the helicity amplitude, since only the *helicity* of the system is important. Now by defining the net helicity flip n as

$$n = |(\lambda_a - \lambda_b) - (\lambda_A - \lambda_B)| = |(\lambda_a - A_A) - (\lambda_b - \lambda_B)| \qquad (4.4.10)$$

the small angle form for the $d^J_{\lambda\mu}$ rotation matrices in the Jacob–Wick partial wave expansion leads to the result

$$T_{(\lambda)} \underset{\theta \to 0}{\to} (\sin \theta/2)^n \qquad (4.4.11)$$

where θ is the c.m. scattering angle. Moreover, by assuming an optical picture in which the projectile scatters off the target at a distance R the small angle approximation for the $d^J_{\lambda\mu}$ suggests that for the physically important region of small θ, each helicity amplitude can be written as

$$T_{(\lambda)} = A_{(\lambda)} J_n(R\sqrt{-t}) \qquad (4.4.12)$$

where t is again the square of the momentum transfer, $t = (p_a - q_A)^2$, and $q_A \cdot R \approx J + \frac{1}{2}$.

The various cross-sections and differential distributions are now calculated from these $T_{(\lambda)}$ given in (4.4.12) together with the functions describing the decay of the dissociated A, B systems. If only $n = 0$ states are important then (4.4.11) implies a peak in the angular distributions in the forward direction and (4.4.12) suggests a dip arising from the zero of the J_0 bessel function. In practice we would expect a dip rather than a zero in the experimental distribution because of other (background) contributions. For a value of R about 1 fermi this dip will be around $-t \approx 0.25$ and in certain reactions such features are indeed observed. Notice however, that as the mass of the dissociated systems, increase, so presumably will the probability of these systems having high spin states. In turn this also implies the presence of more helicity states, and hence the possibility of higher values of n—the net helicity flip. The corresponding helicity amplitudes will now tend to zero in the forward direction like $(\sin \theta/2)^n$ and will give rise to dips at larger values of dips, corresponding to the zeros of the J_n functions ($n > 0$). Thus if some of these high n states are important, i.e. if some of the coefficients $A_{(\lambda)}$ in (4.4.12) are large then the effect of these $n > 0$ amplitudes together with the non-flip amplitude will be to produce a flatter distribution in t. In this way one can see at least qualitatively how the slope-mass correlation could occur. The steep slope for low dissociated masses is created by the occurrence of a sharp dip in the momentum transfer distribution at small values of t. There is already some evidence of such a dip in $pp \to pN^*(1400)$. The flatter slope for higher dissociated masses, on the other hand, is the result of combining several helicity amplitudes with $n = 0, 1, 2$, etc., each having a dip at a larger t-value, which is probably masked by the other amplitudes.

Of course this picture is rather tentative at present. The energy dependence of the helicity amplitudes is presumably contained in the parameter R, and perhaps quantitative fits will also require the coefficients $A_{(\lambda)}$ to have some energy dependence. Also, while the form (4.4.12) is not too bad a description for helicity amplitudes in two-body exchange reactions, the diffractive two-body amplitude apparently requires an additional "central" piece, i.e. an additional e^{At} term. There is reason to believe, however, that for inelastic diffraction reactions only the more peripheral part of the interaction is important and this central piece, (the low partial waves) are absorbed away.

Kane's approach differs from the other models, with a Pomeron pole exchanged, in that his model is an s-channel picture in which the Deck or multi-Regge conditions are "absorbed", i.e. modified by s-channel unitarity effects. Unfortunately, in breaking away from the usual description of diffraction reactions in terms of t-channel pole exchanges, it also has created a great number of unknown parameters, the helicity couplings $A_{(\lambda)}$ and the impact parameter R. It's predictive power, therefore, is limited unless one makes additional assumptions such as the "Pomeron" (vacuum exchange contribution) only populates the $n = 0$ amplitudes. In view of our previous remarks about the observed violation of SCHC, however, it seems more profitable at present to use the model to analyse the currently available data, and deduce the s- and t-dependence of the various helicity states, before speculating on any underlying features of the dynamics.

We must conclude therefore that at present there is no completely satisfactory model for diffraction dissociation processes. Partly this is due to the fact that the nature of the Pomeron singularity is still not understood. A more fundamental problem, however, is that, lacking any concrete theoretical understanding of diffraction dynamics, it is difficult even to know what effects in the experimental distributions are significant, and which questions should we try to answer by further experimental measurements. Perhaps Kane's analysis in terms of helicity amplitudes is the clue to these problems, in which case once again spin will be seen to be an important ingredient of hadron dynamics.

5

Inclusive Reactions

5.1. Particle distributions

In our study of particle production processes so far, which has generally followed the historical development of the subject, we have considered reactions in which the final state is completely specified, i.e. it contains n particles consisting of so many π^-'s, plus so many K^+'s, etc. Such reactions are called *exclusive* since they *exclude* all other n-body processes. On the other hand, reactions in which only a few of the final state particles $(c_1, \ldots, c_m)$ are identified with momentum $(\mathbf{q}_1, \ldots, \mathbf{q}_m)$ respectively are called *inclusive* reactions since they sum over all possible final states which *include* the particles $c_1, \ldots, c_m$. The corresponding experimental measurement is related to the probability of producing at least the particles $c_1(q_1) \ldots c_m(q_m)$ in a two particle collision at a given energy. If no restriction is made as to which particles must be in the final state (i.e. $m = 0$) then this measurement just

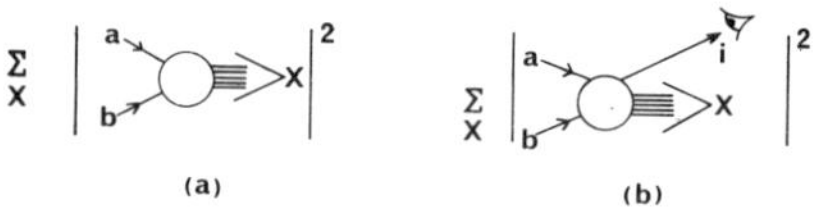

FIG. 39. Illustration of (a) total cross-section, (b) inclusive distribution.

gives the total cross-section. Such a cross-section is shown diagrammatically in Fig. 39a and (see (1.2.1)) can be written as

$$\sigma = \frac{(2\pi)^4}{4p_a^c\sqrt{s}} \sum_n N_n \int \delta^4(p_a + p_b - q_1 \cdots q_n)|\bar{T}_n|^2 \prod_{i=1}^n \frac{d^3\mathbf{q}_i}{2w_i(2\pi)^3} \qquad (5.1.1)$$

where the sum is over all final states (i.e. with all number of possible particles, restricted only by energy-momentum conservation and quantum number selection rules). The integration is over all allowed values of $d^3\mathbf{q}_i$, $i = 1, \ldots, n$ and $\overline{T}_n$ is the spin averaged amplitude for each exclusive n-body reaction contributing to the total cross-section. The normalization factor N_n depends on the identity of the particles $c_1, \ldots, c_n$. If these particles are all identical then to avoid multiple countings of the same final state configuration $K_n = 1/n!$. Similarly, if $c_1, \ldots, c_n$ represents k groups of $m_1, \ldots m_k$ identical particles $N_n = 1/(m_1! m_2! \ldots m_k!)$.

We can also think of holding one or more of the final state particles fixed within a small range dq_i of their values $\mathbf{q}_i$, but still summing and integrating over all other final state particles. In this way we obtain the distributions

$$d\sigma_{1,2,\ldots,m} = f_{1,2,\ldots m}(s, \mathbf{q}_1, \mathbf{q}_2 \ldots, \mathbf{q}_m)\frac{d^3\mathbf{q}_1}{2w_1(2\pi)^3} \cdots \frac{d^3\mathbf{q}_m}{2w_m(2\pi)^3} \qquad (5.1.2)$$

where f is just given by the r.h.s. of (5.1.1) without the factors $d^3\mathbf{q}_i/|2w_i(2\pi)^3|$, $i = 1, \ldots m$; and the normalization factor N_n is modified appropriately. For example, if the m detected particles are m of m_1 identical particles produced in the n-body reaction N_n becomes $1/(m_1 - m)! \, m_2! \ldots m_k!$. The definition of N_n when the m detected particles are not all identical is equally straightforward.

The single particle distribution†

$$\frac{d\sigma_i}{d^3\mathbf{q}_i} = \frac{f_i(s, \mathbf{q}_i)}{2w_i(2\pi)^3} \qquad (5.1.3)$$

is shown diagrammatically in Fig. 39b. This corresponds to the momentum distribution of particle c_i in the process $a + b \to c_i + X$ where X represents all undetected particles in the final state. In the c.m. system the Lorentz invariant momentum distribution of particle c_i, f_i given by (5.1.3), is just a function of s, q_i and $\mathbf{r}_i$ where, as in the last chapter, q_i and $\mathbf{r}_i$ are the longitudinal and transverse components of $\mathbf{q}_i$ respectively. Moreover, if no polarization measurements are made f_i will have no dependence on the angle $\theta_r = \tan^{-1} r_y/r_x$ so that

$$f_i = f_i(s, q_i, r_i) \qquad (5.1.3)$$

where $r_i = |\mathbf{r}_i|$.

Notice that whereas for an n-body exclusive reaction there are $4n - 3$ degrees of freedom, for an inclusive single particle distribution there are only three. This simplification is reflected by the relative ease in which the inclusive

† It is important to notice the difference between "single particle *distribution*" which is an inclusive measurement and "single particle production cross section" which refers specifically to a *three* body exclusive final state.

distribution can be measured experimentally. All one has to do is to monitor the flux of all particles of type c_i into the invariant phase-space volume $d^3\mathbf{q}_i/w_i$ and ignore all other types of particles which may be produced. (Thus the inclusive measurement avoids the very serious problem of trying to specify missing neutral particles). The invariant distribution f_i is then simply proportional to the corresponding rate for the process $a + b \to c_i + X$.

We might mention here that the quantity measured experimentally is not quite f_i but $d\sigma_i/d^3\mathbf{q}_i \propto f_c/w_i$. This is unfortunate because it folds in kinematical effects related to the factor $w_i = (q_i^2 + r_i^2 + m_i^2)^{\frac{1}{2}}$. If m_i is small and f_i is a slowly varying function of q_i the distribution in q_i

$$\frac{d\sigma_i}{dq_i} \propto \frac{f_i}{w_i} r \, dr$$

is likely to be strongly peaked near $q_i = 0$. This peak is a kinematical effect and the dynamics, contained in the function f_i, are found by examining not this gross feature but the more detailed shape of the distribution. To illustrate this we would remind the reader of the example of $K^+ p \to \pi^- X$ discussed by Stone *et al.* (1971). Here the thing of physical interest is not the peak itself, but the fact that the π^- distribution, although not quite symmetric in the $K^+ p$ c.m. system, becomes more so in a Lorentz frame in which $(K^+$ momentum$)/(p$ momentum$) \approx \frac{2}{3}$. That is in their constituent quark-quark c.m. system perhaps?

Kinematical distortions from the energy denominator of course can easily be eliminated by attaching to each measurement of particle c_i as it is recorded a weight factor equal to its energy and taking the sum of these weights Σw_i for all particles with $\mathbf{q}_i$ in $d^3\mathbf{q}_i$. (This can be done satisfactorily with published data provided it is presented in the form of small bins in r_i rather than an averaged distribution over the transverse momentum). In this way the physically meaningful *invariant* distribution is determined by

$$f_i \, d^3\mathbf{q}_i = \sigma_i/N_i \, \Sigma w_i$$

where σ_i is the total cross-section for the production of particle i (i.e. the part of the total cross-section (5.1.1) which gives rise to at least one particle of type c_i in the final state); and N_i is the total number of collisions producing c_i. This weighting procedure is similar to that suggested in the previous chapter to counteract the nonisotropic phase-space effects in the LPS plot.

The simplification implied by the inclusive approach to production dynamics by the lessening of the number of degrees of freedom and the resulting ease in obtaining empirical measurements is clearly very appealing. This is particularly true as one goes to higher and higher incident beam momenta, since the probability of producing a large number of particles in a collision is then increased, and a large part of the cross-section will be built up from exclusive n-body reactions where n is a large number. We stated in the Introduction, however, that one of the interesting features of production dynamics was how *few* particles on average are produced in high energy production reactions, and indeed, the average multiplicity of such produced particles does seem to have only a logarithmic or very small power dependence on s. Thus typically at 300 GeV/c the average number of π^-'s produced in pp collisions is found to be only $\approx 3{\cdot}5$ instead of the ≈ 80 which could in principle be produced. Nevertheless, from the previous chapters it should be quite clear that at present we have difficulty in accurately representing even a three particle final state process with current models, to say nothing of higher multiplicities. Thus when talking about n-body exclusive reactions, for phenomenological purposes n is "large" for $n > 3$ and possibly even for $n = 3$. The recent trend, therefore, away from exclusive reactions and towards the study of inclusive reactions is a very natural progression, particularly with the advent of the new very high energy accelerators such as the Intersecting Storage Rings (ISR) at CERN providing proton-proton collisions at up to 60 GeV c.m. energy and the N.A.L. accelerator providing proton beams at energies up to 500 GeV. However, no matter how appealing is the simplification of the inclusive approach to production dynamics it must be remembered that it is achieved only at the expense of losing a part of the detailed dynamical information.

Of course it is quite true that in principle the two approaches of inclusive and exclusive measurements are entirely equivalent. That is to say, if one measures *all* exclusive cross-sections then one automatically knows *all* the inclusive cross-sections, i.e.

$$\frac{\mathrm{d}\sigma\,(\mathrm{incl})}{\mathrm{d}\phi_1\,\mathrm{d}\phi_2\ldots\mathrm{d}\phi_n} = \sum_{m=0}^{\infty} \frac{1}{m!} \int \frac{\mathrm{d}^{n+m}\,\sigma(\mathrm{excl})}{\mathrm{d}\phi_1\ldots\mathrm{d}\phi_n\,\mathrm{d}\phi_{n+1}\ldots\mathrm{d}\phi_{n+m}}\,\mathrm{d}\phi_{n+1}\ldots\mathrm{d}\phi_{n+m}$$

$$(5.1.4)$$

which is just another way of writing (5.1.2) where $\sigma(\mathrm{excl})$ represents the cross-section for producing exactly $n + m$ particles, and $\sigma(\mathrm{incl})$ is the cross-section for the production of particles $c_1 \ldots c_n$ together with anything else. We have also introduced a short hand form $\mathrm{d}\phi_i$ for the invariant phase-space element

$$\mathrm{d}\phi_i = \frac{\mathrm{d}^3\mathbf{q}_i}{(2\pi)^3 2w_i} \qquad (5.1.5)$$

Conversely, if one measures all inclusive cross-sections then one also knows each exclusive cross-section since

$$\frac{d\sigma(\text{excl})}{d\phi_1 \ldots d\phi_k} = \sum_{j=0}^{\infty} \frac{(-1)^j}{j!} \int \frac{d^{k+j}\,\sigma(\text{incl})}{d\phi_1 \ldots d\phi_k\, d\phi_{k+1} \ldots d\phi_{k+j}}\, d\phi_{k+1} \ldots d\phi_{k+j}$$

$$(5.1.6)$$

This relation may look a little odd at first sight but it can easily be understood as follows. Suppose we wish to isolate that part of a k-body inclusive reaction which in fact has only k particles in it, i.e. the exclusive k-body cross-section. To do so we must first subtract from it the inclusive cross-section where $k + 1$ particles are identified, and hence cannot be part of the k-body process. Now from the $k + 1$ inclusive process we must remove the $k + 2$ particle inclusive contribution and so on so that eventually we are left only with the required k-body exclusive cross-section. This argument is clearly only valid for the case where all the particles are identical, but the generalization to non-identical particles should be fairly obvious.

Although there is this equivalence between the total set of inclusive and exclusive measurements (which incidentally must include the detection of neutral particles) in practice it is difficult to measure any but the low-order inclusive distributions. Thus we must stress that such measurements can only be justified if they, by themselves, are found to lead to interesting and worth-while results. For example, if one has to understand in detail each n-body exclusive reaction contributing to the inclusive distribution then clearly the measurement of one or two low-order inclusive distributions is not going to be a very rewarding proposition.

Fortunately recent work has suggested that the dynamics underlying these inclusive distributions are no more complicated than for exclusive reactions, and in fact, may be somewhat simpler. For instance, the inclusive distribution of order zero, i.e. the total cross-section of (5.1.1), is made up by the sum of the integrated n-body cross-sections which are clearly quite complicated. However, unitarity relates this total cross-section, via the optical theorem, to the forward part of the $a + b$ elastic scattering amplitude. Thus there is no need, $a\ priori$, to have a detailed knowledge of all of the separate n-body reactions contributing to the total cross-section, since it can be calculated instead from a knowledge of the "simple" forward elastic scattering amplitude. In just the same way it has been proposed (Mueller, 1970) that there is a generalized optical theorem which relates the inclusive distributions to the forward amplitude for 3-body $\rightarrow$ 3-body (and in general $m + 2 \rightarrow m + 2$) elastic scattering, where now "forward" implies the out-going particles have the same momenta as the incoming ones.

However, before discussing the implications of this generalized optical theorem which we shall do later in this chapter, it is worth stressing that the inclusive distributions in any case represent some rather useful average properties of multiparticle reactions. For example, from (5.1.1) and (5.1.2) it can be seen that

$$\int f_i(s, \mathbf{q}_i)\, \mathrm{d}\phi_i = \sum_{n_i \geqslant 1} n_i \sigma(n_i) \tag{5.1.7}$$

where $\sigma(n_i)$ denotes the integrated cross-section for the process $a + b \to$ exactly n_i particles of type c_i + anything else. The factor n_i comes from the ratio of $n_i!$ and $(n_i - 1)!$ in the corresponding normalization factors N_n which are used in the definition of $\sigma(n_i)$ and f_i respectively. In other words, an event with n_i particles of type c_i in the final state is counted n_i times in the single-particle inclusive measurement. The right-hand side of (5.1.7) therefore is simply proportional to the average multiplicity, $\langle n_i \rangle$, of particles of type c_i produced in $a + b$ collisions, i.e.

$$\Sigma n_i \sigma(n_i) = \sigma_{ab} \langle n_i \rangle \tag{5.1.8}$$

where σ_{ab} is the total cross-section for the process $a + b \to$ anything. Frequently instead of the total cross-section the average multiplicity $\langle n_i \rangle$ is defined in terms of the total *inelastic* cross-section, i.e.

$$\sigma_{ab}(\mathrm{inel}) = \sigma_{ab}(\mathrm{tot}) - \sigma_{ab}(\mathrm{elastic})$$

This is purely a matter of personal preference. (It arises because it is not clear whether one should consider the elastic amplitude as a production amplitude giving rise to two particles in the final state which just happen to be the same as the initial pair, or whether it truly has a different dynamical origin). In both cases, however, the left hand sides of (5.1.7) and (5.1.8) remain unchanged, and only the definitions of σ_{ab} and $\langle n_i \rangle$ are altered.

Similarly, the two-particle inclusive distribution $f_{ij}(s, \mathbf{q}_i, \mathbf{q}_j)$ is proportional to the average number of particles of type c_i and c_j produced in $a + b$ collisions. More specifically

$$\int f_{ij}(s, \mathbf{q}_i, \mathbf{q}_j)\, \mathrm{d}\phi_i\, \mathrm{d}\phi_j = \sum_{n_i,\, n_j} n_i(n_j - \delta_{ij})\, \sigma(n_i, n_j)$$
$$= \sigma_{ab} \langle n_i(n^j - \delta_i^{\,j}) \rangle \tag{5.1.9}$$

where $\sigma(n_i, n_j)$ is the integrated cross-section for the production of n_i particles of type c_i and n_j particles of type c_j. The delta function δ_{ij} expresses the fact that if c_j is the same type of particle as c_i (i.e. they are both π^-'s, for instance) then in selecting c_i one has only $n_i - 1$ particles of the same type from which to select c_j. The generalization of (5.1.7) (5.1.9) to the case of the m-particle inclusive distribution and the average m-particle multiplicity is straightforward.

Besides the average multiplicity given by (5.1.7) (5.1.8), we can also define the associated average multiplicity by the partial integral of say, $f_{i,\,j}(s, \mathbf{q}_i, \mathbf{q}_j)$, such that

$$\int f_{i,\,j}(s, \mathbf{q}_i, \mathbf{q}_j)\, \mathrm{d}\phi_j \;=\; f_i(s, \mathbf{q}_i)\, \langle n_j - \delta_{ij} \rangle_{q_i} \tag{5.1.10}$$

This associated multiplicity $\langle n_j \rangle_{q_i}$ or $\langle n_i - 1 \rangle_{q_i}$ is therefore the number of particles of type c_j produced in an event where a particle c_i with momentum q_i has already been detected. (For identical particles this is $\langle n_i - 1 \rangle_{q_i}$ because we have already counted one by the first detection). If the two-particle inclusive distribution f_{ij} is simply the product of two one-particle distributions, divided by σ_{ab} to ensure f_{ij} is expressible in the same units as f_i, i.e.

$$f_{ij}(s, \mathbf{q}_1, \mathbf{q}_j) = f_i(s, \mathbf{q}_i)\, f_j(s, \mathbf{q}_j)/\sigma_{ab} \tag{5.1.11}$$

then by integrating (5.1.10) we find

$$\langle n_j - \delta_{ji} \rangle_{q_i} = \langle n_j \rangle \tag{5.1.12}$$

That is to say, the fact that one particle c_i has been detected has no effect on the probability of detecting another particle, c_j. Clearly this is a rather extreme supposition. Even without understanding the underlying dynamics of production processes in any detail, such simple considerations as charge conservation, for instance, suggest that if a positively charged particle is produced, one automatically expects a negatively charged particle should also have been produced. Also at low energies where energy-momentum conservation imposes a strong restriction on the total number of particles which can be formed, the very fact that particle c_i has been detected reduces the probability that c_j can also be produced.

What we are saying, therefore, is that normally there will exist correlations between the production of particles c_i and c_j and more generally the production of any number of $c_i, \ldots, c_m$ particles. In order to obtain a measure of the size of these correlations it is useful to define a set of correlation functions $\rho_{i, \ldots, m}^{(s,\, q_i \cdots q_m)}$ such that

$$\frac{1}{\sigma_{ab}}\, f_i(s, \mathbf{q}_i) = \rho_i(s, \mathbf{q}_i)$$

$$\frac{1}{\sigma_{ab}}\, f_{ij}(s, \mathbf{q}_i, \mathbf{q}_j) = \rho_i(s, \mathbf{q}_i)\, \rho_j(s, \mathbf{q}_i) + \rho_{ij}(s, \mathbf{q}_i \mathbf{q}_j) \tag{5.1.13}$$

$$\frac{1}{\sigma_{ab}}\, f_{ijk}(s, \mathbf{q}_i, \mathbf{q}_j, \mathbf{q}_k) = \rho_i(s, \mathbf{q}_i)\rho_j(s, \mathbf{q}_j)\rho_k(s, \mathbf{q}_k) + \sum_{\text{perms}} \rho_i(s, \mathbf{q}_i)\rho_{jk}(s, \mathbf{q}_j \mathbf{q}_k)$$

$$+ \rho_{ijk}(s, \mathbf{q}_i \mathbf{q}_j \mathbf{q}_k),$$

etc.

If we define the phase-space integrals of these correlation functions by $R^{(s)}_{i,\ldots m}$ then

$$R_i(s) = \frac{1}{\sigma_{ab}} \int \rho_i(s, \mathbf{q}_i)\, \mathrm{d}\phi_i = \langle n_i \rangle \qquad (5.1.14)$$

and

$$R_{ij}(s) = \frac{1}{\sigma_{ab}} \int \rho_{ij}(s, \mathbf{q}_i \mathbf{q}_j)\, \mathrm{d}\phi_i\, \mathrm{d}\phi_j = \langle n_i(n_j - \delta_{ij}) \rangle - \langle n_i \rangle \langle n_j \rangle \qquad (5.1.15)$$

For identical particles

$$\begin{aligned}
R_{ii} &= \langle n(n-1) \rangle - \langle n \rangle^2 \\
&= D^2 - \langle n \rangle \qquad (5.1.16)
\end{aligned}$$

where D is the "dispersion" and gives a measure of the spread of values of n about the average $\langle n \rangle$, i.e.

$$D^2 = \langle (n - \langle n \rangle)^2 \rangle = \langle n^2 \rangle - \langle n \rangle^2. \qquad (5.1.17)$$

Notice that at fairly low energies the number of particles that can be produced will be rather limited and hence we would expect the spread in multiplicity, i.e., D, to be small. From (5.1.16) therefore, we see that R_{ii} is likely to be negative at least for low energies, which is to say, we expect energy-momentum conservation to impose a *negative* correlation effect on the produced particles. Such an effect is indeed, seen in the experimental data, for instance in π^- production from pp collisions ($R_{\pi^- \pi^-} \approx -0{\cdot}25$ at $s = 50\,\mathrm{GeV}^2$), though one of the interesting features of these data, as we shall see later, is that $R_{\pi^- \pi^-}$ becomes positive at about $s = 100\,\mathrm{GeV}^2$ and apparently increases quite rapidly thereafter.

Besides the average multiplicity functions, from a knowledge of the relevant inclusive distributions, we can also calculate other average quantities of multiparticle reactions such as $\langle q_i^\mu \rangle$ the average momentum carried off by all particles of type c_i in an $a + b$ collision. To do this we simply have to note that f_i is the probability of producing a particle c_i with momentum q_i^μ.

Thus the product $q_i^\mu f_i$ integrated over all allowed regions of phase-space must be equal to the average value of q_i^μ obtained in the collision modulated by the size of the ab total cross-section, i.e.

$$\int q_i^\mu f_i(s, \mathbf{q}_i)\, \mathrm{d}\phi_i = \sigma_{ab} \langle q_i^\mu \rangle \qquad (5.1.18)$$

If we now sum over all types of produced particles we obtain, in the c.m. frame,

$$\sum_i \int \mathbf{q}_i f_i(s, \mathbf{q}_i) \, d\phi_i = 0$$

$$\sum_i \int w_i f_i(s, \mathbf{q}_i) \, d\phi_i = \sqrt{s}\sigma_{ab}$$

$$(5.1.19)$$

which just expresses the fact that the c.m. momentum or energy carried off by the produced particles must add up to their respective initial values.

Clearly many other relationships and constraints between the various cross sections, such as those expressing charge conservation, can be derived in a similar way. One rather elegant way in which all of these constraints can be formulated in a systematic fashion has been proposed by Mueller (1971), Biebel and Wolf (1971), Brown (1972) and Koba *et al.* (1972). These authors introduce a *generating functional* which they define as

$$F(s; h_i(\mathbf{q}_i)) = \sum_{n=0}^{\infty} N_n \int \prod_{j=1}^{n} d\phi_j \frac{d^n\sigma(\text{excl})}{d\phi_1 \dots d\phi_n} h_1(\mathbf{q}_1) \dots h_n(\mathbf{q}_n) \quad (5.1.20)$$

where N_n is the usual normalization factor $(= 1/n!$ for identical particles) and $h_i(\mathbf{q}_i)$ represent variable functions defined over $\mathbf{q}_i$. From this definition we can see that

$$\frac{d^n\sigma(\text{excl})}{d\phi_1 \dots d\phi_n} = \frac{\delta F}{\delta h_1(\mathbf{q}_1) \dots \delta h_n(\mathbf{q}_n)}\bigg|_{h_i(\mathbf{q}_i)=0,\,\text{all }i} \quad (5.1.21)$$

Also by comparison with (5.1.4) and (5.1.6) it follows that

$$\frac{d^m\sigma(\text{incl})}{d\phi_1 \dots d\phi_m} = \frac{\delta F}{\delta h_1(\mathbf{q}_1) \dots \delta h_m(\mathbf{q}_m)}\bigg|_{h_i(\mathbf{q}_i)=1} \quad (5.1.22)$$

and hence, for $m = 0$

$$\sigma_{\text{tot}} = F(s; h_i(\mathbf{q}_i))\big|_{h_i(\mathbf{q}_i)=1}$$

(In case the mathematical jargon looks a little off-putting at first sight we would suggest the reader considers the simplest possible case for the functions $h_i(\mathbf{q}_i)$ and simply write $h_i(q_i) = h$ for all i. A few minutes' thought should then be sufficient to convince him of the validity of the above expressions.)

By appropriately choosing the functions $h_i(\mathbf{q}_i)$, all the constraints and relations between the exclusive and inclusive cross-sections can, in fact, be derived. As a simple example of this formalism, let us again deduce the sum-rules (5.1.19) which are implied by energy–momentum conservation. Following the method of Brown (1972) let us consider the transformation

$$h'(\mathbf{q}) = e^{ipx}h(q), \quad (5.1.23)$$

where we drop the indices i and consider for simplicity only one type of particle. Inserting (5.1.23) into (5.1.20) and noting that in each exclusive reaction

$$P \equiv p_a + p_b = q_1 + \ldots + q_n,$$

we see that the generating functional transforms as

$$F(s; h'(\mathbf{q})) = e^{iPx} F(s; h(\mathbf{q})), \qquad (5.1.24)$$

where P is the total momentum of the system. Although (5.1.24) looks as though it provides a continuum of relations depending on the value of x it can be shown (Brown, 1972) that these are in fact redundant. Hence it is sufficient to consider only infinitesimal transformations, and differentiating w.r.t. x_μ at $x_\mu = 0$ we obtain

$$iP_\mu F(s; h(\mathbf{q})) = \int d^3\mathbf{q}' \, \frac{\delta F(s, h(\mathbf{q}))}{\delta h(\mathbf{q}')} \, iq'_\mu h(\mathbf{q}') \qquad (5.1.25)$$

Now evaluating this equation at $h(q) = 1$ gives the result (5.1.19). Similarly, by considering further derivatives of (5.1.24) all other constraints due to energy–momentum of the inclusive cross-sections can be deduced.

It is worth noting that while the meaning of these sum-rules may seem fairly obvious, they do often provide non-trivial constraints on models of inclusive cross-sections. (One can draw a close analogy here with the more familiar unitary relationship of the S-matrix. Although this, too, simply expresses the conservation of probability it is, of course, well known that unitarity puts very severe restrictions and bounds on the form of the scattering amplitude.)

An attractive feature of the generating functional technique is that it can also be used for more complicated processes such as *semi-inclusive reactions*. This has been stressed by Koba *et al.* (1972). See also Bialas *et al.* (1972). The first order semi-inclusive cross-section is defined as

$$r \frac{d^3\sigma_r}{d^3\mathbf{q}} \text{ for the process } a + b \to \underbrace{c(\mathbf{q}) + c + \ldots + c}_{r} + \text{ anything not } c, \quad (5.1.26)$$

i.e. as the cross-section when r particles of type c are produced, together with any number of other particles, but only the momentum of one of the particles c is measured. Such processes, together with higher order semi-inclusive cross-sections (i.e. where the momentum of more than one particle of type c is measured, or where r_1 particles of type c_1 and r_2 particles of type c_2 are produced) have an obvious relevance to experimental situations. Often one knows the number and type of *charged* particles which are produced in a high energy collision, for instance by the tracks they leave in a bubble chamber, and in principle one can measure the momenta of these charged particles. However, one still does not possess sufficient information to deduce the actual n-body exclusive reaction leading to the production of these charged particles since the number and individual momenta of any produced neutral particles are generally incalculable without further empirical measurements.

In order to relate these semi-inclusive distributions to the inclusive (or exclusive cross-sections) we merely have to differentiate the generating functional once and put

$$h(q) = h.$$

Then for the case where only type of particle is detected, from (5.1.20) we obtain

$$\omega \left. \frac{\delta F}{\delta h(\mathbf{q})} \right|_{h(\mathbf{q})=1} = \sum_{r=1}^{\infty} (1 + h)^{r-1} r \frac{\mathrm{d}^3 \sigma_r}{\mathrm{d}\phi_1} \tag{5.1.27}$$

or, in terms of the inclusive distributions, using (5.1.22)

$$\omega \left. \frac{\delta F}{\delta h(\mathbf{q})} \right|_{h(\mathbf{q})=1} = \sum_{r=1}^{\infty} \frac{h^{r-1}}{(r-1)!} \int \ldots \int f_{1,\ldots,r}(s, \mathbf{q}_1, \ldots \mathbf{q}_r) \, \mathrm{d}\phi_2 \ldots \mathrm{d}\phi_r. \tag{5.1.28}$$

By comparing the coefficients of h of (5.1.27) and (5.1.28) we find firstly that

$$\sum_{r=1}^{\infty} r \frac{\mathrm{d}^3 \sigma_r}{\mathrm{d}\phi_1} = f_1(s, \mathbf{q}_1), \tag{5.1.29}$$

which just states that the inclusive cross-section is composed of semi-inclusive cross-sections, corresponding to all possible numbers of particle c that can be produced. Secondly, we have

$$\sum_{r=1}^{\infty} (r-1)r \frac{\mathrm{d}^3 \sigma_r}{\mathrm{d}\phi_1} = \langle r - 1 \rangle_{\mathbf{q}_1} f_1(s, \mathbf{q}_1) = \int f_{1,2}(s, \mathbf{q}_1, \mathbf{q}_2) \, \mathrm{d}\phi_2 \tag{5.1.30}$$

and so on. The generalization of these relations to the situation where more than one kind of particle is detected is fairly straightforward. See Bialas *et al.* (1972) for further details. Also for a fuller discussion of the generating functional approach to inclusive distributions we would refer the interested reader to the very detailed review by Koba (1972).

While the inclusive distributions, as we have discussed above, provide very useful information on the average properties of production phenomena, their recent popularity is due to two further considerations. One of these is the proposal by Mueller that the m-particle inclusive distribution can be related to the analytically continued absorptive part of the $m + 2 \rightarrow m + 2$ forward elastic amplitude. The second stems from the intuitive proposals of Feynman and Yang and collaborators who, noting the nature of high energy total cross-sections and diffraction-dissociation processes, have suggested the scaling and limiting fragmentation hypotheses for high energy inclusive reactions. We shall discuss these hypotheses in some detail. However, as a preliminary to this, it will be useful first of all to consider the kinematics and physical region boundary of the inclusive distributions.

5.1.a. *Kinematics of single particle distributions.* For our purpose it will be sufficient to restrict ourselves to single particle distributions, though hopefully the extension to multi-particle distributions should not be too obscure.

In order to be quite specific let us consider the process $a + b \to c + X$ where X stands for any particles undetected. As usual, we shall assume the positive z-direction to be along the incident beam direction, i.e. along $\mathbf{p}_a$. in the laboratory system. Thus, writing a typical four-momentum as (q_0, q_x, q_y, q_z) where $(q_x, q_y) \equiv \mathbf{r}$ (the component of the momentum perpendicular to the beam direction) we have

$$p_b^L = (m_b, \mathbf{0}, 0)$$

$$p_a^L = \left(\frac{\sqrt{s}}{m_b} E(s, m_b^2, m_a^2) - m_b, \mathbf{0}, \frac{\sqrt{s}}{m_b} P(s, m_b^2, m_a^2) \right) \tag{5.1.31}$$

and therefore

$$p_a^L + p_b^L = \frac{\sqrt{s}}{m_b} (E(s, m_b^2, m_a^2), \mathbf{0}, P(s, m_b^2, m_a^2))$$

$$\equiv \sqrt{s}(\cosh \zeta, \mathbf{0}, \sinh \zeta) \tag{5.1.32}$$

where the functions $P(x, y, z)$, $E(x, y, z)$ are the same as those defined in (2.1.4). Besides the lab. frame it will be necessary also to consider two other frames of reference, namely the centre-of-mass frame (denoted by c) in which $p_a + p_b = 0$, and the projectile frame (P)—the frame in which the beam particle is considered to be at rest, i.e. with $\mathbf{p}_a = 0$. Clearly in this latter frame the four-momenta of particles a and b are given from (5.1.31) simply by interchanging the labels a and b so that

$$p_a^P = (m_a, \mathbf{0}, 0)$$

$$p_b^P = \frac{\sqrt{s}}{m_a} (E(s, m_a^2, m_b^2) - m_a, \mathbf{0}, -\frac{\sqrt{s}}{m_a} P(s, m_a^2, m_b^2)) \tag{5.1.33}$$

and

$$p_a^P + p = \frac{\sqrt{s}}{m_a} (E(s, m_a^2, m_b^2), \mathbf{0}, -P(s, m_a^2, m_b^2))$$

$$\equiv \sqrt{s}(\cosh \zeta', \mathbf{0}, -\sinh \zeta') \tag{5.1.34}$$

F

while in the c.m. frame (see 2.1.6)

$$p_a^c = (E(s, m_a^2, m_b^2), \mathbf{0}, P(s, m_a^2, m_b^2))$$
$$= m_a(\cosh \zeta', \mathbf{0}, \sinh \zeta') \tag{5.1.35}$$
$$p_b^c = (E(s, m_b^2, m_a^2), 0, -P(s, m_b^2, m_a^2))$$
$$= m_b(\cosh \zeta, \mathbf{0}, -\sinh \zeta) \tag{5.1.36}$$

and $p_a^c + p_b^c = (\sqrt{s}, \mathbf{0}, 0)$. As we have already indicated, these frames are related by Lorentz boosts K_z in the beam direction such that

$$q_\mu^L = \exp(i\zeta K_z) q_\mu^c$$
$$q_\mu^P = \exp(-i\zeta' K_z) q_\mu^c \tag{5.1.37}$$

and therefore

$$q_\mu^L = \exp(i(\zeta + \zeta')K_z)q_\mu^P$$

where

$$\exp(\pm \zeta) = [E(s, m_b^2, m_a^2) \pm P(s, m_b^2, m_a^2)]/(2m_b)$$

and

$$\tag{5.1.38}$$

$$\exp(\pm \zeta') = [E(s, m_a^2, m_b^2) \pm P(s, m_b^2, m_a^2)]/(2m_a)$$

If we introduce the variables

$$q^\pm = q_0 \pm q_z$$

which have the property that $q^+ q^- = r^2 + m^2 = m'^2$ the "transverse mass" squared then these transformations take on the particularly simple form

$$(q^\pm)^L = \exp(\pm \zeta)(q^\pm)^c$$
$$(q^\pm)^P = \exp(\pm \zeta')(q^\pm)^c \tag{5.1.39}$$

while

$$\mathbf{r}^L = \mathbf{r}^c = \mathbf{r}^P$$

because the Lorentz boost in the z-direction does not affect the transverse components.

The simple connection between these three frames of reference becomes even more apparent if we define a rapidity variable y by

$$m' \exp(\pm y) = q^\pm.$$

Inverting this relation we have

$$y = \tfrac{1}{2}\log\left(q^+/q^-\right) = \tfrac{1}{2}\log\left(\frac{q_0 + q_z}{q_0 - q_z}\right) = \log\left(\frac{q_0 + q_z}{m'}\right)$$

$$= -\log\left(\frac{q_0 - q_z}{m'}\right) \tag{5.1.40}$$

such that

$$y^L = \tfrac{1}{2}\log\left(q^+/q^-\right)^L = \tfrac{1}{2}\log\left(\frac{e^{\zeta}(q^+)^c}{e^{-\zeta}(q^-)^c}\right) = \tfrac{1}{2}\log\left(q^+/q^-\right)^c + \zeta$$

$$= y^c + \zeta,$$

and similarly

$$y^L = y^c - \zeta'$$

Thus the single particle inclusive distributions of particle c, $w\,d^3\sigma/d^3\mathbf{q}$, considered as functions of this rapidity variable y, i.e. $d^3\sigma/dy\,d^2\mathbf{r}$ will have the same *shape* no matter whether y is calculated in the lab, c.m. or projectile frames. The only difference will be an overall *translation* along the y-axis determined by the boost parameters ζ or ζ'. From (5.1.38) it will be seen that as $s \to \infty$

$$e^{\zeta} \sim \sqrt{s/m_b}, \qquad e^{\zeta'} \sim \sqrt{s/m_a} \tag{5.1.41}$$

and therefore these translations are asymptotically of the order of $\log s$ which, of course, is just the difference in rapidities of the colliding particles a and b.

Returning now to the inclusive distribution $f_1(s, \mathbf{q})$ let us again note that except in polarization experiments, f_1 is, in fact, a function of three variables which we can take as s, q (the longitudinal component of the momentum of particle c_1) and r (the magnitude of the transverse component of this momentum vector). However, these are not the only variables we would choose. For instance, instead of

$$s = (p_a + p_b)^2$$

$$= m_a^2 + m_b^2 + 2m_b E_a^L$$

$$= m_a^2 + m_b^2 + 2m_a E_b^L$$

$$= m_a^2 + m_b^2 + 2m_a m_b \cosh\left(y_a - y_b\right)$$

we might choose E_a^L, E_b^L, $y_a - y_b$ or just simply the incident beam momentum

$$p_a^L = \frac{\sqrt{s}}{m_b}\, P(s, m_a^2, m_b^2)$$

Also, instead of the longitudinal momentum component of particle c_1 defined in any of the three frames L, P or c, i.e., q^L, q^P or q^c it will be useful to introduce a scaled variable, (the so-called Feynman scaling variable) by dividing q by its maximum possible value. Thus in the c.m. frame this scaling variable x is defined as

$$x = \frac{q^c}{q^c_{\max}}$$

(5.1.42)

where $q^c_{\max}$ is obtained by considering the case of only two particles in the final state in which case $q^c_{\max} \sim p^c_a \sim \sqrt{s}/2$ as $s \to \infty$. The advantage of using this scaled variable is that when one plots $f_1(s, q^c, r^c) \propto w(\mathrm{d}^2\sigma/\mathrm{d}q^c\,\mathrm{d}r^2)$ as a function of q^c the distribution extends from $-q^c_{\max}$ to $+q^c_{\max}$, a region which grows like $\sqrt{s}$. By scaling q^c by $q^c_{\max}$ therefore, the distribution has a range from $-1 < x < 1$ irrespective of the value of s. In this way it is possible to compare the *shapes* of the distributions corresponding to different values of the incident beam momenta, the idea being to try to detect features of the distributions which are common to all energies, such as a common asymptotic limit. Notice that if particle c_1 is not the same as one of the initial particles a or b, $q^c_{\max}$ will not in general be p^c_a but $P(s, m^2_c, m^2_d)$ where m_d is the smallest possible mass of the system X, although the difference which is $O(1/\sqrt{s})$ is clearly negligible for large s. Similarly, if the distributions in the longitudinal variable are to be compared for various fixed values of the transverse momenta, then the longitudinal momentum q should be scaled by the maximum value of the *longitudinal* momentum $q^\parallel_{\max}$ rather than the maximum value of the *total* three momentum $|\mathbf{q}|_{\max}$, i.e. we should ideally use the variable

$$x' = \frac{q^c}{\sqrt{(|\mathbf{q}|^2_{\max} - r^2)}},$$

(5.1.42′)

which does have the limits

$$-1 < x' < 1.$$

However, in this case too, it is clear that $x' \sim x$ as $s \to \infty$, so that normally such fine distinctions are overlooked.

In the same way it is possible to introduce a scaled rapidity variable, the "reduced rapidity", where

$$\xi = \frac{y_b - y_c}{y_b - y_a},$$

(5.1.43)

which lies between the limits 0, 1 for any value of s and for *any* reference frame which can be reached by Lorentz boosts in the z (longitudinal) direction. This comes immediately from the definition of the rapidity variable in (5.1.40) and by noting that rapidities in different reference frames are related by simple additive translations.

So far we have considered possible alternatives to the variables s and q in defining the single-particle inclusive distribution. However, there are also one or two commonly used alternatives for the transverse momentum component r. One of these is θ^L, i.e. the angle of particle c relative to the beam direction measured in the laboratory system. This is the natural variable to measure in counter experiments (together with q^L), but it can easily be transformed into a transverse momentum value by noting that

$$\tan \theta^L = \frac{r^L}{q^L}. \tag{5.1.44}$$

Also in theoretical work there is a tendency to work with Lorentz scalars rather than frame-dependent momentum components. Here therefore, instead of q and r let us define the Lorentz invariant quantities

$$X^2 = (p_a + p_b - q_c)^2 \tag{5.1.45}$$

and either

$$t_c = (p_a - q_c)^2$$

or

$$u_c = (p_b - q_c)^2 \tag{5.1.46}$$

where X is the "missing mass" of the system and t_c and u_c are the squares of the momentum transfers from particles a to c and b to c respectively. These three quantities are related to s by

$$s + t_c + u_c = m_a^2 + m_b^2 + m_c^2 + X^2 \tag{5.1.47}$$

so that $f_i(s, q, r)$ can be considered as a function either of (s, X^2, t_c) or (s, X^2, u_c). We leave it as a simple exercise for the reader to show that the minimum value of $(-t_c)$ for large s is given by

$$(-t_c)_{\min} \sim - \frac{X^2}{s}\left(m_b^2 - \frac{m_c^2}{1 - X^2/s}\right) \tag{5.1.48}$$

except in the particular cases when $m_c < m_b$ and $X^2/s < 1 - (m_c/m_b)$ or when $m_c = m_b$ and $X_{\min} < X < m_a$ where

$$(-t_c)_{\min} \sim -(m_b - m_c)^2 \tag{5.1.49}$$

The corresponding minimum value of $-u_c$ is simply given by interchanging m_a and m_b. Notice that in these reactions, unlike elastic scattering, it is often possible, therefore, to reach fairly large positive values of the momentum transfers. The importance of this in the analysis of possible exchange mechanisms has been pointed out by Roberts and Roy (1972).

In Table 4 we summarize these possible choices for the independent variables (besides s) of the single particle inclusive distribution together with

TABLE 4. Variables for Inclusive Distributions

longitudinal variable	transverse variable	phase-space element $d^3\mathbf{q}/2w$
q^c	r	$2\pi r \, dr \, dq^c/2w^c$
q^L or q^P	r^2	$\pi \, dr^2 \, dq^{L,\,P}/2w^{L,\,P}$
$x = q^c/p_a^c$	r^2	$\frac{1}{2}\pi \, dr^2 \, dx \left/ \left(x^2 + \dfrac{m_c^2 + r^2}{(p_a^c)^2} \right)^{\frac{1}{2}} \right.$
y	r^2	$\frac{1}{2}\pi \, dr^2 \, dy$
q^L	θ^L	$\frac{1}{2}\pi (q^L)^2 \, dq^L \, d\cos\theta^L/w^L$
X^2	t (or u)	$\frac{1}{4}dt \, dX^2/(\sqrt{s}\, p_a^c)$

the corresponding expression for the Lorentz invariant phase-space element $d^3\mathbf{q}/2w$.

5.2. Limiting fragmentation and scaling

Armed with this brief introduction to the kinematics of inclusive distributions we now turn our attention to the shape of these distributions, i.e. to the dynamics underlying such inclusive reactions. Remembering that the single particle distribution is a function of three variables, e.g. s, q, r, the questions that come to mind therefore are

(a) How do the distributions vary with s?
(b) For fixed s how do the distributions vary with q and r?

The answer to the second question, as we shall see, depends to some extent on the particular particle identified and the initial particles which collide to produce it. To answer the first question Yang and collaborators and independently Feynman have put forward intuitive arguments to suggest that $f(s, q, r)$ approaches some limiting (finite) distribution as $s \to \infty$.

Specifically Benecke, Chou, Yang and Yen (Benecke 1969) consider either the laboratory or projectile rest frame. For example, in the laboratory frame, their *limiting fragmentation hypothesis* is that $f_c(s, q^L, r)$ approaches an asymptotic limit for large s, i.e.

$$f_c(s, q^L, r) \to \hat{f}_c(q^L, r) \tag{5.2.1}$$

provided q^L is held fixed as $s \to \infty$. Since initially the target particle b is at rest in this frame, while the beam particle (projectile a) has a momentum increasing with s, they consider such detected particles c with finite momentum q^L as representing *fragments* of the target particle. That is to say, $\hat{f}_c(q^L, r)$ reflects the break-up of the target particle b and is independent of the projectile except in determining an overall normalization factor. A handy notation for this process is given by

$$b \overset{a}{\to} c$$

which means the fragmentation of b into a particle c under the impact of particle a. A similar statement to (5.2.1) holds in the projectile rest frame, i.e. for the process $a \overset{b}{\to} c$†

$$f_c(s, q^P, r) \to \tilde{\hat{f}}_c(q^P, r), \tag{5.2.2}$$

when q^P is held fixed as $s \to \infty$. The extension of this limiting fragmentation hypothesis to the case of two- and more-particle distributions is immediate.

The intuitive argument for the limiting fragmentation hypothesis is based on the geometrical (droplet) picture for diffractive scattering considered by Wu and Yang (1965), Byers and Yang (1966) and Chou and Yang (1968). In this model the two colliding particles are interpreted as two spatially extended objects. As the energy increases, the projectile undergoes increasing Lorentz contraction as seen by the target, i.e. in the lab. frame. However, it is also observed that in the case of elastic scattering $d\sigma_{el}/dt$ and therefore also σ_{el}, as well as σ_{tot} apparently all approach asymptotically constant values. Hence, whatever the dynamical mechanism is which controls and transfers momentum and quantum numbers between the matter in the projectile and the matter in the target, it does not seem to change appreciably as the energy increases and the projectile is still further contracted. In the same way, it is argued, one might expect the excitation and break-up of the target also to be asymptotically independent of the energy, and therefore that the single- and many-particle distributions approach limiting values.

Feynman, on the other hand, considers the inclusive processes solely in the centre-of-mass system, but again by mainly intuitive arguments is led to a proposal for the high energy behaviour of the inclusive particle distributions. Specifically *Feynman's scaling hypothesis* for the single particle distribution is that

$$f_c(s, q^c, r) \underset{s \to \infty}{\to} \hat{f}_c(x, r) \tag{5.2.3}$$

where $x = q^c/q^c_{max}$ or more simply, since we are considering the limit $s \to \infty$,

---

† We denote the distribution in the P-system by $\tilde{f}_c$ rather than $\hat{f}_c$ to stress that it may have a different functional form to $\hat{f}_c$. However, from now on we shall not be so pedantic and write $\hat{f}_c$ for the distribution defined in any referance frame.

$x = q^c/\frac{1}{2}\sqrt{s}$. Thus the scaling hypothesis says that if one plots the distributions over the longitudinal momenta, using the scaled variable x rather than q^c so that the distributions lie in the range $-1 < x < 1$ for all s, then these distributions should tend to some limiting shape as $s \to \infty$.

Notice, however, that from (5.1.39) we have

$$q^L = \tfrac{1}{2}(q^+)^L - \tfrac{1}{2}(q^-)^L = \tfrac{1}{2}e^{\zeta}(q^+)^c - \tfrac{1}{2}e^{-\zeta}(q^-)^c$$

$$= \tfrac{1}{2}\sqrt{s}((x^2 + 4m'^2)\sinh\zeta + x\cosh\zeta) \qquad (5.2.4)$$

where $\sinh\zeta = q^c/m_a$ (see (5.1.38)) and $m'^2 = m_c^2 + r^2$. Hence we see that fixed values of $x \gg -2m'\sqrt{s}$ imply fixed (finite) values of q^L. Similarly, fixed values of $x \gg 2m'\sqrt{s}$ imply fixed values of q^P. Therefore, combining these two cases we arrive at the conclusion that, for $x^2 \gg 4m'^2/s$, Feynman's scaling hypothesis is equivalent to the hypothesis of limiting fragmentation. It should be stressed, however, that both are *asymptotic* statements, and at finite energies the variables q^L, q^P and x have a one-to-one correspondence only up to the factors of the order of $1/\sqrt{s}$. Thus when a distribution in the lab. frame, for example, seems to have reached a limiting value, $f(s, x, r)$ could still show some $s^{-\frac{1}{2}}$ dependence.

From our above remarks it is clear that the point $x = 0$ is of special interest. It clearly does not correspond to finite longitudinal momentum in either the lab. or projectile frames, but rather to finite momentum in the c.m. system. This region is sometimes called the pionization region or central region though the latter term often refers not only to $x = 0$ but also to the intermediate regions $4m'^2/s \geqslant x^2 > 0$ which corresponds to values of q^L or $q^P \sim O(s^\beta)$ where $0 < \beta < \frac{1}{2}$. Since there is no uniformity in the literature on these definitions, let us adopt the above convention and refer to the *pionization region* as finite values of q^c for all s (i.e. an asymptotic statement) and the *central region* as $4m'^2/s \geqslant x^2 \geqslant 0$.

Notice now that while Feynman's scaling hypothesis includes the limiting fragmentation hypotheses it is more general in that it also predicts a limiting distribution in the central region. Furthermore, Feynman (1969) suggests that not only does $f(s, x, r)$ become independent of s for large s in this region but that $\hat{f}(x, r)$ should also be independent of x. That is to say, the distribution will be independent of both the target and projectile particles for "wee" values of $|x|$ $(<2m'/\sqrt{s})$ except for a normalization factor. Feynman's suggestion was based on a hadronic bremsstrahlung picture, though in fact, the prediction has been around for a long time. It was first found by Amati *et al.* (1962) as one of the consequences of the multi-peripheral model; and it was subsequently also obtained by Heisenberg (1963) as a result of the Lorentz contraction of the overlap of the target and projectile discs in the c.m. system.

It is interesting to note that one of the consequences of a non-zero scaling function at $x = 0$ is that the average multiplicity $\langle n_c \rangle$ of particles of type c grows like $\log s$. To see this we note from (5.1.7), (5.1.8) that

$$\langle n_c(s) \rangle = \frac{1}{\sigma_{ab}} \int f_c(s, q_c, r) \frac{d^3 \mathbf{q}_c}{w_c}$$

Therefore, if $f_c \to \hat{f}_c(x, r)$, $-1 < x < 1$ this becomes

$$\langle n_c(s) \rangle = \frac{\pi}{\sigma_{ab}} \int \hat{f}_c(x, r) \frac{dx\, dr^2}{[(x^2 + 4(r^2 + m_c^2/s]^{\frac{1}{2}}} \tag{5.2.5}$$

and for large s gives

$$\langle n_c(s) \rangle \approx a_c \log (s/m^2) + b_c$$

where

$$a_c = \frac{\pi}{\sigma_{ab}} \int \hat{f}_c(0, r)\, dr^2$$

and

$$b_c = \frac{\pi}{\sigma_{ab}} \int [\hat{f}_c(x, r) - \hat{f}_c(0, r)] \frac{dx\, dr^2}{x}$$
$$- \frac{\pi}{\sigma_{ab}} \int \hat{f}_c(0, r) \log (1 + r^2/m_c^2)\, dr^2. \tag{5.2.6}$$

Notice that the scaling function evaluated at $x = 0$ as $s \to \infty$ corresponds not only to the distribution of particles in the pionization region, i.e. to particles with finite momenta in the c.m. system, but also to the distribution of particles in the intermediate regions where they have momenta which grow like s^β, $\beta < \frac{1}{2}$. Thus we must be careful to realize that it is the whole of the central region, corresponding to $x = 0$, which determines the behaviour of the average multiplicity in the above scaling argument, and not just the pionization events.

It is also important to stress that while scaling at $x = 0$, as we have seen, implies a logarithmic growth of the average multiplicity with s, this dependence is not in itself a test of scaling. It is quite possible that some non-scaling behaviour for the inclusive distribution $f(s, \mathbf{q}^c)$ at the point $x = 0$ can also lead to a $\log s$ behaviour for $\langle n \rangle$. The statement one can make, however, from (5.2.6) is that if $\langle n \rangle$ is found to have some non-logarithmic dependence, i.e. if $\langle n \rangle \sim s^\delta$, for example, then the inclusive distribution cannot scale at $x = 0$.

At this point, however, it seems appropriate to return to another model which we have already discussed in some detail, i.e. the Chew-Pignotti multi-Regge model, and examine its predictions for the inclusive distribution.

To obtain the single particle distribution from the n-body process we have simply to introduce the appropriate delta function into the De Tar expression (4.2.23), i.e.

$$\frac{d\sigma_{n,i}}{dy}(y, y_b) \propto g^{2n-2} \exp[(2\alpha - 2)y_b] \int \prod_2^n dy_i \delta(y_b - y_n)\delta(y - y_i) \qquad (5.2.7)$$

$$\propto g^{2n-2} \exp\left[(2\alpha - 2)y_b\right] \frac{y^{i-2}}{(i-2)!} \frac{(y_b - y)^{n-1-i}}{(n-1-i)!}, \qquad (5.2.8)$$

where $\delta(y - y_i)$ expresses the fact that we are calculating the probability that the ith particle in the chain has momentum $q_i = m_i' \sinh y_i$. The probability that *any* particle in the chain has this momentum, therefore, is given by

$$\frac{d\sigma_n(y, y_b)}{dy} = \sum_{i=2}^{n-1} \frac{d\sigma_{n,i}(y, y_b)}{dy}$$

$$\propto g^{2n-2} \exp\left[(2\alpha - 2)y_b\right] \cdot \frac{y_b^{n-3}}{(n-3)!}, \qquad (5.2.9)$$

from which we see that the distribution in rapidity y is flat, having no dependence on y at all. The single particle inclusive distribution corresponding to all number of possible particles produced is obtained from (5.2.9) by summing over all n, i.e.

$$\frac{d\sigma}{dy} = \sum_n \frac{d\sigma_n(y, y_b)}{dy} \propto \exp\left[(2\alpha - 2 + g^2)y_b\right], \qquad (5.2.10)$$

which, of course, is also flat in y. Notice that the energy dependence of this distribution $s^{2\alpha - 2 + g^2}$ is the same as for the total cross-section (see 4.2.24). Therefore, we find that if α and g are chosen so that σ_{tot} is constant, then so is the inclusive distribution, i.e. the distribution scales, and as we have already noted in (4.2.27), the average multiplicity is proportional to $\log s$.

Although this is again a very simplified version of the multi-Regge model, it is sufficient to show at least the general features of these models. In fact, Silverman and Tan (1971) have shown that scaling and limiting fragmentation are properties of all such models which are characterized by

(a) damping in the momentum transfer (i.e. in the transverse momentum),
(b) factorization,
(c) the generation of the Pomeron singularity.

Turning briefly now to the question of models and the scaling behaviour at $x = 0$ we note that at first sight it appears the limiting fragmentation hypothesis, based upon the diffraction fragmentation model, suggests that

$\hat{f}(x = 0, r)$ should, in fact, be zero and therefore that $\langle n_c \rangle$ is constant. This comes about as follows. We can write the average multiplicity as[†]

$$\langle n_c \rangle = \frac{1}{\sigma_{ab}} \left\{ \int F(\mathbf{q}^L) \, d^3\mathbf{q}/w + \int \tilde{F}(\mathbf{q}^P) \, d^3\mathbf{q}/w \right\}$$

where F, $\tilde{F}$ are the single particle distributions for target and projectile fragments respectively. Since the diffractive fragmentation model assumes the initial particles are simply excited into fragments, none of which are "slow" in the c.m. frame F and $\tilde{F}$ tend to zero when either q^L or q^P is large respectively. Hence the integrals converge and $\langle n_c \rangle$ is constant. On the other hand, if one modifies and extends the fragmentation model such that F and $\tilde{F}$ tend to some (non-zero) constant for large q^L or q^P then the integrals diverge logarithmically for large s so that once again we find $\langle n_c \rangle \propto \log s$. Models of this type will be discussed in the next chapter.

As we have indicated earlier, the momentum–transfer damping means that it is adequate to consider only the distribution in the longitudinal variable q or, as above, in the laboratory rapidity variable $y = \sinh^{-1}(q^L/m')$ and ignore any dependence on the transverse variables. The factorization property can then be restated as the *short-range correlation condition* for the rapidities, by which we mean that there is no correlation between particles whose rapidities y_i are separated by a distance larger than some correlation length L. (Wilson, 1970). (Fraser *et al.*, 1972a). Thus there will be no correlation between the production of particles c_i and c_j with variables y_i, y_j provided $|y_i - y_j| \gg L$, and similarly any produced particle c_i will have no correlation with the target or projectile so long as $y_i \gg L$ or $y_i \ll y_b - L$ respectively.

In order to see how these conditions lead to the limiting fragmentation and scaling hypotheses let us split the single particle distribution into three parts corresponding to (i) $y < L$ (the target fragmentation region), (ii) $L < y < y_b - L$ (the central region and (iii) $y_b - L$ (the projectile fragmentation region). In (i) the single particle invariant distribution can be considered as a function of the three variables $y_b - y$, y, and r^2 (see Table 4) where we can choose $y_b - y$ instead of s since y is finite (i.e. $< L$) in this region. From the short-range correlation condition, however, this distribution must be independent of $y_b - y$ provided $y_b \gg 2L$ so that $y_b - y \gg L$. Furthermore, since $y_b - y \gg L$ there can be no correlation of the produced particle with the projectile, except for a normalization factor which can be removed by dividing by the total cross-section σ_{ab}. Therefore, in the target fragmentation region, the short-range correlation condition leads to the limiting

---

[†] Here we use the invariant element of phase-space $d^3\mathbf{q}/w$, which differs from that used by Yang and collaborators.

fragmentation hypothesis

$$\frac{w_c}{\sigma_{ab}}\frac{d^3\sigma^c_{ab}}{d^3\mathbf{q}_c} \equiv f^c_{ab}(y_b - y, y, r^2) \simeq \hat{f}^c_a(y, r^2) \quad \text{for} \quad y_b - y \gg L \qquad (5.2.11)$$

Similarly in region (iii), the projectile fragmentation region, the short range correlation condition leads to the result

$$f^c_{ab}(y, y_b - y, r^2) \simeq \hat{f}^c_b(y_b - y, r^2) \quad \text{for} \quad y \gg L \qquad (5.2.12)$$

In the central region, region (ii) where y and $y_b - y$ are both large compared to L, the distribution must be independent of both target and projectile (again up to an overall normalization), hence in this region we recover the scaling behaviour

$$f^c_{ab}(y, y_b - y, r^2) \simeq \hat{f}^c(r^2) \quad \text{for} \quad L \ll y \ll y_b - L \qquad (5.2.13)$$

i.e. a flat distribution in the rapidity y, which depends only on the type of particle produced.

An interesting outcome of this simple short-range correlation picture is that L also provides some indication when we might expect to see a limiting distribution. For example, if s is so small that $y_b < L$ the whole distribution lies within one correlation length so that it is nowhere limiting except perhaps by accident. However, if $y_b \gg L$, then there can exist some values of y such that the conditions for (5.2.11) or (5.2.12) are satisfied, though only if $y_b \gg 2L$. would we expect to see a limiting distribution (i.e. a plateau) in the central region developing. Some theoretical arguments (Abarbanel, 1971) based on Regge phenomenology suggest that $L \approx 1/(1 - \alpha_R)$ where α_R is a typical meson trajectory, i.e. $L \approx 2$. In this case the onset of limiting fragmentation should be observed for $y_b > 2$ which for pp collisions is roughly equivalent to $s = 8\,(\text{GeV})^2$. Notice however that this value of L implies scaling in the central region should not even begin to be seen until $s \gg 64\,(\text{GeV})^2$. This range is now covered by the ISR facility at CERN, and initial measurements of particles in the central region tended to support the presence of a plateau (Sens, 1972). However more recent data indicate a slight (10%) rise of these rapidity distributions through the central region (Foa, 1973).

This brings us to the whole question of the empirical situation in inclusive reactions. It is an indication of the growth of this subject that after only a couple of years' serious investigation it would now require several scores of pages to properly review the current experimental status of limiting fragmentation and scaling. Even if this were feasible, such a review would probably already be out of date by the time of publication. Therefore, rather than attempting such a task let us refer the interested reader to the excellent compilation of inclusive data by Law et al. (1972) and to any of the recent conference reviews of inclusive reactions, e.g., Sens (1972), Ferbel (1972). Here let us merely state that to date it appears that in several reactions there is already good evidence for limiting fragmentation at least within the experimental errors. For example, the q and r distributions of the π^- produced in the reaction $pp \to \pi^- X$ show little energy dependence in the fragmentation region over the five energies considered by Smith (1971) from 13 to 28 GeV. This is still found to be the case even when the energy range is extended to 1500 GeV using the CERN interacting storage rings; (Krisch, 1971; Ratner *et al.* 1971). However, while a limiting distribution seems to have been reached already at fairly low energies in some reactions, such as $pp \to \pi^- X$, it is found that in other reactions there is a much slower approach to limiting fragmentation. One of the intriguing problems of inclusive distributions, therefore, is to try to understand these different rates of approach to a limiting distribution, assuming, of course, such a distribution exists. This problem becomes even more interesting when one looks at the distributions near $x = 0$, i.e. in the central region. Here, although the ISR data tend to have rather large errors, the situation for pp collisions appears to be as follows. The distribution for the process $pp \to pX$ and $pp \to \pi^+ X$ near $x = 0$ show little energy dependence from normal lab. energies ~ 24 GeV to ISR energies > 100 GeV. The distribution for $pp \to \pi^- X$ on the other hand, appears to double in this energy range, (though the experimental errors do allow some room for argument) while the distributions for $pp \to K^- X$ and $pp \to \bar{p}X$ increase by approximately factors of 4 and 10 respectively. (Sens, 1972). This energy dependence of these latter two reactions is particularly noteworthy for the following reason. There have been several proposals to account for the early onset of a limiting fragmentation distribution in certain processes, as we shall discuss later. These proposals all depend on the quantum numbers of the particles concerned. However, if we apply any of these quantum-number criteria to the reactions $pp \to K^- X$ and $pp \to \bar{p}X$ we find that in any case these particular reactions should show an early onset of scaling. The fact that they very markedly do not, therefore, again suggests that the central region is a rather different dynamical region, and is worthy of rather special attention.

With this last point in mind it is relevant to comment briefly on the choice of variables in which to display data. The variables q^L, u_c are best suited for displaying effects in the target fragmentation region, but clearly are not very satisfactory for showing features in the data in the central or projectile fragmentation regions. Using the c.m. variable x or X^2 on the other hand means that we can display limiting fragmentation for both $b \xrightarrow{a} c$ and $a \xrightarrow{b} c$ on the same plot, but the limit will now in general be approached more slowly. Also the central region is confined to the single point $x = 0$ as $s \to \infty$. Alternatively, the rapidity y, or reduced rapidity of (5.1.43), ξ allows us to study the whole of this central region in detail but now contracts the limiting fragmentation regions into small regions at the ends of the plot, i.e. into the points $\xi = 0, 1$ as $s \to \infty$. There is unfortunately no variable which gives rise to finite regions for both the limiting fragmentation regions and the central region. Hence it is necessary to choose the appropriate variable for the corresponding kinematic region; for example y is clearly suitable for particles slow in the c.m. frame.

A further point which is relevant here is that the scaling and limiting fragmentation hypotheses are *asymptotic* statements of inclusive distributions. In this sense, therefore, it does not matter whether one chooses the scaling variable $x = q/q_{max}$ or $q/\frac{1}{2}\sqrt{s}$ or if one normalizes the inclusive distributions whether one uses $\sigma_{ab}(s)$ or $\sigma_{ab}(\infty)$. However, the data are taken at finite values of s, where these choices may show considerable differences. Thus in comparing different sets of data it is important to check that the same variables and normalizations have been assumed.

Finally let us say a few words about the scaling hypothesis as applied to *semi-inclusive reactions*. Rewriting the semi-inclusive cross-section in terms of a distribution function $g_r(s, \mathbf{q}_1)$, i.e.

$$r\frac{d^3\sigma_r}{d\phi_1}(a + b \to c_1(\mathbf{q}_1) + c_2 + \ldots + c_r + \text{anything not } c) = \frac{\sigma_r(s)}{\sigma_{ab}(s)} g_r(s, \mathbf{q}_1),$$

$$(5.2.14)$$

where $\sigma_r(s)$ is the multiplicity distribution of r particles of type c_1 eqn (5.1.29) becomes

$$\sum_{n=1} \sigma_n(s) g_n(s, \mathbf{q}_1) = \sigma_{ab}(s) f_1(s, \mathbf{q}_1) \qquad (5.2.15)$$

where f_1 is the inclusive distribution for the process $a + b \to c_1 + X$. It is fairly easy to see, using the generating functional techniques of the last section that (5.2.15) can be generalized to

$$\sum_{n=q+1}^{\infty} (n - 1)(n - 2)\ldots(n - q)\sigma_n(s) g_n(s, \mathbf{q}_1)$$

$$= \sigma_{ab}(s) \int f_{1,\ldots,(q+1)}(s, \mathbf{q}_1, \mathbf{q}_2 \ldots \mathbf{q}_{q+1}) d^3\phi_1 \ldots d^3\phi_q \qquad (5.2.16)$$

where $f_{1,\ldots,(q+1)}$ is the normalized inclusive cross-section for the process $a + b \to c_1(\mathbf{q}_1) \ldots + c_{q+1}(\mathbf{q}_{q+1})$. The problem now is to deduce the consequences for $g_n(s, \mathbf{q}_1)$ assuming that these inclusive distributions do, in fact, scale. To do this we must be able to "invert" (5.2.16) so that only one semi-inclusive distribution is given in terms of the inclusive distributions. In general this inversion cannot be performed in any simple way. However, Koba *et al.*

(1972) have suggested it is possible to obtain at least the asymptotic result if one assumes Feynman scaling for all orders of inclusive reactions, i.e. that $\hat{f}_{1,\ldots,(q+1)}(x_1, r_1; 0, r_2 \ldots, 0, r_{q+1})$ exist. With this assumption they show that

$$\sum_{n=0}^{\infty} n^q \sigma_n(s) g_n(s, x_1, r_1) = \sigma_{ab}(s)\, \tilde{f}_{1,\ldots,(q+1)}(x_1, r_1; 0, 0; \ldots; 0, 0)\, \log^q(s) \qquad (5.2.17)$$

$$+ \text{ lower order terms}$$

where $\tilde{f}$ is obtained from $\hat{f}$ by integrating over $\mathbf{r}_2, \ldots \mathbf{r}_{q+1}$. Moreover, since (5.2.17), is valid for all $q = 1, 2 \ldots$, they find that the semi-inclusive distribution for $n \gtrsim \langle n \rangle$ and large s must be of the form

$$g_n(s, x_1, r_1) \approx \frac{\sigma_{ab}(s)}{\langle n \rangle \sigma_n(s)} H\left(x_1, r_1; \frac{n}{\langle n \rangle}\right)[1 + O(1/\log s)] \qquad (5.2.18)$$

Now since

$$\int g_n(s, \mathbf{q}_1)\, d\phi_1 = n \qquad (5.2.19)$$

it follows that

$$\frac{\sigma_{ab}(s)}{\sigma_n(s)\langle n \rangle} \log s\, \tilde{H}\left(0; \frac{n}{\langle n \rangle}\right) = n + \text{ lower order terms} \qquad (5.2.20)$$

where

$$\tilde{H}\left(0; \frac{n}{\langle n \rangle}\right) = \int H\left(x_1 = 0, r_1, \frac{n}{\langle n \rangle}\right) r_1\, dr_1$$

and therefore, since $\langle n \rangle \approx f_1(0) \log s$

$$\frac{\sigma_n(s)}{\sigma_{ab}(s)} = \frac{1}{\langle n \rangle} \psi\left(\frac{n}{\langle n \rangle}\right) + O\left(\frac{1}{\langle n \rangle^2}\right) \qquad (5.2.21)$$

with

$$\psi(z) = \tilde{H}(0, z)\, \tilde{f}_1(0)/z.$$

Inserting (5.2.14) and (5.2.21) in (5.2.18) we obtain the "scaling law" for semi-inclusive cross-sections

$$\frac{d\sigma_n}{d\phi_1}(a + b \to c_1(\mathbf{q}_1) + \ldots + c_n + X)$$

$$= \sigma_n(s) h\left(x_1, r_1; \frac{n}{\langle n \rangle}\right)[1 + O(1/\log s)] \qquad (5.2.22)$$

That is to say, the semi-inclusive distributions, normalized by the integrated semi-inclusive cross section $\sigma_n(s)$ is expected to scale for large $\langle n \rangle$; and furthermore, that the scaling distributions depend on n only through the ratio $n/\langle n \rangle$.

The implications of (5.2.21) and (5.2.22) are far-reaching. For example, from (5.2.21) it follows that if one measures the semi-inclusive cross-sections $\sigma_n(s)$ and thus determine $\langle n \rangle$ at one energy s_1, a knowledge of $\langle n \rangle$ at a second energy s_2 is, in principle, sufficient to predict $\sigma_n(s_2)$ for all n. Also, since the empirical evidence tends to suggest that $\langle n(s) \rangle$ is a rather slowly varying function of s for any produced particle c_i, it follows that the ratio $n/\langle n \rangle$ will have only a slight variation over a wide range of energies. Hence (5.2.22) suggests that in this energy range the semi-inclusive cross-section for a given number n of produced particles of type c will also follow, approximately, the scaling hypothesis.

Some initial empirical tests of these results by Chliapnikov *et al.* (1972) tend to support these ideas. However, in view of the very specific assumptions made to derive (5.2.21) and (5.2.22), i.e. scaling satisfied by all orders of inclusive distributions at $x_i = 0$ (practically at all energies), it will be most interesting to see if further tests confirm these conclusions. This is particularly true since Huskins (1972) has demonstrated, by means of counter examples, that the KNO scaling predictions do not follow just from these assumptions: that is to say, there is apparently some further assumption necessary to derive eqns. (5.2.21) (5.2.22) from (5.2.16). (See also Fialkowski and Miettinen (1973) in this connection.)

This concludes our more general discussion of single particle inclusive reactions and the scaling hypotheses. It is somewhat disappointing, though not unexpected, that such distributions have not proved to be a sensitive test for different particle production models. The multiperipheral model, parton model, liquid droplet model (Cheng and Wu, 1969) and diffraction models (Hwa, 1971; Jacob and Slansky, 1972) as well as some versions of the Hagedorn thermodynamic model (Hagedorn and Ranft, 1968; Hagedorn, 1972) all give similar predictions for the inclusive limiting distributions. To really test model predictions we must also look at *multiparticle* inclusive reactions and look for possible correlations between the distributions. This is just the same point as we stressed in the earlier chapter on exclusive reactions; unless we are very fortunate, a distribution in terms of a single variable or particle is just not sufficient to fully determine the production dynamics. This is not to say that single particle distributions are no use at all. For instance, the total cross-section has been a valuable tool in providing information of various resonances, the existence of the Pomeron (whatever it is), and more recently with the advent of duality, on the connection between "exotic" channels and exchange degenerate Regge poles. In a similar way single particle distributions should throw considerable light on these topics and probably on others too. How we can go about extracting this information is the subject of the next section.

5.3. Generalized optical theorem and Regge phenomenology of inclusive distributions

Besides the introduction of the scaling and limiting fragmentation hypotheses, the other great impetus to the study of inclusive phenomena has been the suggestion of a generalized optical theorem for inclusive reactions. The existence of such a theorem was first suggested by Mueller (1970) who proposed that the inclusive distribution for the process $a + b \rightarrow c + X$ could be related to a discontinuity in the analytically continued forward amplitude

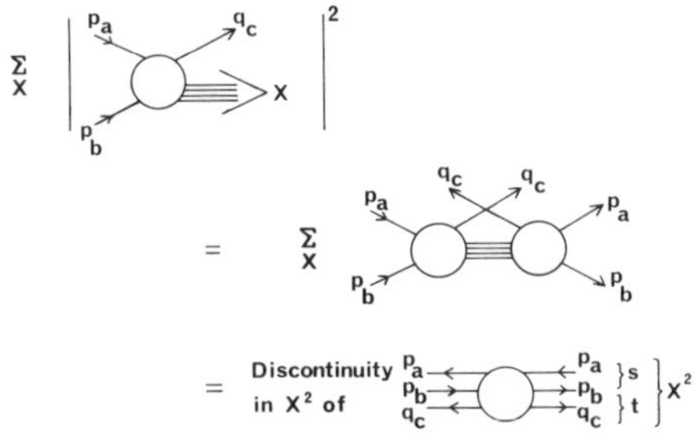

FIG. 40. Relation between the single particle inclusive distribution and the forward six-point amplitude.

for the process $a + b + \bar{c} \rightarrow a + b + \bar{c}$ (where "forward" in this context means the outgoing particles have the same momenta as the incoming ones), i.e.

$$f_i(s, \mathbf{q}_i) = \frac{1}{2p_a^c \sqrt{s}} \mathrm{Im}_{X^2} \langle ab\bar{c} | T | ab\bar{c} \rangle. \tag{5.3.1}$$

This Mueller hypothesis is shown diagramatically in Fig. 40 where the invariants s, t, u, X^2 are defined as in (5.1.45), (5.1.46) by

$$s = (p_a + p_b)^2$$
$$t_c = (p_a - q_c)^2$$
$$u_c = (p_b - q_c)^2 \tag{5.3.2}$$
$$X^2 = (p_a + p_b - q_c)^2 = (\text{mass})^2 \text{ of } X \text{ system.}$$

Assuming the validity of this generalized optical theorem we see that in order to determine the inclusive distribution we have only to specify the value of the invariants in (5.3.2) at which the discontinuity in the six point amplitude is to be evaluated and consider which diagrams are likely to best approximate the amplitude in this kinematic region. This is in exact analogy with the optical theorem for the "inclusive distribution of order zero", i.e. for the total cross-section. This theorem, of course, relates the total $a + b$ cross-section to the imaginary part of the elastic scattering amplitude. Thus, rather than evaluating explicitly each n-body cross section which contributes to σ_{ab}^{tot}, we may instead consider the structure of the forward elastic amplitude, for example, in terms of possible Regge exchanges. However, the obvious question arises: is it possible to generalize this optical theorem to the case of inclusive distributions?

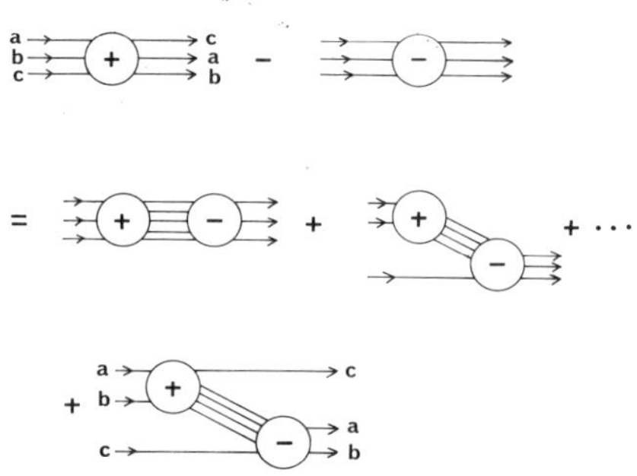

FIG. 41. Unitarity for the forward 3-body → 3-body amplitude.

It has been our understanding that, in general, discontinuities of scattering amplitudes arise from the unitarity condition and the formation of various intermediate states. This is, in fact, the basis of the optical theorem. If we now apply unitarity, however, to the *three* body amplitude (see Eden, *et al.* 1966) we find that the corresponding total discontinuity is given by several terms, only one of which resembles the single particle inclusive distribution. This is shown in Fig. 41 where the "+" and "−" indicate the corresponding functions are to be evaluated just above, or just below their cuts, i.e. $\ominus$ is obtained from $\oplus$ by continuing round all the singularities of the amplitude. In Fig. 41 only the last term resembles the single particle distribution, which therefore is not given *simply* from unitarity. Of course, if we were to analytically continue this unitarity condition to some region where the other energy variables s_{ab}, s_{bc}, s_{ca} $(s_{bc} = (p_b + q_c)^2$, $s_{ac} = (p_a + q_c)^2$ are below their respective thresholds such that only $X^2 = (p_a + p_b - q_c)^2$ is physical, then perhaps only this last term remains and would then correspond to the now extended unitarity discontinuity. Even so, the necessary analytic continuation is by no means straightforward. Remember, for instance, when there are more than five particles involved in a reaction there exist quadratic relationships between the invariants as well as linear ones like those in (2.1.3). This means when we continue the energy variables below their thresholds we must be careful to avoid (if possible) singularities associated with related variables and also not to cross any of the kinematic cuts arising from the quadratic nature of these relations. Assuming this can be done we can relate the inclusive distributions to the "Mueller" discontinuity in the six point function by analytically continuing back to the physical values of the variables Now, however, we must ensure that the continuation is such that we do not cross any of the singularities in the energy variables. We thus arrive at the relation shown in Fig. 42 where it should be noted that because of the "+" and "−" signs in the same bubble the inclusive distribution is given by a discontinuity of the six point amplitude evaluated on an *unphysical* sheet, not the physical one.

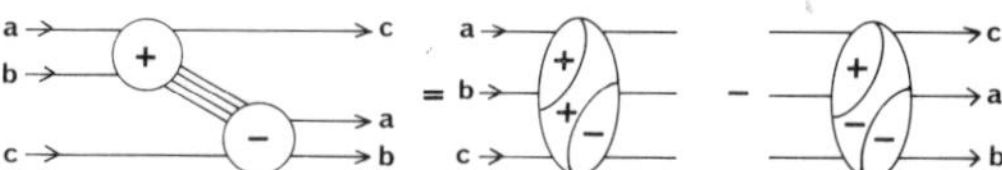

FIG. 42. The inclusive distribution related to a discontinuity in the 3- to -3-body amplitude through extended unitary.

There is still a further difficulty we might mention. On the right hand side of Fig. 42 the many particle intermediate state involves an integration over the phase-space elements

$$\prod_{k=1}^{n} d^3\mathbf{q}_k, \qquad n = 2, 3, 4, \ldots$$

Because of the analytic continuations we have outlined in order to write the expression shown in Fig. 42, we should also check that the integration contour has not been irrevocably disturbed in the process, perhaps on to an unphysical sheet of the $a + b \rightarrow c + X$ amplitude.

These problems are clearly very complex and no *simple* solution is likely to be forthcoming though Stapp (1971) has presented some very detailed justification of the Mueller hypothesis (see also Polkinghorne, 1972). Clearly it is going to be even more complicated to obtain a rigorous proof of the generalized optical theorem which relates the m-particle distributions $a + b \rightarrow c_1 + \ldots + c_m + X$ to a discontinuity in the forward elastic $m + 2 \rightarrow m + 2$ scattering amplitude. However, in view of the arguments presented by Stapp we shall from now on assume the validity of these relations.

Let us again stress that the beauty of the Mueller hypothesis is that we do not need to explicitly evaluate each of the n-particle exclusive reactions contained in the inclusive distribution (even if we had a reliable model for these exclusive reactions!). Such a task has implicitly been performed for us in relating the distribution to an amplitude discontinuity. This does not mean that the effect of each n-particle exclusive reaction has somehow been lost. It is still present in the six point amplitude, giving rise to a branch cut in the appropriate "missing mass" variable X. Of course, as in two body reactions such a combination of many-particle intermediate state singularities is often approximated at high energies by Regge contributions or at low energies by resonances. It is clear that these must be only approximations since generally they do not contain branch cuts corresponding to the various intermediate states. However, from two-body phenomenology we see that they can often provide a very good description of the amplitude.

With this in mind let us now consider how such ideas may be incorporated into a phenomenology of inclusive reactions.

5.3a. *Fragmentation region and the single Regge limit.* Since, as we believe, the inclusive approach to particle production becomes increasingly more necessary and relevant as we go to higher and higher energies, we are interested

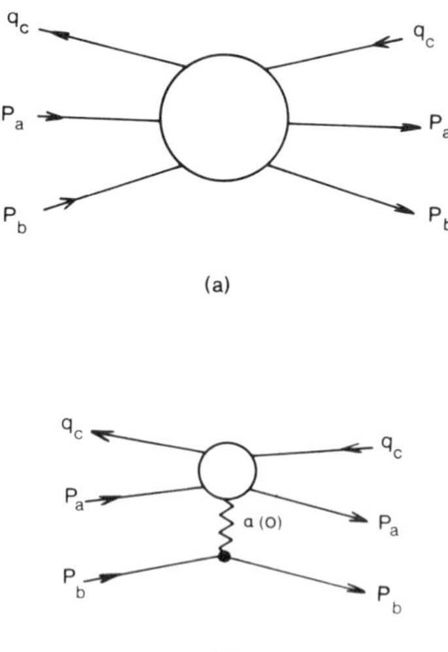

Fig. 43. (a) Forward elastic three-body amplitude. (b) Single Regge pole graph for the forward three-body amplitude.

in determining the various Regge asymptotic limits of the six point amplitude $T_{a\bar{b}c \to a\bar{b}c}$ shown in Fig. 43a. The first of these, the single Regge limit, was discussed by Mueller (1970) using the Toller group theoretic approach (Toller, 1965). Although this is a very elegant technique we do not wish to digress from our immediate goal (which is simply to approximate the six point amplitude) by outlining the necessary mathematics of this approach. Here, therefore, let us merely give a non-rigorous description of the results.

If we consider the particular kinematic region in which t_c is held fixed as $s \to \infty, X^2 \to \infty, u_c \to -\infty$ then the $a\bar{c}$ system in the $3 \to 3$ amplitude can be considered in some sense as a pseudo-particle and the amplitude as a pseudo $2 \to 2$ amplitude describing the process $(a\bar{c}) + b \to (a\bar{c}) + b$. In this case the single Regge exchange diagram shown in Fig. 43b should be the appropriate approximation to the six point amplitude. Notice, however, that these constraints on s, t, u, X^2 correspond to the fragmentation region of particle a. To see this let us consider the rest frame of particle a and write

$$p_a = (m_a, 0, 0, 0)$$

$$p_b = (m_b \cosh y_b, 0, 0, m_b \sinh y_b) \tag{5.3.3}$$

$$q_c = (m' \cosh y, \mathbf{r}, m' \sinh y)$$

since s is large, $s \approx m_a m_b \exp(y_b)$, y_b is large,

$$u = m_b^2 + m^2 - 2m_b m' \cosh (y_b - y) \tag{5.3.4}$$

while

$$t = m_a^2 + m^2 - 2m_a m' \cosh y$$

Thus fixed t implies fixed y (and fixed $\mathbf{r}$) and therefore a fixed value of the longitudinal momentum if particle c in the rest frame of particle a.

The single Regge limit applicable in this fragmentation region can be written as

$$f_c(s, \mathbf{q}_c) = \frac{1}{s} u^{\alpha(0)} \tilde{g}^c_{ab}(\mathbf{q}_c) \tag{5.3.5}$$

where $\alpha(0)$ is the intercept of the exchanged Reggeon, and we have approximated the flux factor by its high energy form $1/s$. Although we have used u in (5.3.5), from (5.3.4) it is clear that

$$u \approx -s \frac{m'}{m_a} e^{-y},$$

and therefore since y is fixed, we can recast (5.3.5) in the form

$$f_c(s, \mathbf{q}_c) = s^{\alpha(0)-1} g^c_{ab}(\mathbf{q}_c) \tag{5.3.6}$$

where the factors $-(m'/m)e^{-y}$ have been incorporated into g^c_{ab}. From (5.3.6) we see that the statement of limiting fragmentation (i.e. that $f_c(s, \mathbf{q}_c)$ is independent of s) now becomes the statement that the leading Regge trajectory exchanged in the $3 \to 3$ body amplitude has unit intercept—i.e. presumably the Pomeron. Moreover, assuming this Regge exchange factorizes we can write g^c_{ab} in the form

$$g^c_{ab}(\mathbf{q}_c) = \beta_b \gamma^c_a(\mathbf{q}_c) \tag{5.3.7}$$

where β_b is simply a normalization factor which describes the coupling of the Reggeon to the particle b. Provided the same Reggeon also dominates the elastic $a + b$ two-body amplitude we have that

$$\sigma^{\text{tot}}_{ab} \approx \beta_a \beta_b \tag{5.3.8}$$

if $\alpha(0) = 1$, and hence the inclusive distribution divided by this total cross section in the fragmentation region of particle a, should have no dependence on particle b, even as regards the normalization of the distribution.

Notice that we have relied heavily here on the assumption of a pure Regge pole exchange in order to invoke factorization and the $s^{\alpha(0)}$ dependence in the amplitude. If it turns out that the leading Pomeron-like Regge exchange is not a pole but some form of cut (say), this would introduce extra logarithmic factors into the energy dependence of the amplitude. Thus the scaling and limiting fragmentation hypotheses would only be satisfied up to terms of the order of log s. This would also be reflected in a similar "non-scaling" behaviour of the inclusive cross-section of order zero, i.e. the total cross-section. Moreover, if the Regge exchange is not a pole there is no reason *a priori* to expect that its dependence on b and the $(a\bar{c})$ system factorizes. In fact this factorization property of the leading Regge exchange has been considered by Chen *et al.* (1971) who examined the processes

$$\pi^+ p \to \pi^- X$$

$$K^+ p \to \pi^- X$$

$$pp \to \pi^- X$$

$$\pi^- p \to \pi^- X$$

over a range of beam momenta from 7 to 28·5 GeV/c. They found that the π^- inclusive distributions for the first three reactions divided by the corresponding total cross-sections do apparently agree with each other in the proton fragmentation region (at least within the experimental errors). The π^- distribution for the fourth reaction, $\pi^- p \to \pi^- X$, on the other hand, did not agree so well with the others, though this may well be because this distribution has not yet reached a scaling limit. Since this initial investigation of factorization and limiting fragmentation further evidence of Pomeron factorization has also been found. See for example Miettinen 1973.

Of course, the question of whether the Pomeron contribution factorizes or not has already been asked for several years in two-body phenomenology. Unfortunately, with the limited number of possible target and beam particles available to us it has not been possible to examine this question in any great detail. For example, since it is not possible to measure the total $\pi\pi$ cross-section directly such obvious tests as

$$(\sigma_{\pi N}(\text{tot}))^2 = \sigma_{\pi\pi}(\text{tot}) \cdot \sigma_{NN}(\text{tot})$$

has not been seriously confronted with experiment. Only by examining diffraction dissociation processes has it been possible to obtain any direct evidence for factorization (see Section 4.4). Now, on the other hand, by measuring the inclusive distributions and using the generalized optical theorem, many more tests of factorization are open to us. From Fig. 43b it is clear that in the fragmentation regions we can define many more "targets" or "projectiles" with a wide range of quantum numbers by suitably choosing the particles a and c which make up the $(a\bar{c})$ pseudo particle. However, it has to be admitted that unlike the situation in diffraction dissociation, inclusive reactions and the optical theorem only allow us to test factorization at one isolated point, i.e. in the forward direction. It could therefore be that even though the Regge exchanges do not, in general, factorize, they can conspire to have this property at this one isolated point $t = 0$.

Having noted this possibility we shall assume from now on that the Pomeron is after all a simple pole with $\alpha(0) = 1$. Therefore we shall expect it to factorize, and the inclusive distributions eventually to exhibit the property of limiting fragmentation. The question now arises—can we say anything about how, and at what rate, is the limit approached? Returning to Fig. 43b and the single Regge exchange mechanism it seems natural to try to answer this question by the inclusion of a secondary trajectory α_M. The inclusive distribution can then simply be written in the form

$$f_c(s, \mathbf{q}_c) = g_{ab}^P(\mathbf{q}_c) + s^{\alpha_M(0)-1} g_{ab}^M(\mathbf{q}_c) \tag{5.3.9}$$

where the superscripts P and M denote the Pomeron and secondary (meson) trajectories. Typically $\alpha_M(0) \approx 0.5$ corresponding to the intercept of the exchange degenerate $\rho - f_0 - \omega - A_2$ trajectory so that (5.3.9) says that in general the inclusive distribution has a non-scaling part which dies away like $1/\sqrt{s}$ as $s \to \infty$. Notice one can make exactly the same statement about the energy-dependent part of the total cross-sections for the same reason. Also, as in the case of total cross-sections, it is possible to isolate this non-scaling term in inclusive distributions by taking the difference of two measurements corresponding to particle-antiparticle processes. For example, the inclusive distributions for production of particle c in the processes

$$\pi^{\pm} p \to cX$$

in the target fragmentation region can be written as

$$g^P_{\pi^+ p}(\mathbf{q}_c) + \frac{1}{\sqrt{s}} g^{\rho}_{\pi^+ p}(\mathbf{q}_c)$$

and

$$g^P_{\pi^- p}(\mathbf{q}_c) + \frac{1}{\sqrt{s}} g^{\rho}_{\pi^- p}(\mathbf{q}_c)$$

respectively. However, because of the charge conjugation properties of the $(\pi^{\pm}\pi^{\pm}P)$ and $(\pi^{\pm}\pi^{\pm}\rho)$ vertices $g^P_{\pi^+ p} = g^P_{\pi^- p}$ and $g^{\rho}_{\pi^+ p} = -g^{\rho}_{\pi^- p}$, and hence

$$f^{\pi^+ P}_c(s, \mathbf{q}_c) - f^{\pi^- P}_c(s, \mathbf{q}_c) = \frac{1}{\sqrt{s}} g^{\rho}_{\pi^+ p}(\mathbf{q}_c) \tag{5.3.10}$$

Having isolated these non-scaling meson-exchange contributions in this manner Meittinen (1972) found that these Regge exchanges also factorize, at least as far as one can tell within the experimental errors.

Although (5.3.9) tells us that in general the dominant non-scaling contribution to the inclusive distribution falls off like $1/\sqrt{s}$ this is not really too helpful unless one knows the relative strengths of the scaling and non-scaling parts of the distribution. In two body collisions, for instance, the same arguments which lead to (5.3.9) also suggest that

$$\sigma^{\text{tot}}_{ab} = \sigma^P_{ab} + \frac{1}{\sqrt{s}} \sigma^M_{ab}$$

However, this by itself does not explain why some total cross-sections are apparently flat† from very low energies right up to the highest energies yet measured, while others exhibit a very strong energy dependence over the whole of this range. Nevertheless, it is observed that those reactions which have a constant cross-section are those reactions which have exotic quantum numbers, i.e. they do not correspond to baryon states which can be formed out of three quarks or meson states formed by a quark and an anti-quark. Such reactions, therefore, do not possess any of the known resonances. In order to account for this situation in terms of Regge exchanges, duality suggests that except for the Pomeron all other exchanges occur in exchange degenerate pairs so that their contribution to the imaginary part of the exotic elastic amplitude, and hence also to the total cross-section vanishes.

Following Chan *et al.* (1971) let us apply the same reasoning to the energy dependence of inclusive distributions. First of all let us notice that now we

† At least up to logarithmic factors of s.

are not talking about particle a + particle b scattering but rather to particle b + system $(a\bar{c})$ scattering.† However, if the ("mass") of the $a\bar{c}$ system, i.e. t_c, is not too large, it may well be possible to consider it as a pseudo-particle, at least for these purposes. With this assumption, duality now tells us that in the fragmentation regions, the non-scaling term in (5.3.9) is absent, (or negligibly small) whenever $b + (a\bar{c})$ or respectively $(b\bar{c}) + a$ is exotic. That is to say, the inclusive distributions reach an early scaling limit whenever the "missing mass" system $ab\bar{c}$ has exotic quantum numbers and therefore as far as we know does not couple to any known resonances.

This simple $ab\bar{c}$ exotic criterion for the onset of early scaling seems to be supported by much of the present data on inclusive reactions, though nothing like a rigorous or exhaustive study is yet possible. It would certainly be useful to see if the requirement of the $b\bar{c}$ (or $a\bar{c}$) mass being kept small is necessary, or whether the same simple exotic criterion applies also to the central region where t_c and u_c are both large. Moreover, in two-body collisions all particles involved in the reactions are obviously non-exotic. However, in the case of the $ab\bar{c}$ forward amplitude it may well be that the $b\bar{c}$ (or $a\bar{c}$) system is already exotic by itself. In such situations it would be interesting to see if the $ab\bar{c}$-exotic rule which comes from our experience of duality for two-body reactions also applies here.

There is yet another question about the derivation of the $ab\bar{c}$-exotic criterion which has been stressed by Ellis et al. (1971). They pointed out, quite rightly, that the process $a + (b\bar{c}) \rightarrow a + (b\bar{c})$ differs from a proper two-body reaction in one important respect; namely, that here the $a + b$ system is well above threshold. Thus besides considering possible resonances in the $ab\bar{c}$ missing-mass channel one should also take into account possible resonance effects in the ab channel. From this they conclude that necessary and sufficient conditions for the onset of early scaling should be that both the $ab\bar{c}$ and ab channels are exotic. Chan and Hoyer (1971) have subsequently argued with this result and shown that at least in terms of the B_6 dual-resonance model the $ab\bar{c}$-exotic condition is alone sufficient to determine early scaling.

(In view of this it is worth mentioning a "no-exotic-no-go theorem" which has been proposed by Kugler et al. (1972) based upon the assumed factorization properties of Regge and Pomeron exchanges. A consequence of this theorem is that for meson-meson-baryon systems, if $ab\bar{c}$-exotic is indeed a sufficient condition for early scaling then so is the condition $b\bar{c}$-exotic in the fragmentation region of a).

Since these initial suggestions many other criteria for the onset of an early approach to a scaling limit have been suggested. Of these the proposal of

---

† Clearly the argument can apply equally well to fragmentation of particle a.

Einhorn *et al.* (1972) is of special interest since it introduces a strong form of the Harari–Freund ansatz. Specifically, by examining the quark content of the amplitude they propose that

"only the Pomeron, but no secondary Regge pole, is exchanged when quark lines are connected across in a given channel".

From this rule they deduce in the case of the six point amplitude, that for the reaction $a \overset{b}{\to} c$ to have little or no energy dependence both the channels (ab) and $(b\bar{c})$ must be exotic. Similarly, for the reaction $b \overset{a}{\to} c$, (ab) and $(a\bar{c})$ must be exotic. Although this strong form of the Harari–Freund ansatz is not valid in the dual resonance model, where Regge contributions are found even when quark lines are connected across in a given channel, this does not, of course, mean that it must automatically be wrong. No doubt it is possible to construct an alternative form of dual model which would satisfy this condition. The interesting thing about the result of Einhorn *et al.* is that it allows a different approach to a scaling limit in the target fragmentation region and the projectile fragmentation region, unlike the global conditions suggested by Chan *et al.* and Ellis *et al.* Somewhat similar quark-model ideas have been used by Lo and Phua (1972) who conclude that the process $a \overset{b}{\to} c$ rapidly approaches a scaling limit provided the transition $a \to c$ does not permit any through-going (spectator) quarks.

The Harari–Freund ansatz has also been the starting point for an investigation by Gordon and Veneziano (1971) (see also Veneziano, 1971) which was continued in more detail by Tye and Veneziano (1972). However, these authors accepted the more usual version of the Harari–Freund conjecture, i.e. that a total cross-section is composed of two components, one corresponding to direct channel resonances which is dual to ordinary Regge exchanges, and the other corresponding to non-resonant background which is dual to Pomeron exchange. Gordon and Veneziano proposed that in the case of single-particle inclusive distributions one must now consider *seven* components corresponding to possible resonant/non-resonant effects in the various parts of the forward 3-body to 3-body amplitude. By a very careful study of these seven terms Tye and Veneziano conclude that the inclusive distribution has an early approach to its scaling limit if and only if *all* channels, ab, $ab\bar{c}$, $a\bar{c}$, $b\bar{c}$ are exotic. Moreover, with the addition of certain other assumptions, i.e. if necessary cancellations which take place in an energy–momentum sum-rule in fact takes place locally at each point x (the Feynman scaling variable), they conclude that:

"if ab and $ab\bar{c}$ are exotic and $a\bar{c}$ and/or $b\bar{c}$ is not, the scaling limit will be reached from *below* in the corresponding non-exotic fragmentation region(s)".

Thus unlike the situation for total cross-sections where each contribution is positive definite, they find that the meson–Regge contribution in (5.3.9), $g_{ab}^M(\mathbf{q}_c)$, can sometimes be negative for certain values of $\mathbf{q}_c$.

It may seem a little surprising that there can be so many differing suggestions for the onset of early scaling in the single-particle inclusive distributions. However, it implies that the application of what appears to be a rather simple idea—i.e. duality—to the six point function is by no means as straightforward as it might look. It will indeed be interesting to see which criterion the data favour, not only for the single-particle distributions but for the m-particle inclusive distributions.

5.3b. *Triple-Regge limit.* In the above we have written the residues of the single Regge pole contributions to the forward $3 \to 3$ body amplitude as functions of $\mathbf{q}_c$, i.e. $g_{ab}^c(\mathbf{q}_c)$ in (5.3.6). In fact, a group theoretical analysis shows that these functions are actually functions of t_c, the invariant (mass)^2 of the $a\bar{c}$ cluster, and a Toller angle variable s/X^2 which describes the orientation in space of this $a\bar{c}$ cluster relative to the parricle b. (Similarly, for the process $b \xrightarrow{a} c$, g_{ab}^c is a function of u_c and s/X^2.) It is a simple matter of kinematics to show that this Toller angle variable s/X^2 is related to the scaling variable $x = q/\frac{1}{2}\sqrt{s}$ in the fragmentation region by

$$s/X^2 = s/(s + m_c^2 - 2\sqrt{s}\sqrt{(q^2 + r^2 + m_c^2)}) \underset{s \to \infty}{\approx} (1 - |x|)^{-1} \quad (5.3.11)$$

Let us reiterate that the single Regge limit was attained by letting s and X^2 tend to infinity while keeping t_c and s/X^2 constant. Therefore, from (5.3.11) we see that this limit corresponds to some fixed value of $|x|$, which in general will lie *between* the extremus 0 and 1. Now the question arises: what are the corresponding approximations to the six point amplitude in the cases when $|x| = 1$ or 0? To try to answer the first part of this question it is useful to note that while we are still considering the limits $s \to \infty$, $X^2 \to \infty$ the point $|x| \approx 1$ corresponds to the situation where $s \to \infty$ much more rapidly than X^2. In this case a rather heuristic argument suggests that each n-body amplitude contributing to the inclusive process can be represented by a Regge pole exchange between the $a\bar{c}$ vertex and the bX^2 vertex as shown in Fig. 44a. (Notice this is only meaningful if X^2 is very much less than s, since the Regge propagator is of the form $(s/X^2)^{\alpha(t)}$. In two $\to$ two-body reactions where all the masses are fixed X^2 is replaced by a free parameter which may be a typical (mass)^2 in the system or may just be a term factored out of the residue function to define a scale. (Remember we stressed that one of the successes of the B_4-dual model was that it predicted this scale factor to be the inverse of the universal slope of the Regge trajectories.)

Assuming the above Regge pole approximation to the exclusive reactions, the inclusive reaction can now be calculated by "sewing together" and summing over all the intermediate states X. In this way we see that the inclusive

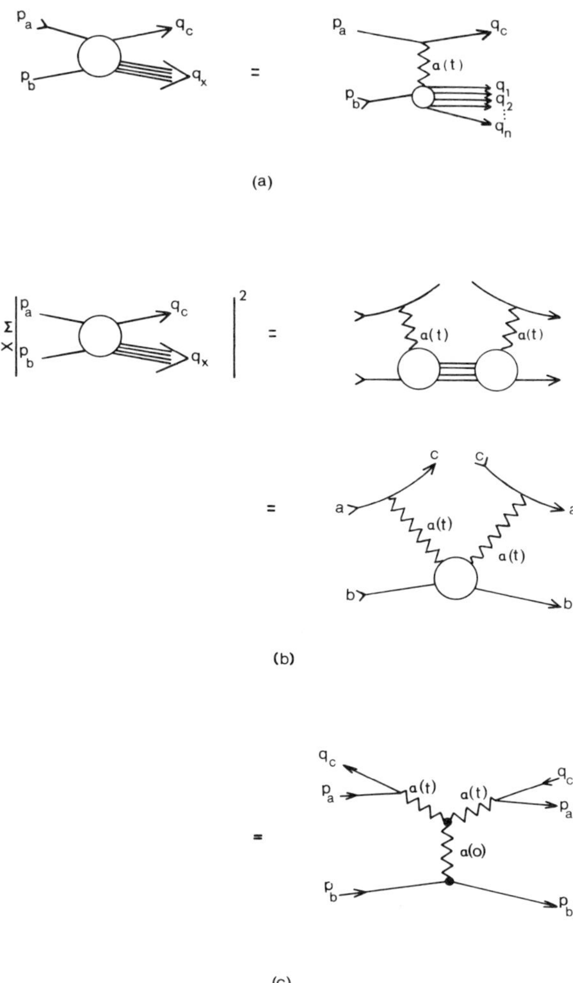

FIG. 44. Derivation of the triple Regge pole limit in single particle inclusive processes.

distribution involves the total cross-section for Reggeon $\alpha(t)$ + particle b scattering, which by a naive use of the optical theorem can be written in terms of the imaginary part of the forward elastic $\alpha(t) + b$ scattering amplitude, as indicated in Fig. 44b. Now provided the energy of this $\alpha(t) + b$ reaction is sufficiently large, i.e. if X^2 is sufficiently large, it would seem plausible to approximate the forward scattering amplitude by its leading Regge exchange, i.e. by Pomeron exchange. In this way we obtain the triple-Regge graph shown in Fig. 44c.

Repeating this argument in terms of algebraic functions we obtain from Fig. 44b that

$$\frac{d\sigma}{d\phi_c} \approx \frac{1}{s}\beta^2(t_c)(s/X^2)^{2\alpha(t_c)}X^2\sigma^{tot}_{Rb}(X^2,t_c) \qquad (5.3.12)$$

where $\beta(t_c)$ is the form factor describing the Reggeon $\alpha(t)$ coupling to particles a and c and σ^{tot}_{Rb} is the total cross-section for Reggeon $+ b$ scattering. Applying the optical theorem to this process we find†

$$\sigma^{t\,ot}_{R\,b}(X^2,t_c) \approx (X^2)^{\alpha_P(0)-1}\gamma_{PRR}(t_c) \qquad (5.3.13)$$

where $\gamma_{PRR}(t)$ is some form factor describing the $\sigma(tot)$ dependence on the Reggeon mass; i.e. it is the Pomeron–Regge–Regge coupling function; and $\alpha_p(0)$ is the Pomeron trajectory which at $t = 0$ is, by definition, very close to unity. Combining (5.3.12) and (5.3.13), the triple Regge contribution to the single particle inclusive distribution is given by

$$\frac{d\sigma}{d\phi_c} = 8\pi^2 s\frac{d\sigma}{dt_c\,dX^2} = \frac{1}{s}\beta^2(t_c)\gamma_{PRR}(t_c)(s/X^2)^{2\alpha(t_c)}(X^2)^{\alpha_p(0)} \qquad (5.3.14)$$

Although this is by no means a rigorous derivation, it does indicate the main steps in the argument. As one may imagine, the exact mathematical derivation is far from simple and as shown by De Tar and Weis (1971) it involves taking a mixed Regge–helicity–pole limit. That is to say, besides considering the pole structure of the amplitude considered as a function of complex angular momentum we must also consider the amplitude as a function of complex *helicities* and angular momentum. (The usefulness of the helicity pole approach is that it allows us to obtain the asymptotic behaviour of amplitudes in limits which are not strictly Regge limits in the normal sense. For example, if we look closely at the top left hand part of the triple Regge graph in Fig. 44c, it appears that the (energy)^2 variable corresponding to $\alpha_R(t)$ exchange is given by $(p_a + k_p)^2$ where k_p is the four-momentum carried by the Pomeron. However, this is just $P_b - P_b = 0$ so that the (energy)^2 variable is m_a^2 which is not exactly an asymptotic limit! Nevertheless, the helicity pole analysis is applicable, and often agrees with the Regge limit. (See for example, Goddard and White (1971), White (1972), Weis (1972), Abarbanel and Schwimmer (1972), who use the helicity pole approach to obtain the correct singularity structure of some low order exclusive reactions).

The simplicity of the form of the distribution (5.3.14) makes it a very appealing tool to analyse inclusive data near the kinematical boundary

---

† It is interesting to note that to verify the optical theorem for this case one really has to consider a six-point function, contracting two pairs of external lines to form the Reggeons and thus is equivalent to justifying the generalized optical theorem of Mueller.

$x \approx 1$. It should be noticed, however, that its simplicity is due to two basic assumptions:

(i) All Regge poles have factorizable residues

(ii) The Pomeron is an isolated Regge Pole, with intercept close to 1.

Also, since we are assuming Regge pole dominance for the amplitude it would seem that (5.3.14) is only applicable when the ac system is non-exotic so that there does indeed exist some trajectory $\alpha(t_c)$ which couples to the $a\bar{c}$ cluster. Nevertheless, allowing for these assumptions, there is presently considerable interest in testing this formula (5.3.14) against experiment in view of the keen theoretical interest in the size of the triple-Regge coupling functions, which we shall now try to summarize.

Firstly, it was noted that under the conditions

(i) $\alpha_p(0) = 1$

(ii) $\left. \dfrac{d\alpha_p(t)}{dt} \right|_{t=0} < \infty$

(iii) Pomeron is a factorizable Regge pole

the triple Pomeron term which should dominate the inclusive distributions of the form $a + b \rightarrow a + X$ near $x = 1$ must vanish at $t = 0$, i.e. when all three Pomerons have zero mass. (De Tar and Weis, 1971; Henyey and Zakrzewski, 1972; Abarbanel et $al.$, 1971).

To get some idea why this must happen let us note that the invariant inclusive distribution near $x = 1$ from (5.3.14) can be written as

$$\hat{f}(x, r) = \beta^2(t_c)\gamma_{ppp}(t_c)(1 - x)^{1 - 2\alpha(t_c)}, \qquad x \approx 1 \qquad (5.3.15)$$

where $t_c \approx -r^2$ when $x \approx 1$. Since each Reggeon is supposed to be an isolated pole the form $(1 - x)^{1 - 2\alpha(t_c)}$ must dominate over a finite region of x near $x = 1$. Thus in any inclusive reaction sum-rule of the form (5.1.18) (say), integrating over this region of x when $t_c = 0$ gives a divergent contribution if $\alpha(0) = 1$. On the other hand, if $\alpha(0) = 1 - \varepsilon$ the integral converges and γ_{ppp} may again be non-zero.

From this condition that $\gamma_{ppp}(0) = O$ Jones et $al.$ (1972) and, by a somewhat different method, Abarbanel et $al.$ (1972) deduced that certain Reggeon–Reggeon–Pomeron couplings must also be zero when the Pomeron has zero mass. Observing that the Reggeon legs in the Reggeon–Reggeon–Pomeron vertex could presumably be extrapolated to physical particle masses, Brower and Weis (1972) argued that this meant that the Pomeron must also decouple from two-body total cross-sections. Thus we have the paradoxical situation that the Pomeron which was introduced in the first instance only to account for the approximate constancy of the high energy total cross-section is not now expected to contribute to two-body cross-sections.

In fact the situation is somewhat more fluid than indicated above. For instance, it has been shown by Moen and White (1972) that in the helicity pole approach the particular part of the Reggeon–Reggeon–Pomeron coupling function which is required to vanish by the arguments of Abarbanel *et al.* (1972) does not automatically lead to the result that the Pomeron decouples from particles. Moreover, in a subsequent paper by Moen and Zakrzewski (1972) in which they compare the helicity pole and Regge pole limits for the 8-point amplitude, they suggest that the arguments of Jones *et al.* (1972) also may not necessarily imply that the Pomeron decouples from total cross-sections. However, whatever the outcome of these theoretical speculations it is worth stressing that the starting point for these considerations was that the Pomeron was a simple isolated Regge pole which, of course, is by no means established at present.

The phenomenological situation, unfortunately, is also undecided. Although many triple-Regge fits have been made, e.g. Chliapnikov *et al.* (1972), Peccei and Pignotti (1971), Ting and Yesian (1971), Wang and Wang (1971), only recently have the necessary high energy data become available to ensure that one is actually in a triple-Regge region. For most of this data, e.g. Albrow *et al.* (1973) there is no evidence of a vanishing triple-Pomeron coupling at $t = 0$, although this is contradicted by the experiment of Childress *et al.* (1974). In any case all current phenomenological analyses (wrongly) assume that there are no interference effects between the various triple-Regge diagrams. Thus all conclusions on the triple Pomeron couplings are as yet only tentative. We would refer the interested reader to a recent study by Roberts and Roy (1974) for a further discussion of the problem.

5.3c. *Finite Mass Sum Rules.* In view of the practical difficulties associated with obtaining data which are truly in the triple Regge region it is worth mentioning the work by Einhorn (1972), Sanda (1972), Oleson (1971), Jen *et al.* (1971), Kwiecinski (1972), Einhorn *et al.* (1972) and Moen (1972) in deriving sum-rules to relate the triple-Regge couplings to integrals over low missing masses in inclusive reactions. Basically, the idea is to relate the high energy (or rather large X^2) limit of the forward three to three body amplitude to its low X^2 structure in just the same way that finite *energy* sum rules relate high and low energy two-body scattering. However, in the case of finite *mass* sum rules (FMSR) the situation is more complicated because the analytic structure of the six-point amplitude is obviously much more complex than the analytic structure of the four-point amplitude. Nevertheless, by examining the analytic structure of the six-point amplitude in terms of various models it would appear that it may well be possible to write the analogous FMSR's for inclusive reactions.

In order to describe these sum rules let us first of all define an appropriate set of variables which are symmetric or antisymmetric under the crossing operation $p_a \leftrightarrow -q_c$. Besides the invariant momentum transfer variable

$$t_c = (p_a - q_c)^2$$

let us introduce the variables

$$v = p_b \cdot (p_a - q_c) = \tfrac{1}{2}(X^2 - t_c - m_b^2) \tag{5.3.16}$$

and

$$\eta = p_b \cdot (p_a + q_c) = \tfrac{1}{2}(s - s_{bc} - m_a^2 + m_c^2)$$

where $s_{bc} = (p_b + q_c)^2$, to replace X^2 and s respectively. The generalized optical theorem can then be written as

$$\frac{\mathrm{d}\sigma}{\mathrm{d}\phi_c} = \frac{1}{p_a\sqrt{s}} \operatorname*{Disc}_{v>0} A(\eta, v, t_c) \tag{5.3.17}$$

where $A(\eta, v, t_c)$ is the forward 3-body amplitude, and the discontinuity of $A(\eta, v, t_c)$ is taken across the branch cut in v for $v > 0$, keeping the variable $s = (p_a + p_b)^2$ just above and just below its normal threshold cut on the left and right of three-to-three body diagram respectively. (See Fig. 42). In terms of this variable v, however, we see that besides the discontinuity in v for $v > 0$ corresponding to the states coupling to the $ab\bar{c}$ system, the amplitude $A(\eta, v, t_c)$ also has a discontinuity in v for $v < 0$ corresponding to the physical missing mass states coupling to the bca system. For values of $\eta \gg v$, or $|t_c|$, the amplitude has the asymptotic behaviour

$$A \approx \sum_{i,j} \eta^{\alpha_i(t_c) + \alpha_j(t_c)} f_{ij}(v, t_c) + \tilde{A}(\eta, v, t_c) \tag{5.3.18}$$

where α_i and α_j are Reggeons coupling to the $a\bar{c}$ cluster. (5.3.18) is the generalization of (5.3.12) when more than one trajectory is considered, see Fig. 44b). The remainder $\tilde{A}(\eta, v, t_c)$ contains the non-leading Regge contributions and therefore provided η is sufficiently large $\tilde{A}$ should not contribute to the v discontinuity—the assumption being, of course, that there actually is a leading Regge *pole* contribution. The residue function $f_{ij}(v, t_c)$ can be thought of as the forward amplitude for Reggeon-particle scattering and is related to the inclusive distributions by (5.3.7), i.e.

$$\frac{\mathrm{d}\sigma_{ab}}{\mathrm{d}\phi_c} \approx \sum_{i,j} \eta^{\alpha_i(t_c) + \alpha_j(t_c) - 1} \operatorname*{Disc}_{v>0} f_{ij}(v, t_c) \tag{5.3.19}$$

and similarly

$$\frac{\mathrm{d}\sigma_{bc}}{\mathrm{d}\phi_a} \approx \sum_{i,j} \eta^{\alpha_i(t_c) + \alpha_j(t_c) - 1} \operatorname*{Disc}_{v<0} f_{ij}(v, t_c)$$

The positive and negative discontinuities of this function f_{ij} correspond to the physical intermediate states in the processes $\alpha_i(t_c) + b \to \alpha_j(t_c) + b$ and $\bar{\alpha}_j(t_c) + b \to \bar{\alpha}_i(t_c) + b$ respectively. However, since the (mass)^2 of the Reggeons α_1, α_j is $t_c < 0$, the question arises whether $f_{ij}(v, t_c)$ has any other singularities. There seems to be no firm answer to this, but the evidence from the dual resonance model and, more importantly, from perturbation theory and Gribov's Reggeon calculus suggests that $f_{ij}(v, t_c)$ does *not* have any further singularities.

At this point it should be clear where we are heading. We have obtained a Reggeon-particle scattering amplitude which we believe has only the normal threshold singularities of a typical particle–particle amplitude. Moreover, from (5.3.19) we have related the right and left hand discontinuities of this amplitude to single-particle inclusive distributions, which, while they must be evaluated at large η (i.e. large s) correspond to all values of the missing masses. Hence, if we can now write a finite energy sum-rule for this Reggeon-particle amplitude we can relate the low missing-mass data to the high missing-mass limit, i.e. the triple Regge limit. In fact, the asymptotic behaviour of $f_{ij}(v, t_c)$ as $v \to \infty$ is given by

$$\eta^{\alpha_i(t_c) + \alpha_j(t_c)} f_{ij}(v, t_c) \underset{v \to \infty}{\sim} \sum_k (\eta/v)^{\alpha_i(t_c) + \alpha_j(t_c)} v^{\alpha_k(0)}$$

$$\times \left\{ \frac{\tau_i \tau_j \tau_k + \exp\left[-i\pi(\alpha_k(0) - \alpha_i(t_c) - \alpha_j(t_c))\right]}{\sin \pi(\alpha_k(0) - \alpha_i(t_c) - \alpha_j(t_c))} \right\} \xi_i(t_c) \xi_j^*(t_c) \qquad (5.3.20)$$

$$\times \beta^i_{a\bar{c}}(t_c) \, \beta^{j*}_{a\bar{c}}(t_c) \, \beta^k_{b\bar{b}}(0) \, \gamma_{ijk}(t_c)$$

where $\beta^i_{a\bar{c}}(t_c)$ are the Reggeon couplings to particles $a\bar{c}$, $\xi_i(t_c)$ is the usual Regge phase-factor

$$\xi_i(t_c) = \frac{\tau_i + \exp\left[-i\pi\alpha_i(t_c)\right]}{\sin \pi\alpha_i(t_c)},$$

τ_i is the signature of the α_i trajectory, and $\gamma_{ijk}(t_c)$ is the triple-Regge coupling function. The only unusual feature of (5.3.20) is the triple-Regge phase-factor in the curly brackets. Although it is not immediately obvious how this factor comes about it can be verified by considering all six-point diagrams contributing to this limit (just as one considers the (s, t) and (u, t) terms in two-body scattering), or equivalently, by drawing all the relevent quark diagrams contributing in the dual resonance model.

Having obtained the asymptotic form for f_{ij} the FMSR follows in the normal manner by integrating $v^n \eta^{\alpha_i + \alpha_j} f_{ij}$ round a contour which envelops the right and left hand cuts and is closed by two semi-circles of radius $|v| = N$. The circle integration is performed by assuming f_{ij} is given by this asymptotic form (5.3.20). In this way we obtain a relationship between the triple Regge coupling functions and integrals over the large s, low X^2 inclusive distributions of the form

$$\int_0^N dv\, v^n \left[\frac{d\sigma_{ab}}{d\phi_c} + (-1)^{n+1} \frac{d\sigma_{cb}}{d\phi_a} \right]$$

$$= \sum_{i,j,k} (1 + (-1)^{n+1}\, \tau_i\tau_j\tau_k)\, (\eta/N)^{\alpha_i(t_c)+\alpha_j(t_c)-1}\, N^{\alpha_k(0)+n} \qquad (5.3.21)$$

$$\times\, \frac{\xi_i(t_c)\xi_j^*(t_c)\beta_{a\bar c}^i(t_c)\beta_{a\bar c}^{j*}(t_c)\gamma_{ijk}(t_c)\beta_{b\bar b}^k(0)}{\alpha_k(0) + n + 1 - \alpha_i(t_c) - \alpha_j(t_c)}$$

and similarly a Schwarz-like sum rule, (Schwarz 1967) of the form

$$\int_0^N dv\, v^n \left[\frac{d\sigma_{ab}}{d\phi_c} + (-1)^n \frac{d\sigma_{cb}}{d\phi_a} \right]$$

$$= \sum_{i,j} \eta^{\alpha_i(t_c)+\alpha_j(t_c)-1}\, \xi_i(t_c)\xi_j^*(t_c)\beta_{a\bar c}^i(t_c)\beta_{a\bar c}^j(t_c) \left[R_{i,j}^{(n)}(t_c) + \sum_k (1 + (-1)^n\tau_i\tau_j\tau_k) \right.$$

$$\left. \times\, \frac{\gamma_{ijk}(t_c)\beta_{b\bar b}^k(0)\, N^{\alpha_k(0)+n+1-\alpha_i(t_c)-\alpha_j(t_c)}}{\alpha_k(0) + n + 1 - \alpha_i(t_c) - \alpha_j(t_c)} \right] \qquad (5.3.22)$$

where $R_{i,j}^{(n)}(t_c)$ is a nonsense wrong signature fixed pole residue.

One of the obvious applications of these finite-mass sum-rules is to estimate the triple Reggoen vertices $\gamma_{ijk}(t_c)$ by integrating over data at relatively low values of the missing mass. Notice that the two-body processes $a + b \to c + d$ and $b + c \to a + d'$ are included in these sum-rules and probably represent a sizeable part of the low-mass integrals.

Several other applications of (5.3.21) and (5.3.22) come to mind. For example, it would be most interesting to investigate whether the Harari-Freund conjecture does in fact generalize to the case of Reggeon particle scattering as was discussed in the last section when we considered possible criteria for the onset of early scaling. Also by evaluating the functions $\gamma_{ijk}(t_c)$ from (5.3.21) and substituting them into (5.3.22) the nonsense wrong signature fixed pole residue $R_{ij}^{(u)}(t_c)$ can be calculated. Such quantities are interesting since they set the scale of Reggeon–Reggeon cut contributions to two-body scattering amplitudes. Unfortunately, for such a project to be practical very accurate data is required over a range of momentum transfers t_c. Otherwise, only very crude estimates for the size of these Reggeon–Reggeon cuts are possible. However, initial results by Roy and Roberts (1972) seem to be quite encouraging though the inclusive data they use is at too low an energy for the sum-rule evaluation to be very reliable.

5.3d. *The central region and the double Regge limit*. The final kinematic limit for the forward three-body amplitude we wish to consider corresponds to the central region ($x \approx 0$) in which particles are not slow either in the laboratory or the projectile frame. Reverting now to our original variables s, t_c and u_c this limit corresponds to $t_c \to \infty$, $u_c \to \infty$ and the ratio $K = (t_c u_c)/s$ fixed. To see how this correspondence comes about it is useful to introduce Sudakov variables, which for our particular situation means we write

$$C = xA + yB + R, \tag{5.3.23}$$

where A, B, C are the four momenta of particles a, b and c in any Lorentz frame and R is the transverse momentum of particle c, i.e.

$$R.A = R.B = 0.$$

The parameters x and y lie between 0 and 1, and since by squaring (5.3.23) such that

$$C^2 = m_c^2 = xys + m_a^2(x^2 - xy) + m_b^2(y^2 - xy) + R^2$$

we see that quite generally

$$x = O(s^{-\gamma})$$
$$y = O(s^{\gamma - 1}) \tag{5.3.24}$$

where $0 < \gamma < 1$, provided $|R|$ is finite.

The fragmentation region of particle a corresponds to $A = O(1)$ and $C = O(1)$ while $B = O(s)$ and therefore $y = O(1/s)$ with x finite. Similarly, the fragmentation region of particle b is given simply by changing A for B and x for y. Thus the two fragmentation regions correspond to the two limiting values of γ in (5.3.24), i.e. 0 and 1 respectively; while the region not corresponding to either target or projectile fragmentation, (the central region) corresponds to all other possible values of γ, $0 < \gamma < 1$. (Notice the *pionization* region as we defined it earlier is given by γ taking the value $\frac{1}{2}$, i.e. $x = O(s^{-\frac{1}{2}})$, $y = O(s^{-\frac{1}{2}})$). However, via (5.3.23) and (5.3.24) it is a simple matter to show that to leading order in s

$$t_c = (A - C)^2 \approx (x - 1)ys \approx O(s^\gamma)$$
$$u_c = (B - C)^2 \approx x(y - 1)s \approx O(s^{1 - \gamma})$$

and hence

$$|t_c| \to \infty, |u_c| \to \infty \quad \text{with} \quad (t_c u_c)/s = O(1).$$

The corresponding Regge diagram corresponding to this kinematic limit is the double Regge diagram shown in Fig. 45. As discussed by Mueller (1970), the generalized optical theorem relates this graph to the inclusive distribution such that

$$f_c(s, \mathbf{q}_c) = f_c(t_c, u_c, K) = \sum_{i,j} \beta_{ij}(K) |u_c|^{\alpha_i(0) - 1} |t_c|^{\alpha_j(0) - 1} \tag{5.3.25}$$

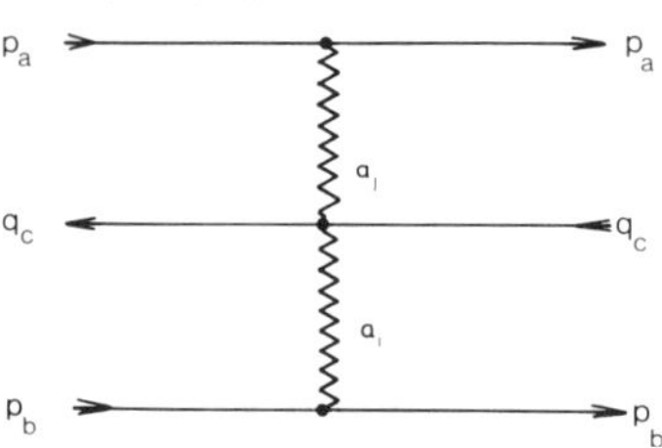

FIG. 45. Double Regge pole graph for the forward three-body amplitude.

where α_i, α_j are the possible Regge exchanges. (We again restrict our discussion to purely Regge *pole* exchanges, though we should note that this is more for convenience than for any strong physical argument). Since it is possible for both Regge exchanges i and j to carry vacuum quantum numbers it would seem very likely that the leading contribution to the sum in (5.3.25) will be given by double Pomeron exchange with $\alpha_i(0) = 1$, $\alpha_j(0) = 1$. In this situation we see that the asymptotic limit of the inclusive distribution is expected to be purely a function of $K = t_c u_c / s$. However, provided t_c, u_c and s are all much larger than the transverse momentum component and any of the masses involved we find from (5.1.39) that

$$t_c \approx - \sqrt{s}(q^+)^c$$

and

$$u_c \approx = \sqrt{s}(q^-)^c$$

where $(q^\pm)^c = (w_c^c \pm q_c^c)$. Hence we have that

$$K = r^2 + m^2 = m'^2$$

and the rapidity of particle c, is given by

$$y = \tfrac{1}{2} \log (u_c/t_c).$$

Thus the asymptotic limit of the inclusive distribution is again seen to satisfy the scaling hypothesis: the double Regge pole analysis gives a limiting distribution which is independent of the rapidity y and is merely a function of the "transverse" mass of particle c. This is equivalent to the prediction of a central plateau in y resulting from the short range correlation hypothesis in (5.2.13). Furthermore, if we assume that the Pomeron exchanges factorize such that

$$f_c^{ab}(r) = \beta_a \beta_b \gamma^c(r) \tag{5.3.26}$$

where β_a, β_b are again the couplings of the Pomeron to the particles a and b respectively (as in (5.3.7)) and $\gamma^c(r)$ describes the central vertex, we obtain the prediction that $f^{ab}_c(r)/\sigma_{ab}(\text{tot})$ should be independent of both target and projectile particles a, b and is determined merely by the type of particle produced.

In order to see how this limiting distribution is approached it is necessary to consider the secondary exchanges which can occur in the double Regge region. If we denote a typical secondary trajectory by α_R then (5.3.25) gives

$$f(t_c, u_c, K) = \beta_{pp}(K) + \beta_{PR}(K)\,|t_c|^{\alpha_R(0)-1} + \beta_{RP}(K)\,|u_c|^{\alpha_R(0)-1} \qquad (5.3.27)$$

$$+ \;(\text{lower order terms in } |t_c|, |u_c|)$$

where β_{PR} represents the Pomeron–Reggeon–c–$\bar{c}$ vertex function. Notice that the limiting distributions $\beta_{pp}(K)$ is approached from above if β_{PR}, β_{RP} are positive and from below if they are negative. Also, the rate of approach to this limit depends on the size of these functions and the intercept of α_R. Typically we might assume α_R to be the trajectory for the leading meson exchange, i.e. the degenerate ρ-f-ω-A_2 system. Although neither t_c or u_c is small enough to consider the process as a pseudo two-body reaction $(a\bar{c}) + b \to (a\bar{c}) + b$ or $(b\bar{c}) + a \to (b\bar{c}) + a$ as we did in the fragmentation regions, it is difficult to see, *a priori*, what other exchanges might be important besides these meson exchanges. With this assumption, we can write $\alpha_R(0) \approx 0.5$ in (5.3.27) and thus obtain the prediction that the non-scaling contribution to the central region is proportional to $|t_c|^{-\frac{1}{2}}$ and $|u_c|^{-\frac{1}{2}}$. Since t_c and u_c are both less than s this implies that the approach to scaling is somewhat less rapid than in the fragmentation. In particular in the pionization region $|t_c| \approx |u_c|$ and the approach to the scaling limit is proportional to $s^{-\frac{1}{4}}$. Hence provided the Reggeon coupling strengths β_{PR} are roughly the same order of magnitude as the corresponding Reggeon couplings in the fragmentation regions we would expect the limiting fragmentation distributions to be reached long before a limiting behaviour is observed in the central region, which again is predicted by the short range correlation hypothesis discussed in Section 5.2. Moreover, it is interesting to note that, assuming $\beta_{PR} = \beta_{RP}$ (for instance when $a = b$), the non-scaling contribution to $f(t, u, K)$ in the central region is proportional to

$$|t_c|^{-\frac{1}{2}} + |u_c|^{-\frac{1}{2}} \approx s^{-\frac{1}{4}}(s^{\frac{1}{4}-\frac{1}{2}\gamma} + s^{-\frac{1}{4}+\frac{1}{2}\gamma}) \qquad (5.3.28)$$

which is always greater than $2s^{-\frac{1}{4}}$, its value in the pionization region. This means if β_{PR} is positive so that the scaling limit is approached from above, the rapidity distribution will tend to be concave at finite values of s in the central region; similarly, if β_{PR} is negative, so that the limit is approached from below, the rapidity distribution will tend to be convex. In terms of the distribution in x this convexity or concavity corresponds to a cusp at $x = 0$.

The question of whether the scaling limit is in fact approached from above

G*

or below in the central region cannot be answered at present. Comparing the ISR results of π's, K's, p's and $\bar{p}$'s produced in proton–proton collisions to data obtained at much lower energies it would appear that the scaling limit is actually approached from below implying that β_{PR}, β_{RP} are negative. However, as pointed out by Chan *et al.* (1972), this result is apparently in contradiction to many of our theoretical ideas about duality and inclusive distributions. For example, if we consider the sum of the distributions for the processes $pp \rightarrow \pi^+ x$ and $pp \rightarrow \pi^- x$ the non-scaling contribution should be given only by f_0 exchange. Thus the data suggests that in particular β_{Pf_0} and β_{f_0P} are negative. Assuming factorization the same central vertex also occurs in $\pi^+ p \rightarrow \pi^+ x$, but now since the $a\bar{c}$ channel is exotic for the process $\pi^+ p \rightarrow \pi^- x$, exchange degeneracy requires $\beta_{f_0P} = \beta_{\rho P}$. A simple application of charge conjugation then leads to the prediction

$$f_{\pi^+ p - \pi^+ X}(x = 0) < f_{\pi^+ p - \pi^- X}(x = 0)$$

i.e. that more π^-'s are produced in $\pi^+ p$ collisions than π^+'s. This prediction does not sound at all plausible, and is certainly not supported by the data. Moreover, if one looks at the distributions in terms of dual models we find that quite generally the coupling functions β_{PR} and β_{RP} should be greater than or equal to zero (Tye and Veneziano 1972, Einhorn *et al.* 1971). In particular, for the process $pp \rightarrow \bar{p}X$ all channels are exotic and hence by any exoticity (duality) criterion for early approach to a scaling limit this process should exhibit early scaling. In fact, of course, this process shows the biggest energy dependence from 28 GeV to 1500 GeV of any process yet measured.

In view of these apparent contradictions between theory and experiment it is worth pointing out that the double Regge limit on which the analysis of the central region is based is purely a mathematical limit. It is presumably going to be the correct diagram at some sufficiently large values of t_c and u_c. However, if the double Regge formula (5.3.25) is applied to describe data corresponding to much lower values of t_c and u_c it is possible that the Pomeron and ρ-f-ω-A_2 trajectory no longer dominate the amplitude and one must also consider other Regge exchanges.† The energy at which this becomes necessary depends on the structure of the forward three-body amplitude and probably therefore on the type and mass of particle produced. It is a fairly simple matter, for instance, to construct models which suggest that the energy dependence of the inclusive spectra in the central region is a kinematical or "threshold" effect even up to ISR energies (e.g. Humble, 1972a). Thus even though such models may not be the final answer to the problem of the central region dynamics they do provide an appealing solution to our

---

† It is worthwhile noting that while it is plausible that the same Regge exchanges which describe the fragmentation regions also dominate the central region it is purely an assumption at present which, like factorization, has still got to be rigorously tested.

difficulties. That is to say, they suggest that in the central region up to ISR energies we are in some sense in a low-energy regime where threshold or kinematical effects mask the dynamical situation. Therefore, only above these energies should we expect the analysis of the double-Regge diagram to be applicable where, for instance, the inclusive distributions may decrease towards their asymptotic limit.

This completes our discussion of the generalized optical theorem and the various Regge limits of the forward three-body amplitude. Clearly, assuming the validity of the optical theorem for *two-particle* inclusive spectra, we can similarly relate such spectra to the various Regge limits of the forward four-body amplitude. For example, if we consider the process $a+b\rightarrow c_1+c_2+X$ where q_{c_1} is in the fragmentation region of particle a and q_{c_2} is in the fragmentation region of particle b, the Mueller hypothesis relates the distribution of particles c_1 and c_2 to the X^2 discontinuity in the single Regge limit of the $ab\bar{c}_1\bar{c}_2$ amplitude shown in Fig. 46. Notice that the bubbles connected by the Regge exchange in this figure are the same as the bubbles

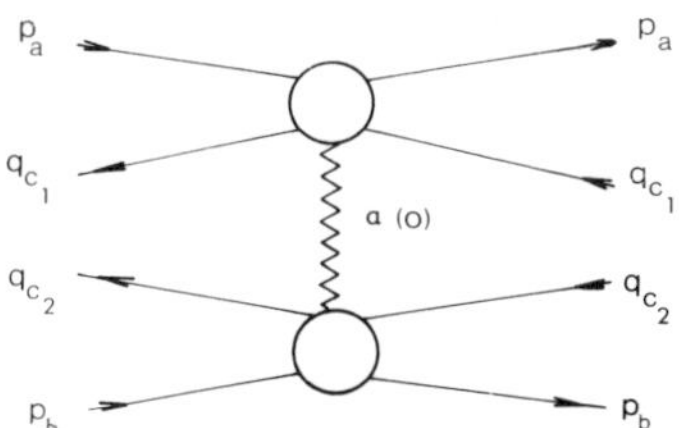

FIG. 46. A single Regge pole diagram for the forward four-body amplitude.

obtained in the single Regge limits of the single-particle inclusive diagrams for the a and b fragmentation regions respectively. Thus, assuming factorization, the generalized optical theorem immediately predicts the limiting distribution for the process $a + b \rightarrow c_1 + c_2 + X$ in this kinematical region from the single particle distributions. Furthermore, assuming factorization also holds for the non-leading Regge exchanges, the approach of this two-particle distribution to a scaling limit is similarly predicted by the way in which the single particle spectra approach their limit.

There is, of course, no reason to restrict ourselves to fragmentation regions, and many predictions for the two-particle distributions in all regions of phase-space are possible. See, for instance Abarbanel (1971b) for a discussion of the two-particle distribution when both c_1 and c_2 are in the central region; and Freedman *et al.* (1971) who consider the situation where c_1 and c_2 are close to the kinematic boundary, i.e. in the so-called "double triple-Regge region". Moreover, there is probably considerable merit in calculating the forward three-body and four-body amplitudes not only in terms of simple Regge limits but also by using B_6 and B_8 dual model approximations to these amplitudes. In this way it is possible to calculate how the distributions might be expected to go over from one well-defined kinematic region to another. (See for example Kang and Shen, 1972; Jen *et al.* 1972). However, as always in dual model analysis, there is necessarily going to be the introduction of some rather ad hoc features when one tries to include the Pomeron into the dual scheme. Nevertheless, dual models do have the very interesting property that they very naturally give rise to the damping in the transverse momenta; a feature which must frequently be added quite arbitrarily to other models.

6

Hadrodynamics

6.1. Prong cross-section and multiperipheralism versus diffraction dissociation

In the previous chapter a generalized optical theorem was used to relate the inclusive distributions to a discontinuity in certain forward elastic scattering amplitudes. By this means a phenomenological analysis of inclusive reactions was proposed which required only a knowledge of the structure of these forward elastic amplitudes in certain kinematic regions. Consequently, we did not need to concern ourselves with the problem of describing the production dynamics leading to these distributions, but simply noted that whatever the individual exclusive cross-sections looked like they must somehow conspire to satisfy this (extended) unitarity requirement. While this is obviously an attractive proposition when we are trying to correlate and predict the characteristic features of inclusive spectra it is none the less desirable to consider what the specific production mechanisms must be which give rise to these features.

In this final chapter therefore we wish to concentrate on this aspect of hadron physics which is sometimes referred to simply as "hadrodynamics" and discuss some of the many models that have been proposed to describe high energy production phenomena. Clearly, when the number of particles in the final state becomes large it will be practically impossible to accurately describe *all* the fine details of the reaction and instead we must look at the system as some kind of statistical ensemble. In this sense the models presented in this chapter are statistical models, i.e. models which are constructed with the aim of describing multiparticle dynamics in some sort of gross average manner.

Most of the models currently under active consideration fall into one of two classes which we can loosely term *multiperipheral* or diffractive. As we have seen in Chapter 4, multiperipheral models are based on the idea of factorization between the various links of the multiperipheral chain. Thus although neighbouring particles emitted along the chain may interact with one another, the interaction between particles widely separated along the chain is negligibly small. For such widely separated particles, therefore, there can be little or no correlation between their production, except perhaps for kinematical, phase-space, constraints. In Section 4.2 we found, typically, that these models give a value of the integrated n-body cross-section $\sigma_n(s)$ which depends on the type of exchange along the chain and the coupling parameter g. However, by assuming that

$$\Sigma \sigma_n(s) = \sigma_{\text{tot}} = \text{constant} \tag{6.1.1}$$

we deduced that $\sigma_n(s)$ follows a Poisson distribution in n of the form

$$\sigma_n(s) = c\, \frac{\langle n \rangle^n \exp\left(-\langle n \rangle\right)}{n!} \tag{6.1.2}$$

with a mean given by

$$\langle n \rangle \propto g^2 \log s. \tag{6.1.3}$$

In practice, of course, there can be more than one type of particle produced unlike the simple idealized example given in Section 4.2. Even assuming that most produced particles are pions, and neglecting kaons and proton–anti-proton pair production, we still have to consider the three possible charge states of the pions, i.e. $Q = 1, 0, -1$. Now the easiest and, in most cases, the only way of measuring $\sigma_n(s)$ is by looking at the number of tracks (prongs) made by charged particles in a bubble chamber or some similar device. Thus, practically all measurements of $\sigma_n(s)$ are, in fact, measurements for the prong cross-sections $\sigma_{n_{\text{ch}}}(s)$ which include the production of all possible numbers of undetected neutral particles. Nevertheless, if we make the hypothesis that the number of neutral particles produced is the same as the number of charged pairs produced (with $Q = \pm 1$), we should expect the multiperipheral mechanism to lead approximately to a Poisson distribution for $\sigma_{n_{\text{ch}}}(s)$ or rather for $\sigma_{n_2}(s)$—the cross-section for the production of n_2 charged pairs (Wang 1969). In fact, the essential ingredient leading to a type of Poisson distribution is the lack of correlation either between produced particles, or between produced charged pairs, or more generally, between any produced clusters. Thus we would expect distributions of the form (6.1.2) to hold more generally than in some simple types of multiperipheral models. Conversely, when one does allow finite correlations between the various produced clusters, e.g. as considered by Ball and Marchesini (1970) and Mueller (1971), such models do not necessarily give rise to Poisson distributions.

Notice that one of the consequences of the multiperipheral view of production dynamics is that $\sigma_n(s)$ goes to zero as $s \to \infty$. Thus a typical multiperipheral cross-section grows from threshold up to some maximum value, $\sigma_n^{\max}(s)$, from which it decreases to zero like $s^{-g^2}(g^2 \log s)^n/n$! However, in Chapter 4 we noted that there was apparently a class of production reactions which may not go to zero as $s \to \infty$, but like the elastic cross-sections approach a non-zero constant limit. In order to reconcile these diffractive phenomena with the observed† $\log s$ behaviour for $\langle n \rangle$, Hwa (1971) suggested that while

$$\sigma_n(s) \approx \sigma_n \qquad (6.1.4)$$

where σ_n is independent of s for large s,

$$\sigma_n \propto 1/n^2 \qquad (6.1.5)$$

In this case $\Sigma n \sigma_n(s)$ would diverge logarithmically except that at any value of s, only a finite number of particles can be produced, the number depending on $\sqrt{s}$, the total energy available. Thus since

$$\langle n \rangle = \frac{\Sigma n \sigma_n}{\Sigma \sigma_n}$$

(6.1.4) and (6.1.5) again produce a logarithmic dependence on s.

A simple, but hopefully illuminating, way of visualizing diffractive dissociation processes is to consider the two colliding particles becoming excited as they pass through the field of influence of the other particle. Since the process is assumed to be diffractive the cross-section for the production of these excited states (which are often referred to as "fireballs" or "nova") will be taken as constant at high energy. Further, the subsequent break-up of the fireballs or nova is assumed to take place in a definite manner which, again by the diffractive nature of the process, is assumed to be independent of the presence of any other fireballs in the process—except, of course, for energy-momentum constraints. A consequence of this assumption is that each fireball will give a definite multiplicity distribution of the secondaries, and thus give rise to constant values for σ_n.

We shall consider such diffractive processes in more detail in Sections 6.3 and 6.4. However, it is worth stressing that both production mechanisms, multiperipheralism and diffractive dissociation, are somewhat ad hoc with the parameters of the models being chosen to mirror the observed features

† In fact, although the average charge multiplicity is usually assumed to have a logarithmic dependence on s it may be wise to remember that the data at present are equally well fitted by a power dependence, i.e. $\langle n_{\mathrm{ch}} \rangle \sim s^a$ with $a \approx 0.25$ (Ganguli and Malhotra, 1972).

of the data. For instance, while a log s behaviour for $\langle n \rangle$ is a fairly natural consequence of multiperipheral models, a constant total cross-section is only achieved for certain values of the exchange trajectory and its associated coupling constant. On the other hand, practically by definition, diffraction models lead to constant cross-sections; but a log s behaviour for $\langle n \rangle$ is only obtained by choosing a suitable dependence of σ_n on n or, as we shall see, a suitable dependence of the fireball density distribution on its mass M.

Nevertheless the two mechanisms are distinguishable. For example, in a multiperipheral model the Poisson-type of distribution of σ_n in n has a peak $\sigma_n^{\max}(s)$ which increases with s. On the other hand, in diffraction models, since σ_n is constant for fixed n as s increases, the average multiplicity increases simply because the high n "tail" of the σ_n versus n distribution extends as the energy increases. Thus the two types of distribution may be distinguished even if the high-n cross-sections are too small for us to differentiate between a $1/n!$ and $1/n^2$ behaviour for σ_n.

Moreover, the two mechanisms suggest very different forms for the two-particle inclusive distributions which are related to the effects of correlations between produced particles. For instance, the factorization hypothesis of multiperipheral dynamics suggests the way one particle is produced has practically no correlation with the way in which another particle is produced. Thus the two-particle inclusive distribution should, to some approximation, be simply the product of the single particle inclusive distributions, i.e.

$$f_{1,2}(s, \mathbf{q}_1, \mathbf{q}_2) \approx f_1(s, \mathbf{q}_1) f_2(s, \mathbf{q}_2) \tag{6.1.6}$$

Hence, integrating over these distributions we find

$$\langle n(n-1) \rangle = \int\!\!\int d\phi_1 d\phi_2 \, f_{1,2}(s, \mathbf{q}_1, \mathbf{q}_2)$$

$$\approx \int d\phi_1 f_1(s, \mathbf{q}_1) \int d\phi_2 f_2(s, \mathbf{q}_2) \tag{6.1.7}$$

$$= \langle n \rangle^2$$

$$\propto (\log s)^2.$$

However, while this factorization hypothesis holds in diffraction models for particles produced from different fireballs, there will, in general, be correlations between particles produced from the same fireball. This will be reflected in the very different form for the scaling limit of the two-particle inclusive distribution, $\hat{f}_{1,2}(x_1, x_2)$ predicted by the diffraction model as compared with the multiperipheral model, i.e. (6.1.6) (Koba, 1972). Unfortunately, when working at the relatively low energies presently available to us and taking into account the fact that the production mechanism may in reality be a sum of these two models, it is difficult to decide whether these differences should yet be apparent. Perhaps the simplest way to test the predictions for the two-particle inclusive distributions is to consider their integrated values. From (6.1.7) we observed that $\langle n(n-1) \rangle \propto (\log s)^2$ in a typical multiperipheral picture. However, assuming the n-dependence for σ_n as given in (6.1.5), we see that the diffraction mechanism suggests that

$$\langle n(n-1) \rangle = \sum_n n(n-1)\sigma_n/\sigma_{\text{tot}} \propto s^{\frac{1}{2}} \qquad (6.1.8)$$

since we must again truncate the sum to those values of n sith $mn \lesssim s^{\frac{1}{2}}$. Therefore, a measurement of the energy dependence of the quantity $\langle n(n-1) \rangle$ gives us directly an indication of the operative production mechanisms.

Of course, in practice, the two predictions (6.1.7) and (6.1.8) may be modified by starting from slightly different assumptions, or by adding more parameters to the model. In view of this we devote the rest of this chapter to a more detailed discussion of some particular versions of these models which, while not covering the field completely, allow us to investigate the consequences of certain particular assumptions.

6.2 Statistical phase-space models

The simplest statistical model for multiparticle production was suggested by Fermi (1950). He noted that the probability, $p^{(n)}$, of producing n particles with specified momenta within a small volume element dR_n of the n-particle momentum phase-space is given by

$$p^{(n)} = \left| T^{(n)} \right|^2 dR_n, \qquad (6.2.1)$$

where $T^{(n)}$ is the quantum mechanical transition amplitude $\langle \mathbf{q}_1, \ldots, \mathbf{q}_n | T | \mathbf{p}_a, \mathbf{p}_b \rangle$ for the process $a + b \rightarrow c_1 + c_2 + \ldots + c_n$. In the absence of any contrary dynamical evidence Fermi assumed that this probability was proportional to the probability that the n particles simultaneously go into a small region in *configuration space* which is defined by the range of the strong interaction. This implies that $\left| T^{(n)} \right|^2$ is just proportional c^n where the constant c should be the order of an inverse pion mass. (Another way of obtaining this result is to assume an analogy with a free gas and take the particles to be equally distributed in momentum space). Of course, since n-body phase-space increases with n like s^n this definition for $\left| T^{(n)} \right|^2$ does not automatically ensure the conservation of probability unless one suitably chooses the proportionality factor. This introduces an s-dependent term into $\left| T^{(n)} \right|$ which may be chosen, for instance, to produce a constant total cross-section. Thus Fermi's model for the n-body cross-section is simply

$$\sigma_n(s) = \Omega_n(s)/\sum_n \Omega_n(s), \qquad (6.2.2)$$

where†

$$\Omega_n(s) = \frac{c^n}{n!} \int_0^\infty \delta^4(p_a + p_b - q_1 - \cdots - q_n) \prod_{i=1}^n \mathrm{d}\phi_i, \tag{6.2.3}$$

and $\mathrm{d}\phi_i$ is the Lorentz invariant phase-space element for particle i (eqn (5.1.5)). In other words, except for the normalization factor $\sum_n \Omega_n(s)$, Fermi's model is just the phase-space volume for n-particles modulated by the necessary dimensional term c^n. In a sense it is the necessary first step in a study of hadrodynamics since only by comparing the phase-space (i.e. kinematical) prediction for $\sigma_n(s)$ can we deduce the dynamical content of these reactions. Unfortunately, as we pointed out in previous chapters, the phase-space integrals cannot be performed analytically for $n > 3$. However, techniques have been devised to estimate these integrals for all n at large s. We shall discuss these techniques later; but for the moment, let us merely state the main results. From (6.2.2) and (6.2.3) we obtain (Satz 1965)

$$\langle n \rangle = \tfrac{2}{3} (\pi m c s)^{\frac{1}{4}} \left\{ 1 - \tfrac{1}{3} \left(\frac{m^2}{s} \right)^{\frac{1}{4}} \frac{\pi m^3 c - 1}{(\pi m^3 c)^{\frac{1}{4}}} \right\} \tag{6.2.4}$$

and a Poisson distribution for the n-body cross-section

$$\sigma_n(s) = \frac{\langle n \rangle^n \exp\left(-\langle n \rangle\right)}{n!}. \tag{6.2.5}$$

Although these results are purely asymptotic results they do tend to differ sufficiently from the present trend of the data to suggest that one must indeed insert some dynamical information into the model. For example the distribution of σ_n tends to be narrower than a Poisson distribution at small s but becomes wider than (6.2.5) at higher energies. Also while the current measurements of $\langle n \rangle$ do not rule out a possible s^ν behaviour at large s, an exponent of $\nu = \tfrac{1}{3}$ would seem altogether too large (Ganguli and Malhotra, 1972). (See however Wolfendale and Wdowczyk, 1973 for the possible situation at extremely high cosmic ray energies). Moreover a similar $s^{\frac{1}{4}}$ is predicted for $\langle r^2 \rangle$, the mean square transverse momentum (Humble et al. 1973), which is also in conflict with the experimental evidence.

In view of these discrepancies, let us consider how we might modify Fermi's model by the introduction of some dynamical input. Notice that the basic assumption of this model was the equi-distribution of particles in momentum space. This implies that in the c.m. system a typical reaction looks like that shown in Fig. 47a, i.e. where the produced particles are distributed randomly about the collision region, and hence retain no

† The $1/n!$ term is introduced because we are assuming for simplicity that the produced particles are identical.

"memory" of the direction of the incident particle. In practice, of course, there appears to be a strong correlation between the initial particle direction and the way in which the produced particles are distributed. This correlation (peripherality condition) restricts the particles to be produced predominantly in the same direction as the initial particles and strongly suppresses the production of particles with large transverse momentum. Thus instead of Fig. 47a a more physical picture of high energy particle production is that

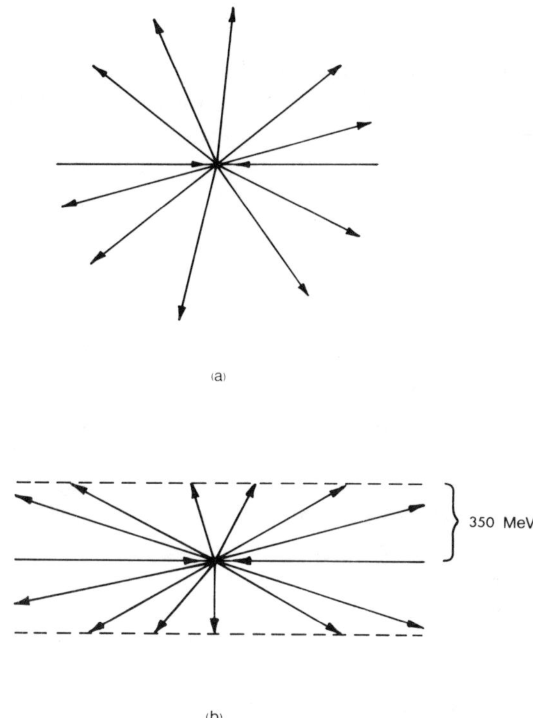

FIG. 47. (a) Typical equi-distribution of particles in momentum space. (b) Typical peripheral distribution of particles indicating the sharp cut-off in transverse momenta.

shown in Fig. 47b, i.e. where the average transverse momentum is typically about 350 MeV.

Probably the fundamental problem of hadrodynamics is to try to account for the limitation of the transverse momenta. However, even without knowing what causes this limitation, it is interesting to introduce some structure into the n-body scattering which reflects this effect and deduce the consequences of this peripherality condition for quantities such as σ_n and $\langle n \rangle$. Therefore, instead of assuming the n-body matrix element $T^{(n)}$ is a constant, c^n, which is independent of the momenta of the particles let us assume in (6.2.2) and (6.2.3) that

$$|T^{(n)}|^2 = \prod_{i=1}^{n} f_i(r_i), \qquad (6.2.6)$$

where r_i is the transverse momentum of the ith produced particle and f_i is some suitable monotonically decreasing function, e.g.

$$f_i(r_i) = c \exp(-a_i r_i^2). \qquad (6.2.7)$$

This parameterization of the production amplitude is perhaps the simplest and most obvious modification of Fermi's model. Depending on what one assumes for the factor f_i, or for the parameters a_i in (6.2.7) it covers all types of factorizable models. For example, if a_i is a linear function of $\log s_{i, i+1}$ where $s_{i, i+1}$ is the square of the invariant mass of the ith and $(i + 1)$th particles (6.2.6) is just the multi-Regge amplitude discussed in Chapter 4. However, since the dependence on $s_{i, i+1}$ was found to be of little importance except at the ends of the chain, here let us simply assume the form (6.2.7) and take the parameters a_i to be constants. This approximation is often referred to as the "Uncorrelated Jet Model", or "Independent Emission Model" since the particles are still produced independently of one another (i.e. have an equi-distribution) but now in a suitably restricted region of phase-space, i.e. as "jets" in the longitudinal direction. See, for example, Van Hove (1963), (1964), Krzywicki (1964) and more recently Ruijgrok (1972), de Groot (1972).

Not surprisingly perhaps the uncorrelated jet model (UJM) has many of the average features of the multi-Regge model. In particular, it has a logarithmic energy dependence for the average multiplicity (instead of the power dependence of Fermi's model) and a Poisson-like distribution for the high energy n-body cross-sections. Recently several modifications of the UJM have been proposed which attempt to include further aspects of production reactions such as quantum-number conservation laws and leading particle effects, i.e. small average inelasticity. (Chao, 1972; Sivers and Thomas, 1972). In this way it is hoped to be able to deduce which features of the data, if any, require the introduction of new dynamical assumptions and which features are simply kinematic or phase-space reflections of these very basic dynamical ideas. However, before we can discuss such modifications it is necessary once again to consider how we can perform the various phase-space integrations. As we mentioned in the previous chapters, one way of estimating the phase-space integrals is by the use of Monte Carlo techniques. Unfortunately, such numerical estimates often obscure the dependence of the results on the initial input and, of course, it is just this dependence that the UJM programme is designed to study. Therefore in the following we shall describe the rather elegant analytic method of approximating the high energy n-body phase-space integrals, due to Lurcat and Mazur (1964) and extended by Krzywicki (1965).

6.2a. *Approximate analytic evaluation of the UJM.* The basic problem is to evaluate the n-particle production rate which typically is given by the form

$$\Omega_n(P) = \frac{\lambda(s)c^n}{n!} \int \prod_{i=1}^{n} \left\{ \exp(-ar_i^2) \frac{d^3\mathbf{q}_i}{\omega_i} \right\} \delta^4(P - \Sigma q_i) \qquad (6.2.8)$$

where $P = (\sqrt{s}, 0, 0, 0)$ in the c.m. system, $\lambda(s)$ is a normalization factor independent of n, and for simplicity we have assumed the production of n identical particles. Notice that the reason (6.2.8) is difficult to evaluate is because of the four delta functions expressing energy momentum conservation. Without these functions the integrations would indeed be quite trivial. Therefore the suggestion of Lurcat and Mazur is to remove the delta functions by taking the Laplace transform of $\Omega_n(P)$, i.e. by defining a new function $\Phi(z^\mu)$ such that

$$\Phi_n(z^\mu) = \frac{1}{\lambda(s)} \int d^4P \exp(-z^\mu P_\mu) \Omega_n(P), \qquad (6.2.9)$$

where the parametric four-vector $z_\mu(z_0, z_x, z_y, z_L)$ has $z_0 > 0$ and $z_0^2 - z_L^2 - (z_x^2 + z_y^2) > 0$. The d^4P integration can be performed immediately because of the delta functions in $\Omega_n(P)$ to yield the fully factorizable form

$$\begin{aligned}
\Phi_n(z^\mu) &= \frac{1}{n!} \int \prod_{i=1}^{n} \left\{ c \exp(-ar_i^2 - z^\mu q_\mu) \frac{d^3\mathbf{q}_i}{\omega_i} \right\} \\
&= \frac{1}{n!} [\phi(z^\mu)]^n.
\end{aligned} \qquad (6.2.9')$$

The integration over the longitudinal component of $\mathbf{q}_i$ can be performed analytically to give

$$\phi(z^\mu) = \pi c \int_{\mu^2}^{\infty} dr^2 \exp[-a(r^2 - \mu^2)] I_0 [z_T \sqrt{(r^2 - \mu^2)}] K_0(zr), \qquad (6.2.10)$$

where $z^2 = z_0^2 - z_L^2, z_T^2 = z_x^2 + z_y^2, I_0(x)$ and $K_0(x)$ are modified Bessel functions and the factor π arises from a trivial integration over the transverse momentum polar angle. Although the remaining integral in (6.2.10) is not so simple $\phi(z^\mu)$ can be evaluated from (6.2.10) by any number of standard techniques depending on the particular choice of the parameters z and z_T and also on the value of a which determines the damping in the transverse momenta. For example, provided z and z_T are sufficiently small we can expand $K_0(x)$, $I_0(x)$ in a power series about $x = 0$ such that

$$\phi(z^\mu) \approx \frac{\pi c}{a^2}\left(-\tfrac{1}{2}\gamma - \log\frac{z}{2a} + O(z^2 \log z) + O(\mu^2 \log \mu)\right) \qquad (6.2.10')$$

provided that the mass μ is also small. (In 6.2.10') γ is Euler's constant, i.e. $\gamma = 0{\cdot}577215\ldots$.

Having obtained the function $\phi(z^\mu)$ and hence also from (6.2.9) the Laplace transform $\Phi_n(z^\mu)$, the rate Ω_n can now be determined by applying the inverse transform to $\Phi_n(z^\mu)$, i.e.

$$\frac{\Omega_n(P)\exp(-zP)}{\Phi_n(z^\mu)} = \int_{-\infty}^{\infty} \frac{\mathrm{d}^4 t}{(2\pi)^4} \exp(-iPt)\frac{\Phi_n(z_\mu - it_\mu)}{\Phi_n(z_\mu)} \qquad (6.2.11)$$

which is valid for all positive time like z_μ. Of course, if we could perform this inverse transformation exactly we would have completely solved the n-body phase-space problem, and it is likely that we need never have constructed the Laplace transform in the first place. However, in general we cannot perform the t-integrations exactly (except perhaps in the limit in which the particle mass μ tends to zero) and so we must resort to approximation techniques. For instance, Lurcat and Mazur suggested (6.2.11) could be estimated using the method of steepest descents. Thus expanding

$$\log\left\{\frac{\Phi_n(z_\mu - it_\mu)}{\Phi_n(z_\mu)}\right\}$$

as a power series in t_μ such that

$$\log\left\{\frac{\Phi_n(z_\mu - it_\mu)}{\Phi_n(z_\mu)}\right\} = -iA_n^\mu t_\mu - \tfrac{1}{2}B_n^{\mu\nu}t_\mu t_\nu \qquad (6.2.12)$$

the condition

$$-A_n^\mu = P_\mu \qquad (6.2.13)$$

ensures that the integrand in (6.2.11) is approximately gaussian. By analogy with the central limit theorem in mathematical statistics this condition gives an estimate for the integral which will be valid provided the number of particles n is fairly large. See, for instance Khinchin (1949).

Notice that the conditions (6.2.13), the so-called temperature equations, allow us to calculate the optimum values for the parameters z_μ. Specifically in the centre of mass system in which $P_L = 0, |P_T| = \sqrt{(P_x^2 + P_y^2)} = 0$ and $P_0 = \sqrt{s}$ we find

$$z_T = 0$$

$$z_L = 0 \qquad (6.2.14)$$

and

$$P_0 = -\frac{\partial}{\partial z} \log \Phi_n(z, z_T)\Big|_{z_T = 0},$$

where the solution to this latter equation, i.e. $z = \zeta$, is referred to as the inverse temperature. Having obtained these values of $(z, z_T) = (\zeta, 0)$ we must now evaluate the coefficients $B_n^{\mu\nu}$ of the quadratic term in the expansion (6.2.12). In this particular frame it turns out that only the diagonal elements of $B_n^{\mu\nu}$ are non-zero (since the off-diagonal elements are proportional to z_T or z_L) and these are given from (6.2.12) by

$$B_n^{00} = \frac{\partial^2}{\partial \zeta^2} \log \Phi_n(\zeta, z_T)\Big|_{z_T = 0} \qquad (6.2.15)$$

$$B_n^{LL} = (\sqrt{s})/\zeta$$

and

$$B_n^{xx} = B_n^{yy} = B_n^{TT} = \frac{1}{z_T} \frac{\partial}{\partial z_T} \log \Phi_n(\zeta, z_T)$$

with

$$\det B_n = B_n^{00} B_n^{LL} (B_n^{TT})^2$$

The resulting estimate for the rate $\Omega_n(P)$ then becomes

$$\Omega_n(P) = \frac{\lambda(s) \exp(\zeta P_0)\, \Phi_n(\zeta, 0)}{(2\pi)^2 (\det B_n)^{\frac{1}{2}}} \exp(-P_T^2/2B_n^{TT}) \{1 + R_n(\zeta, P_T)\} \qquad (6.2.16)$$

where $R_n(\zeta, P_T)$ contains the corrections to the gaussian form arising from the higher powers of t_u in the expression (6.2.12). However, this approximation scheme ensures that at any energy the correction term is $O(n^{-\frac{1}{2}})$ for large n, and hence may be neglected beyond some large (but undefined) value of n. We would refer the reader to the original papers of Lurcat and Mazur (1964) and Krzywicki 1965) and also to the work of Sivers and Thomas (1972) for a further discussion of this correction term.

It is important to note that quite often in evaluating the integrals in (6.2.11) we must work in a frame where $P_L = 0, P_T = 0$ is not the same as the overall centre of mass system. For instance, in the above example we performed the phase space integrals for all n particles. However, in determining the exclusive rates

$$d^k\Omega_n\Big/\left(\frac{d^3\mathbf{q}_1}{\omega_1} \cdots \frac{d^3\mathbf{q}_k}{\omega_k}\right)$$

only $n - k$ of the integrations are performed and the resulting rate depends on $Q_\mu = q_1 + \ldots + q_k$ where $\mathbf{Q}$ is not zero in the overall centre of mass system. Thus, in general, the determination of the parameters z_μ from the temperature equations (6.2.13) may not be as simple as indicated by (6.2.14). Nevertheless, for our example of the production of n identical particles (without leading particle effects) we can use the recursion relation

$$\Omega_n(P) = \int \frac{d^3\mathbf{q}_1}{\omega_1} \exp\left(-ar_1^2\right) \frac{\lambda(s)}{\lambda(M_1^2)} \Omega_{n-1}(P - q_1), \qquad (6.2.17)$$

where $M_1^2 = (P - q_1)^2$ to evaluate in a straightforward manner the exclusive distributions.

The Lurcat and Mazur technique of removing the energy-momentum conserving delta functions from the phase-space integrals by taking the Laplace transform, and then estimating the inverse transformations, is clearly very useful for many calculations in multiparticle physics. However, if we were to calculate the total inelastic rate, or the inclusive distributions, by adding up all these approximate estimates for the n-body exclusive distributions it is likely that the accumulative error in this procedure would invalidate the result. As stressed by Sivers and Thomas (1972), for example, this summation of the approximate exclusive reactions is equivalent to inappropriately interchanging the infinite sum over n and taking the high energy limit. Therefore in order to obtain the total inelastic rate $\Omega(P)$, where $\Omega(P) = \Sigma_n \Omega_n(P)$, or the inclusive distributions $f(s, \mathbf{q}_i)$ let us perform a summation over n in (6.2.9) before any approximations are made and introduce the function $\Phi(z_\mu)$ where

$$\Phi(z_\mu) = \sum_{n=2}^{\infty} \Phi_n(z_\mu) = \sum_{n=2}^{\infty} \frac{[\Phi(z_\mu)]^n}{n!} \qquad (6.2.18)$$

$$= \exp\left[\phi(z_\mu)\right] - \phi(z_\mu) - 1.$$

The two terms 1 and $\phi(z_\mu)$ in (6.2.15) correspond to 0 and 1 particles respectively in the final state and hence will give a vanishing contribution to $\Omega(P)$ if s is above threshold. This can also be seen by inverting the Laplace transform and regaining the delta functions. Therefore, these two terms can be ignored and we can write

$$\Phi(z_\mu) = \exp[\phi(z_\mu)] \qquad (6.2.18')$$

Now the total inelastic rate can be calculated in exactly the same way as the exclusive rate by replacing $\Phi_n(z_\mu)$ by $\Phi(z_\mu)$ in (6.2.11) to (6.2.16). In particular, it is interesting to note that the temperature equation (6.2.14) becomes

$$\sqrt{s} = -\left.\frac{\partial \phi(z^\mu)}{\partial z}\right|_{z_T = 0} \qquad (6.2.19)$$

which, using (6.2.10') gives

$$z \approx \frac{\pi c}{a^2 \sqrt{s}} \tag{6.2.20}$$

Thus, provided s is large enough or that $a^2 \gg c$ we see *a posteriori* that z is small enough for the power series expansion of the Bessel functions in (6.2.10) to be meaningful.

We shall not present the expression for the total inelastic rate here because, although its calculation is straightforward, it is a little tedious algebraically, and probably not very illuminating. (It is given explicitly for several choices of the parameter a in the work of Humble *et al.*, 1973). We might mention here, however, that the normalization function $\lambda(s)$ is arbitrary within the philosophy of the model, and presumably can be chosen therefore to give any required energy dependence to the total inelastic cross-section. This is similar to the situation in the multi-Regge model where the trajectory and coupling parameter are chosen to reproduce the desired energy dependence of the total cross-section. If, for instance, $\lambda(s)$ is chosen to give a constant total inelastic cross-section, the n-body cross-sections are then found to have a Poisson distribution, i.e.

$$\sigma_n(s) \approx \frac{1}{s^2} \frac{(\bar{c} \log s)^{n-1}}{(n-1)!},$$

where $\bar{c}$ is some constant, provided the parameter a is independent of s, and hence

$$\langle n \rangle \approx \bar{c} \log s.$$

This again is similar to the predictions of the multi-Regge model. Nevertheless, there are differences between the uncorrelated jet model the multi-peripheral description of hadrodynamics. For example, the details of the peripherality of produced particles is different in the two models, and while (in the strong ordering limit) we found that the two end particles in the multi-peripheral chain carried away practically all of the available energy this is not the case in the UJM. In this model all the particles have been treated on exactly the same footing, and therefore, not surprisingly, each particle carries with it its own fraction of the available energy.

In practice, however, it is found that in most multi-particle reactions two of the particles in the final state carry with them a large fraction of the available c.m. energy. Because of this, these particles are often referred to as *leading particles*, or, since they can usually be identified with the initial particles, they are also known as the through-going particles. This phenomenon of small inelasticity for through-going particles means that, on the average, there is only a fraction of the c.m. energy available to the other final state particles. Although this fraction (typically about $\frac{1}{2}$) is not usually as small as suggested by the strong ordering limit in the multi-Regge model calculation it does not appear to change rapidly with the momentum of the incident beam. In order to incorporate this leading particle effect into UJM calculation it has been suggested by Chao (1972) and Sivers and Thomas (1972) that the functions f_i which define the production amplitude, as in (6.2.6), should be modified for these particles so that

$$f_{\text{lp}}(q) = \exp\left[\frac{2\lambda q_{\text{lp}}}{\sqrt{s}} - a_{\text{lp}} r_{\text{lp}}^2\right] \qquad (6.2.21)$$

where lp denotes a leading particle, and λ^μ is a four vector which in the c.m. system has only a time component λ^0. The transverse momentum cut-off parameter a_{lp} may or may not be the same as the cut-off parameter for non-leading particles; but in any case such parameters (and λ^0) are chosen purely to describe the physically observed particle distributions. With the inclusion of f_{lp} given by (6.2.21) the Laplace transform of the phase-space integral for the process $pp \to pp + (n\pi)$ becomes

$$\tilde{\Phi}_n(z_\mu) = \frac{1}{n!} \left[\phi(z^\mu)\right]^n \left[\psi(z^\mu)\right]^2$$

where $\phi(z^\mu)$ is given as before by (6.2.10) and $\psi(z^\mu)$ is given by

$$\psi(z^\mu) = \pi c' \int_{m^2}^{\infty} dr^2 \exp\left[-a_{\text{lp}}(r^2 - m^2)\right] I_0[z_T\sqrt{(r^2 - m^2)}] K_0(\tilde{z}r), \qquad (6.2.22)$$

where

$$\tilde{z} = \left[(z_0 - 2\lambda_0/\sqrt{s})^2 - (z_L - 2\lambda_L/\sqrt{s})^2\right]^{\frac{1}{2}}$$

and m is the nucleon mass. The evaluation of the n-body cross-sections or the total inelastic rate follows through exactly as before though, of course, because of the presence of the ψ functions the solution of the temperature equation is rather more complicated (Chao, 1972; Sivers and Thomas, 1972).

Although we have stressed the importance of the Lurcat and Mazur technique for the evaluation of phase-space integrals (which is natural in view of its historical importance, we should point out that it is not the only method we could have used. There are several ways in which the energy-momentum conserving delta functions can be written in terms of a mathematical transform, besides the Laplace transform considered by Lurcat and Mazur. For instance, it has recently been suggested by de Groot (1972) that the Fourier–Laplace transform is a particularly favourable way of rewriting the delta-functions, so that $\Omega_n(P)$ becomes

$$\Omega_n(P) = \frac{1}{n!}\frac{1}{(2\pi i)^4}\int_{\varepsilon-i\infty}^{\varepsilon+i\infty} \exp\left[Pz + n\int_{-\infty}^{\infty}\frac{\mathrm{d}^3\mathbf{q}_i}{\omega_i}f_i(r_i)\,\mathrm{e}^{-q_i z}\right] \qquad (6.2.23)$$

De Groot's method has the advantage that $\Omega_n(P)$ can be calculated analytically for *all* $n \geq 2$ provided the transverse-momenta of the produced particles is fairly small. Since this is one of the basic assumptions of the model it is not a very severe restriction on the method. Notice that its strength lies in its ability to estimate even the low multiplicity processes, unlike the Lurcat and Mazur technique, which requires n to be fairly large. Thus it represents a considerable improvement over such previous methods for calculating the low-order exclusive cross-sections, and is therefore worthy of much serious attention. However, it is unlikely to make a significant difference to the calculation of the gross average features of the model such as σ_{tot}, $\langle n \rangle$ or $\langle r^2 \rangle$ etc.

Finally in this class of independent emission models, whether they are truly uncorrelated jet models, or the multiperipheral models of Chapter 4, we must consider the production of particles with different quantum numbers. So far we have been rather simple-minded and considered only the production of some fictitious universal scalar particle. However, in reality, nature is not so simple and even if we assume that, to some approximation, only pions are produced in a high energy collision, they come in three sorts, π^+, π^- or π^0. In order to describe the relative distributions of these three charge modes in the multiperipheral model, for example, Chew and Pignotti (1968) suggested that pions were produced by alternating exchanges of an isospin $I = 0$ and $I = 1$ trajectory along the multiperipheral chain. Subsequently Caneschi and Schwimmer (1970) considered a multiperipheral model leading to $I = 1$ ("ρ") pion pair production to account for the relative production rates of charged and neutral pions. More recently this type of approach has been extended by Berger *et al.* (1973) who also considered the production of $I = 0$ ("σ") pion pairs, not only according to Poisson (multiperipheral) distribution but also to an inverse square (limiting fragmentation) distribution.

Alternatively, isospin conservation between the initial and final states can be imposed in an average way as was first suggested by Fermi. For example, in the particular example of $pp \to pp + n(\pi)$ the initial state has isospin $I = 1$. The most simple statistical assumption to make about the final state is that all ways of forming a corresponding $I = 1$ final state from the produced particles are equally probable. Thus, one has to consider all possible final states with n_+, n_-, n_0 number of π^+'s, π^-'s, π^0's respectively where $n_+ + n_- + n_0 = n$, and determine a suitable isospin weighting factor $W(n_+, n_-, n_0)$ for each particular choice of the numbers n_+, n_-, n_0.

Fortunately these isospin weighting factors have been given in a closed form by Cerulus (1961), (see also Shapiro, 1960) as

$$W(n_+, n_-, n_0) = \frac{n!}{n_+! n_-! n_0!} \frac{3}{2^{n_+ + n_- + 3}} \int_{-1}^{1} dx (1 - x)^{n_+ + n_- + 2} x^{n_0} \qquad (6.2.24)$$

$$= 0 \quad \text{if} \quad n_+ < n_- \quad \text{or} \quad n_+ > n_- + 2$$

(A somewhat similar result to (6.2.24) for the random production of pions when charge conservation is imposed was obtained by Horn and Silver (1970)). Consequently the total inelastic rate becomes

$$\Omega(P) = \sum_{n=0}^{\infty} \left(\sum_{n_+ + n_- + n_0 = n} W(n_+, n_-, n_0) \right) \Omega_n(P)$$

where $\Omega_n(P)$ is the rate for the production of two protons and n identical pions. In other words, the identity of the produced pions has been factored out of the production mechanism in this case and is contained solely in the statistical weight factor W. We would refer the reader to the paper of Berger *et al.* (1973) for a detailed examination of combining isospin conservation using this method and the dynamical (charge conservation) assumption that pions are produced via "ρ" or "σ" correlations.

6.3. Diffraction fragmentation models

Besides the multiperipheral or independent emission model the other type of production mechanism which seems to be relevant to a study of hadrodynamics is the diffractive fragmentation model. As we have already outlined at the beginning of this chapter such a model is characterized by an assumed energy independence of the integrated cross-sections which is associated, as the term "diffractive" implies, to vacuum quantum number exchange. In order to make this statement meaningful we must, of course, define between which particles, or system of particles, such a vacuum quantum number exchange is supposed to operate. This brings us to the "fragmentation" part of the model. It is proposed that one or both of the initial state particles are excited into a state of some higher mass M which subsequently decays, or fragments, by the emission of π's, K's, p's and $\bar{p}$'s, etc. The diffractive nature of such a process suggests that these excited states are, in fact, identical to the initial particles except in the very important respect of their mass. Thus the vacuum quantum number exchange is between the initial particles and the decay products corresponding to each excited state. To make matters more concrete it is helpful to picture the diffractive fragmentation hypothesis as shown in Fig. 48. The "jets" produced from the excited states are often referred to variously as fireballs, jets, clusters or more recently, novas, though in practice they all mean just about the same thing. Therefore for convenience we shall follow the language of Jacob and Slansky (1971, 1972) and call them novas. Notice that Fig. 48 suggests that a produced particle is either a decay product of one or other of the nova or else is a leading particle, i.e. one of the initial particles which has not been excited in the collision. A leading particle will presumably be fairly easily identified since it will tend to

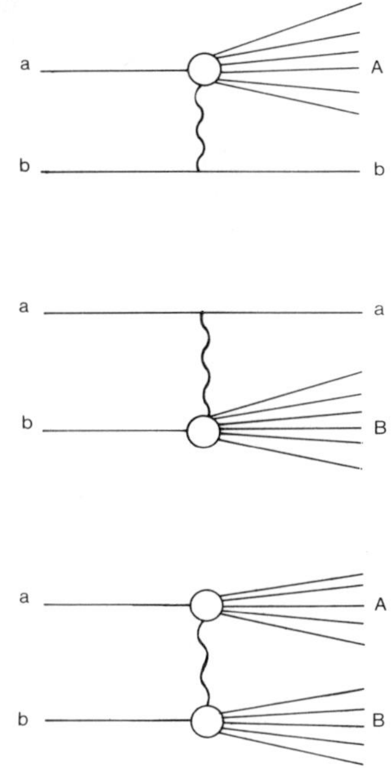

FIG. 48. Single and double nova production.

be travelling in the same direction with roughly the same momenta as the initial particle. Non-leading particles, however, coming from the nova decay will have a much more complicated dependence on the initial state. The purpose of particular diffractive fragmentation models, therefore, is to account for the distribution of non-leading particles by making specific statements about the way in which the nova fragment into their decay products.

Not surprisingly, therefore, there are currently several slightly different versions of such models under consideration. Hwa (1972) has termed his model a diffractive excitation model which deals with the asymptotic behaviour of the mechanism, while the diffraction dissociation model of Adair (1968, 1972) concentrates on the numerical aspects of the problem.

In this section, however, we wish to discuss only the general assumptions of this class of model and we are not concerned too closely with some of the particular aspects of the numerical investigations. Therefore, let us concentrate on only one of these models, i.e. the so-called nova model of Jacob and Slansky (1972). In this way we can now state the basic features of the diffractive excitation mechanism by means of a simple and concrete example. We would only mention that the other diffractive fragmentation models are fairly similar in their general philosophy to the nova model while their technical differences can be fairly easily deduced from an examination of the original papers.

Let us start our discussion of the nova model by considering the nova itself. This is characterized by three quantities: its excitation spectrum $\rho(M)$, the mean number $n(M)$ of decaying particles and the average decay distribution dD/d^3k in the nova rest-frame. (Berger, 1972a; Berger, Jacob and Slanksy, 1972a). For simplicity we shall consider here the production of pions in which case $n(M)$ becomes the mean number of pions decaying from the nova. (Notice that in general we should also consider $n_k(M), n_p(M)$ and so on corresponding to kaon decays, proton decays, etc.). Even so, $n(M)$ must be separated into $n^+(M)\, n^-(M)\, n^0(M)$ if we are to describe the ratio of π^+, π^-, π^0 production.

Having characterized the nova by the introduction of these three basic features we must now choose suitable algebraic forms for these functions in order to flesh-out the diffractive fragmentation hypothesis into a phenomenological model. The simplest assumption to make about the average pion decay distribution $dD/d^3\mathbf{k}$ is that it has a symmetrical form in the nova rest frame, i.e. the decaying particles have no preferred direction in this frame. If they did, then the simplest explanation would be that the nova carried some memory of the way in which it was formed, in particular of the other particle in the initial state. However unless some care is exercised this could be contrary to the factorization and limiting fragmentation hypotheses of diffractive reactions. Thus a suitable parametric form for dD/d^3k would appear to be

$$w\frac{dD}{d^3\mathbf{k}} = \frac{a}{R^2}\exp\left(-k^2/R^2\right) \tag{6.3.1}$$

where $w = (k^2 + \mu^2)$ is the energy of the pion in the nova rest frame $k^2 = |\mathbf{k}|^2$ and a is a normalization constant so that

$$\int \frac{dD}{d^3\mathbf{k}}\, d^3\mathbf{k} = 1$$

and the parameter R is introduced to provide the damping in the transverse momentum of the produced pion. The reason this parameter determines the transverse momentum distribution even in the c.m. system is that when a Lorentz boost is performed on (6.3.1) to obtain the c.m. distribution the boost is found to be predominantly in the longitudinal direction thus leaving the transverse momentum distribution virtually unchanged. Clearly other symmetric forms for $dD/d^3\mathbf{k}$ could be chosen† but (6.3.1) would seem to be one of the simplest giving rise to this desired transverse momenta cut-off.

Another feature of an isotropic decay distribution is that the mean energy release per decay is a constant, which incidentally is also determined by the parameter R. Hence the average number of decaying pions depends linearly on the mass of M of the nova, i.e.

$$n(M) = \lambda(M - M_0), \tag{6.3.2}$$

where M_0 is the mass of the nova's ground state.

Having thus considered two of three characteristics of the nova, namely $dD/d^3\mathbf{k}$ and $n(M)$ we now come to the third function, the nova's excitation spectrum, $\rho(M)$. In order to see what form $\rho(M)$ must take if the diffractive fragmentation model is to be a viable description of the data, let us for the moment concentrate only on single nova excitation and note that

$$\sigma_{\text{inel}}^{ab} = \beta \int_{M_0}^{(\sqrt{s}) - m_b} \rho(M)\, dM \tag{6.3.3}$$

$$\sigma_{\text{inel}}^{ab} \langle n \rangle = \beta \int_{M_0}^{(\sqrt{s}) - m_b} n(M)\, \rho(M)\, dM. \tag{6.3.4}$$

Of course, in general the right-hand side should include single nova contributions from both a and b initial particles and also a double nova contribution. However, let us assume that each of these contributions gives the same energy dependence for σ_{inel} and $\sigma_{\text{inel}} \langle n \rangle$. In this case, therefore, (as we previously pointed out in our discussion of (6.1.5)), σ_{inel} will tend to a constant at high energy and $\langle n \rangle$ will have a log s behaviour provided $\rho(M) \propto M^{-2}$ (i.e. $\propto n(M)^{-2}$) for large M. The low M behaviour of this function, on the other hand, is rather more arbitrary although it should presumably fall to zero at threshold, i.e. at $M = M_0$, and probably peak around some value $M_0 + 1$ GeV‡ before decreasing asymptotically like $1/M^2$. A typical parameterization of $\rho(M)$ which incorporates these features is

$$\rho(M) = (M_0 + \gamma M)^{-2} \exp\left(-\delta/(M - M_0)\right). \tag{6.3.5}$$

where γ is chosen to be about $6\ \text{GeV}^{-1}$ and $\delta \approx 2$ GeV. It should be stressed,

† Adair, for example, uses instead a simple exponential form $\exp\left(-k/R\right)$.

‡ This is perhaps not obvious until one compares the model predictions to the data at moderate values of s.

however, that except for the M^{-2} asymptotic behaviour the parameterization of $\rho(M)$ is determined only by detailed comparison with the data.

Having discussed some of the features of the nova with regard to the most basic measurements of a production process, namely σ_{inel} and $\langle n \rangle$ let us now consider the predictions of the model for single and double particle distributions, i.e. the one particle and two particle inclusive reactions. To do this let us rewrite the average particle decay distribution (6.3.1) in terms of rapidity variables, i.e.

$$\frac{\mathrm{d}D}{\mathrm{d}y_0 \mathrm{d}k_T^2} = \frac{\pi c}{R^2} \exp\left(-\frac{k_T^2}{R^2}\right) \exp\left\{-\left(\frac{m' \sinh y_0}{R}\right)^2\right\}, \tag{6.3.6}$$

where k_T is the transverse component of $\mathbf{k}$ and m' is the transverse mass, $m' = \sqrt{(k_T^2 + \mu^2)}$. y_0 is the rapidity of the decaying particle in the nova rest frame such that $w = m' \cosh y_0$. Since the transverse momenta are practically unaltered in Lorentz transforming this distribution into the c.m. system we consider only the transformation in the rapidity variable and write

$$y = y_0 + Y_a(s, M)$$

or
$$\tag{6.3.7}$$

$$y = y_0 - Y_b(s, M)$$

depending on whether the nova is produced from the beam particle a or the target particle b. As we saw in the last chapter the rapidity translations Y_a, Y_b are simply given by the relations

$$\sinh Y_a = P(s, M^2, m_b^2)/M$$

or
$$\tag{6.3.8}$$

$$\sinh Y_b = P(s, M^2, m_a^2)/M$$

Integrating (6.3.6) over k_T the nova decay rapidity distribution is given by

$$\frac{\mathrm{d}D}{\mathrm{d}y_0} \approx \frac{c}{\cosh^2 y_0} \exp\left\{-\left(\frac{\mu \sinh y_0}{R}\right)^2\right\} \tag{6.3.9}$$

so from (6.3.7) the nova decay distribution in rapidity in the c.m. system becomes

$$A(M, y) = \frac{\mathrm{d}D(y \mp Y)}{\mathrm{d}y_0} \tag{6.3.10}$$

Notice that (6.3.9) is not an exact expression since at finite energy the integration cannot be continued right out to $k_T = \infty$. However, since the k_T

distribution anyway is strongly damped the error incurred by this assumption will be rather small.

Now we are in a position to calculate the c.m. rapidity distribution of pions produced from a single nova. These are given simply by including $A(M, y)$ of (6.3.10) into the integrals of (6.3.4), i.e.

$$\frac{d\sigma}{dy} = \int_{M_0}^{(\sqrt{s})-m_b} n(M)\,\rho(M)\,A(M, y)\,dM \tag{6.3.11}$$

for the single pion inclusive distribution and

$$\frac{d^2\sigma}{dy, dy_2} = \int_{M_0}^{(\sqrt{s})-m_b} \langle n(M)(n(M) - 1)\rangle \rho(M)\,A(M, y_1)\,A(M, y_2)\,dM \tag{6.3.12}$$

for the inclusive distribution of two identical particles. The function $\langle n(M)(n(M) - 1)\rangle$ is twice the average number of pairs of pions produced from the decay of a nova. Again, in general, the total inclusive distributions are obtained by including effects from the single excitation of the other nova and from double nova excitation.

One of the most interesting features of these diffractive fragmentation models can be seen by integrating (6.3.12) over dy_1 and dy_2 to obtain the two particle expectation value, i.e.

$$\langle n(n - 1)\rangle \sigma_{incl} = \beta \int \langle n(M)(n(M) - 1)\rangle \rho(M)\,dM \tag{6.3.13}$$

where the functions $A(M, y_1)$, $A(M, y_2)$ drop out because dD/d^3k from which they were derived was normalized to unity. Since $\rho(M) \sim M^{-2}$ for large M and $n(M).\,(n(M) - 1) \propto M^2$, (6.3.13) shows that these models predict an $s^{\frac{1}{2}}$ asymptotic behaviour for $\langle n(n - 1)\rangle$ which is much larger than the multi-peripheral-cluster-model prediction of log s, suggested by Mueller (1971). Moreover, as shown in Fig. 49, the nova model prediction for $\langle n(n - 1)\rangle$ (Berger, 1972b) indicates that this asymptotic behaviour does not set in until quite large energies. In fact, the model can account quite well for the negative correlation effect in $R_{\pi^-\pi^-} = \langle n_{\pi^-}(n_{\pi^-} - 1)\rangle - \langle n_{\pi^-}\rangle^2$ from pp collisions at lower energies which, as we discussed previously, is probably a result of phase-space limitations.

This marked difference between the diffractive fragmentation model prediction for the asymptotic behaviour of $\langle n(n - 1)\rangle$ and the multi-peripheral model prediction is often stated as a definitive test of these types of models. However, it is only a test of one class of diffractive models, i.e. those which have $\rho(M) \approx M^{-2}$ and an *isotropic* average particle decay distribution dD/d^3k. If instead of the assumption that $\langle k_L\rangle \approx \langle k_T\rangle$ in the nova rest frame (where L and T refer to the longitudinal and transverse

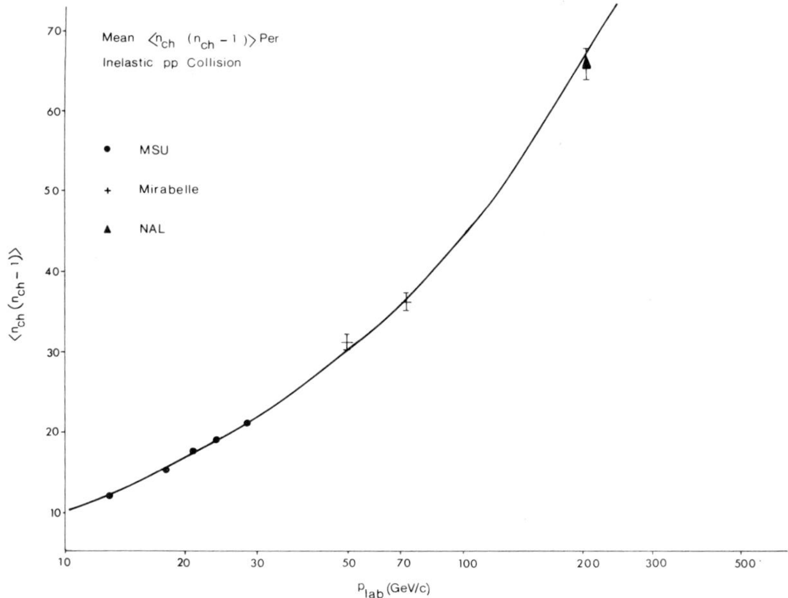

FIG. 49. Recent measurements of $\langle n_{ch}(n_{ch} - 1) \rangle$ observed in pp collisions.

components of the momentum) we make the alternative hypothesis that $\langle k_L \rangle \gg \langle k_T \rangle$ even in this Lorentz frame then $n(M)$ will no longer be proportional to M at large M. Instead, for this "one dimensional" distribution we would have $n(M) \propto \log M$ and in order that $\langle n \rangle \propto \log s$ we must take $\rho(M) \propto (M \log M)^{-1}$ for large M. In this case we find that

$$\sigma_{\text{incl}} \underset{s \to \infty}{\to} \log (\log s) \qquad (6.3.14)$$

$$\langle n \rangle \underset{s \to \infty}{\to} \log s/\sigma_{\text{incl}} \qquad (6.3.15)$$

and

$$\langle n(n - 1) \rangle \underset{s \to \infty}{\to} (\log s)^2/\sigma_{\text{incl}} \qquad (6.3.16)$$

so that the large $s^{\frac{1}{2}}$ behaviour for $\langle n(n - 1) \rangle$ is not a feature of this version of the fragmentation model.

In fact Chan and Tsou (1971) have shown that $n(M) \propto \log M$ is precisely the relation which follows from a study of the cluster properties of dual models. Of course, whether the dual model is a fragmentation or multiperipheral model is somewhat difficult to decide, though probably the best guess would be that it is somehow a mixture of both. It is therefore interesting to note that several authors have found (6.3.15) and (6.3.16) also result from a combination of the diffractive and multiperipheral mechanisms. See, for example, Wilson (1970) and Le Bellac (1971).

So far we have only discussed the possibility of one of the initial particles becoming excited into a nova which subsequently decays to produce the multiparticle final state. However, in principle there is nothing to prevent both particles being excited into nova, and indeed, as the energy of the system rises we might expect this to become an increasingly relevant production mechanism. (We might note, however, that at normal laboratory energies there is little evidence for double diffraction dissociation.)

In order to allow for this double nova possibility we have only to specify the relative importance of double nova excitation with respect to the excitation of either single nova. To do this a function $R(M_a, M_b)$ is introduced into the model such that for double nova production

$$\sigma_{\text{incl}} = \int \rho_a(M_a)\rho_b(M_b)R(M_a, M_b)\, \mathrm{d}M_a\, \mathrm{d}M_b \qquad (6.3.17)$$

$$\sigma_{\text{incl}}\langle n\rangle = \int \rho_a(M_a)\rho_b(M_b)R(M_a, M_b)(n_a(M_a) + n_b(M_b))\, \mathrm{d}M_a\, \mathrm{d}M_b \quad (6.3.18)$$

$$\frac{\mathrm{d}\sigma}{\mathrm{d}y} = \int \rho_a(M_a)\rho_b(M_b)R(M_a, M_b)$$

$$\times\ (n_a(M_a)A_a(M_a, y) + n_b(M_b)A_b(M_b, y))\, \mathrm{d}M_a\, \mathrm{d}M_b \qquad (6.3.19)$$

$$\frac{\mathrm{d}^2\sigma}{\mathrm{d}y, \mathrm{d}y_2} = \int \rho_a(M_a)\rho_b(M_b)R(M_a, M_b)\, \mathrm{d}M_a\, \mathrm{d}M_b$$

$$\times\ \{n_a(M_a)\,n_b(M_b)(A_a(M_a, y_1)A_b(M_b, y_2) + A_a(M_a, y_2)A_b(M_b, y_1)$$

$$+\ \langle n_a(n_a - 1)\rangle_{M_a}A_a(M_a, y_1)A_a(M_a, y_2)$$

$$+\ \langle n_b(n_b - 1)\rangle_{M_b}A_b(M_b, y_1)A_b(M_b, y_2)\} \qquad (6.3.20)$$

where $A_i(M, y)$ as before are given by (6.3.10) except the rapidity translation Y is determined from (6.3.8) with M_a or M_b replacing m_a, m_b wherever appropriate. Notice that in (6.3.20), for instance, the three terms in curly brackets correspond to the three possibilities of one particle coming from each nova, or both coming from one or the other nova.

The function $R(M_a, M_b)$ must reflect the relative ease with which two nova of masses M_a and M_b can be produced. Hence besides M_a, M_b, $R(M_a, M_b)$ will probably also be a function of the total energy of the system $\sqrt{s}$ and the momentum transfer t in the reaction where $t = t(M_a^2, M_b^2, s)$. In fact, in some phenomenological fits (Berger $et\ al.$, 1972a) it has been found adequate to take $R(M_a^2, M_b^2)$ to be a simple θ-function in t, i.e. typically

$$R(M_a, M_b) = R \cdot \theta(t + 1) \qquad (6.3.21)$$

where R is a constant to be determined by comparison with the data. In numerical fits to $f_{\pi^- \pi^-}$, the two-particle inclusive distribution obtained from $\pi^+ p$ reactions at $18 \cdot 5$ GeV/c with this model the double nova contribution is found to be 30% of the total inelastic cross-section (Berger *et al.*, 1972a). A similar result was also found from a study of $f_{\pi^- \pi^-}$ obtained from pp collisions at 21 GeV/c (Berger *et al.*, 1972b).

Although it is probably possible to live with a double nova contribution of 30%, such a large fraction does seem to be surprising in view of our comment that there is very little evidence for double-diffraction dissociation in many exclusive processes at these energies. It is perhaps wise, therefore, to examine these diffractive fragmentation models in more detail. Clearly they cannot be expected to reproduce all the features of the data. They cannot for example, give a precise description of low energy resonance production effects where the spin of the resonance is likely to produce an anisotropic decay distribution. More importantly, the model as outlined above does not specify the exchange mechanism in detail. This is reflected in two ways. Firstly, it assumes only vacuum exchange (diffraction scattering) which at moderate energies clearly only forms a part of the dynamics. There is abundant evidence for non-diffractive contributions to the cross-sections, but how to separate diffractive and non-diffractive effects in a meaningful way is a difficult task. However, unless one does separate these effects, a diffractive model fit to the whole data must necessarily incur some error. In this connection it is relevant to mention the work of Gottfried and Koefoed-Hansen (1972). They take the very reasonable view that the inelastic cross-section can only be equated to the cross-section calculated by the diffraction model in the limit that $s \to \infty$. By then extrapolating the model down to lower energies they deduce that 29% of the inelastic cross-section must be due to non-diffractive effects even at ISR energies, $s \approx 440$ GeV^2 while at still lower energies $s \approx 50$ GeV^2, the diffractive component is found to be yet much smaller.

Secondly, the s-dependence of the diffraction model in this extrapolation comes not only from the shape of the nova spectra and the allowed range of M_a, M_b, but also from the assumed momentum transfer dependence of the process. So far we have neglected any t-dependence in the simple description of the nova model given above. However, it becomes very clear when one considers, for instance, the distribution of a through-going, non-excited, particle that one requires a rather definite statement about the momentum transfer dependence. In particular, Berger *et al.* (1972a) make the additional assumption that for a leading particle its rapidity distribution is given by the form

$$\frac{d\sigma}{dy} = k\sqrt{s} \int \frac{1}{M^*} \frac{d\rho(M^*)}{dt} dr^2 \tag{6.3.22}$$

where k is the c.m. momentum, r is the transverse momentum and M^* is the mass of the recoil nova such that

$$M^{*2} = s - 2\sqrt{(s)}m' \sinh y + \mu^2.$$

The t-dependence of the exchange mechanism is parameterized by choosing the form

$$\frac{d\rho(M^*)}{dr^2} = \rho(M^*)B \exp(-Br^2). \tag{6.3.23}$$

However, even without introducing any specific t-dependence in the production amplitude it should be recognized that the kinematical constraints which restrict the minimum value of $|t|$ to a value depending on M_a, M_b and s can already produce considerable changes into the model. (See, for instance, Hwa (1971).) This point has also been taken up recently by Abarbanel and Kane (1973), who introduce $|t|_{\min}$ effects simply by multiplying each integrated cross-section σ_n (which in the diffraction picture is proportional to $1/n^2$) by an $\exp(-b|t|_{\min})$ factor. Then by noting that for single nova production

$$|t_{\min}| \approx M_a^4 m_b^2/s^2 \tag{6.3.24}$$

while for double production

$$|t_{\min}| \approx M_a^2 M_b^2/s \tag{6.3.25}$$

and choosing $M_a \approx M_b$ with $M_{a,b}$ proportional to the number of particles produced

$$M_{a,b} = cn$$

they suggest that

$$\sigma_n \approx \frac{1}{n^2} \exp(-\alpha n^4/s^2) \qquad \text{(for single nova production)} \tag{6.3.26}$$

$$\sigma_n \approx \frac{1}{n^2} \exp(-\beta n^4/s) \qquad \text{(for double nova production)} \tag{6.3.27}$$

From a comparison of these $|t|_{\min}$ modified diffraction model predictions with data on pp, $\pi^- p$ and $k^- p$ collisions they conclude that *in order to fit the σ_n data* the double nova production must be far more important than single nova excitation. (In this situation they also show that while one still has

$\langle n \rangle \sim \log s$, $\langle n^r \rangle \sim s^{(r-1)/4}$, $r > 1$ so that $\langle n^2 \rangle$ now goes like $s^{\frac{1}{4}}$ instead of $s^{\frac{1}{2}}$ as in the unmodified model.) However, this predicted dominance of double nova production even in fairly low energy regions is contrary to the empirical evidence from exclusive reactions which is against any sizeable double diffraction dissociation effects. This strongly suggests therefore that the diffraction mechanism must account for only a part of the hadrodynamics, or that diffraction models have to be drastically changed from their simple form given here.

6.4. Thermodynamical and hydrodynamical models

In the introduction to this chapter we noted the statistical nature of the models we were going to consider, but we did not enlarge on the differences between these statistical pictures. However, to understand the motivation for the thermodynamical models of this section it is perhaps relevant to do so now. In Section 6.2, for instance, the statistical nature of particle production was invoked to permit an approximate method of calculation using quantum mechanical, phase-space formulae. In Section 6.3, on the other hand, a statistical description was invoked to describe in a simple manner particle production phenomena in terms of average particle decay distributions, etc., in certain Lorentz frames. Now in the present section the statistical hypothesis is that there are a sufficient number of particles produced which interact with another to form a system in statistical equilibrium. This hypothesis enables us to consider the production reaction to some extent along classical lines as a process developing in space and time. Of course, depending on the assumptions we make concerning this development we can obtain several different models of production phenomena. Here we shall concentrate principally on three models due to Fermi, Landau and Hagedorn respectively, though from time to time we shall also touch on other models. In any case, if our treatment of a rather difficult subject is as simple minded as we intend it to be, the reader shall become aware that numerous other variations of these three basic models may be possible if physical intuition or empirical evidence suggest the need for them.

6.4a. *Fermi's black body radiation model.* The first model we consider is due to Fermi (1950, 1951). The basic postulates of this model are most easily stated in the following way: (Belenkji and Landau, 1956):

(i) When two very fast nucleons collide the energy in the c.m. system is released in a very small volume V. As the nuclear interaction is very great and the volume small the energy distribution will be determined by statistical laws so that the high energy collision may be examined without regard to any particular theory of nuclear interactions.

(ii) The volume V into which the energy is released is determined by the dimensions of the nucleon meson cloud whose radius is of the order of $\hbar/\mu c$, where μ is the pion mass. However, since the nucleons move at high velocity the meson cloud surrounding them undergoes Lorentz contraction in the longitudinal direction. Thus we have

$$V \approx \frac{4\pi}{3}\left(\frac{\hbar}{\mu c}\right)^3 \frac{2mc^2}{\sqrt{s}} \tag{6.4.1}$$

(iii) The particles are created in the volume V at the initial moment of collision, according to the laws of statistical equilibrium. *After this initial creation the particles no longer interact with one another and escape from the volume in a "frozen" state.*

This third postulate is really the crux of the Fermi model so let us discuss it a little more fully. The assumption of statistical equilibrium is certainly an extreme assumption though as we shall see it does permit us to simplify the calculations very considerably. [It is interesting to compare it, for example, with Heisenberg's earlier descriptions of very high energy collisions of two nucleons in terms of some sort of turbulent motion of the pion cloud which is produced by the impact. Heisenberg (1949).]

The statistical equilibrium condition, of course, must be qualified so that additive quantum numbers such as charge are conserved and that energy momentum conservation is satisfied. Also, as pointed out by Fermi (1950) only those states which are easily reachable from the initial state may actually attain statistical equilibrium. Thus radiation phenomena in which photons are created will not have time to develop and the model must be restricted to a description of the production of strongly interacting particles only, e.g. $\pi^{\pm}$, π^0's and $N\overline{N}$ pairs. Nevertheless, the statistical equilibrium condition and the assumption that all produced particles are extremely relativistic allowed Fermi to consider a thermodynamical approach to the problem, as we shall now indicate. This is to be contrasted with the statistical computation of the probabilities of the various final state configurations which we discussed earlier.

Firstly, Fermi noted that for extremely relativistic particles it would appear that we could apply a version of Stefan's law so that the energy density is proportional to the fourth power of the temperature. More specifically, he pointed out that pions like photons obey Bose–Einstein statistics; and further, *if the temperature is very high*, the rest mass of the particles is negligible so that the energy momentum relationship will be approximately the same as for photons. Hence the Stefan's law for pions should be similar to Stefan's law for black body radiation; the only difference being a statistical weight factor. The statistical weight for photons is 2 corresponding to the two polarization states of the photon, whereas for pions the weight factor will be 3 corresponding to the three charge states of the pion. Consequently Fermi postulated a Stefan's law for "black-body" radiation of pions as

$$\varepsilon_\pi = \frac{3}{2}\frac{6\cdot494\,(kT)^4}{2\pi^2\hbar^3 c^3} \tag{6.4.2}$$

where ε_π is the pion energy density, T is the temperature, k is the Boltzman constant, and $6\cdot494 \approx \pi^4/15 = \Sigma(1/n^4)$. He further postulated a Stefan's law for nucleon and anti-nucleon radiation in which case

$$\varepsilon_N = \frac{8}{2}\frac{5\cdot682(kT)^4}{\pi^2\hbar^3 c^3}, \tag{6.4.3}$$

where 8 represents the various spin and isospin states of the nucleons and anti-nucleons and the factor $5\cdot682 \approx 6\Sigma(-1)^{n+1}/n^4$ arises because the particles now obey Pauli statistics. Using (6.4.2) and (6.4.3) the temperature is now easily obtained by equating the total energy of the system, $\sqrt{s}$, to the volume V (given by (6.4.1)), multiplied by the sum of the two energy densities, i.e.

$$\sqrt{s} = V(\varepsilon_\pi + \varepsilon_N). \tag{6.4.4}$$

Hence we have

$$(kT)^4 = 0\cdot152\frac{3c^4 s}{4m\pi}. \tag{6.4.5}$$

Fermi was also able to use the statistical nature of the theory to calculate the density of the various particles. According to the standard procedures of statistical mechanics he observed that in the extreme relativistic case these densities would be proportional to $(kT)^3$. Specifically he obtained

$$\rho_\pi = 0\cdot367\frac{(kT)^3}{\hbar^3 c^3}, \quad \rho_N = 0\cdot855\frac{(kT)^3}{\hbar^3 c^3}. \tag{6.4.6}$$

Therefore the number of pions and nucleons can be calculated by multiplying these densities by the volume V and substituting the expression given by (6.4.5) for the temperature. After applying corrections for angular momentum conservation Fermi predicted that the number of pions and the number of nucleons and anti-nucleons produced in very high energy collisions would be

$$\langle n_\pi \rangle = 0\cdot54\,(\sqrt{s}/mc^2)^{\frac{1}{2}}$$
$$\langle n_N \rangle = 1\cdot3\,(\sqrt{s}/mc^2)^{\frac{1}{2}} \tag{6.4.7}$$

That is to say, when relativistic conditions are achieved for all particles, whether pions or nucleons, he predicted that nucleons and anti-nucleons would be more numerous than pions!

This rather startling result is a direct consequence of the assumption of a fixed volume into which larger and larger amounts of energy are dumped during the collision. Consequently the temperature increases without limit and the masses of the produced particles become negligible, i.e. the Boltzman factor $\exp(-m/T) \to 1$ for all values of m. Thus the number of particles produced depend, for the most part, only on their internal degrees of freedom. Hence as we have seen, Fermi arrives at the approximate result

$$\langle n_\pi \rangle : \langle n_N \rangle = 3 : 8$$

whereas even in cosmic rays the ratio is found to be

$$\langle n_\pi \rangle : \langle n_N \rangle \approx 1000 : 1$$

Fermi's model, as we have stated it, applies only to central collisions, i.e. to collisions with impact parameter equal to zero. Hence this model would suggest an isotropic angular distribution of particles (see Fig. 47a) which is also not observed experimentally. Although this model was subsequently extended (Fermi, 1951) to describe collisions with non-zero impact parameter this extension has been severely criticized by Belenkji and Landau (1956). In any case it cannot compensate in any dramatic way for the relative abundance of nucleons suggested by the theory. To do this it would seem necessary to restrict the temperature T to a finite value by allowing the volume V to increase.

However, before we discuss this possibility, it is interesting to note an alternative hypothesis suggested by Koba (1965a, 1965b). Instead of the physical particles reaching thermodynamical equilibrium in the fixed volume V, Koba assumes that it is the constituent quarks which reach equilibrium. Then assuming $SU(6)$ symmetry, Koba calculates that the ratio of produced pions to kaons to nucleons, etc., does after all, closely resemble the empirically observed ratios.

Nevertheless, in spite of Koba's achievement, it does not seem possible to consider a large number of particles in a small volume V as a perfect gas. As stressed by Pomeranchuk (1951) such particles must interact and undergo numerous transformations until the distances between them are all of the order of the range of the forces, i.e. $\hbar/\mu c$. That is to say, one cannot assume that the particles escape from V in a "frozen state". In general, therefore, the initial volume V into which the total energy of the collision is assumed to be dumped must expand until

$$V_{\text{new}} \equiv V_p = \alpha \langle n \rangle V \gg V, \tag{6.4.8}$$

where α is some constant (≈ 1). Hence this new volume V_p is a function of the number of particles produced. However, if we assume that we can still apply the standard procedures of statistical mechanics we still have the conditions (6.4.6), i.e. that

$$\langle n \rangle \approx V_p \frac{(kT)^3}{\hbar^3 c^3}, \tag{6.4.9}$$

where both $\langle n \rangle$ and $(kT)^3$ are proportional to the entropy S of the system. In this case, therefore, we can combine (6.4.8) and (6.4.9) to estimate the decay temperature T_d as

$$kT_d \approx (\alpha V)^{-\frac{1}{3}} \approx \mu = \text{constant} \tag{6.4.10}$$

However, while we now have a decay temperature T_d which is constant and low enough to give the correct ratio of pions to nucleons this model predicts a rather large energy dependence for $\langle n \rangle$. To see this we note that as before Stefan's law suggests

$$\sqrt{s} \approx V_p \frac{(kT)^4}{\hbar^3 c^3},$$

but now T is constant while V_p is proportional to $\langle n \rangle$. Therefore we obtain the result

$$\langle n \rangle \propto \sqrt{s}. \tag{6.4.11}$$

This energy dependence is practically the largest allowed by energy conservation, and although there is *some* evidence for such a rapid increase in $\langle n \rangle$ with $\sqrt{s}$ at extremely high cosmic ray energies (see Wolfendale and Wdowczyk, 1973) it is certainly not observed below 10^3 GeV. In order to understand where the Fermi and Pomeranchuk models break down it is interesting to consider a rather more appealing version of their models proposed by Landau (1953).

6.4b. *Landau's hydrodynamical model.* Following the paper of Belenkji and Landau (1956) we can enumerate the basic features of this model as follows:
   (i) The collision of the two initial particles (nucleons) creates a compound system and energy is released into a small, Lorentz contracted, volume V. At the moment of collision a large number of "particles" is created, the collision mean free path in the system is small compared with the dimensions of the volume V, and a statistical equilibrium is established.
   (ii) There is now a second stage to the collision during which the system expands. Considering the expansion as the motion of an ideal (non-viscous and non-heat-conducting) fluid this stage should be examined hydro-dynamically. The mean free path during this stage remains small compared

to the dimensions of the system which justifies the application of hydro-dynamics. In the first and second stage of the collision, because of the high energy density and strong interactions in the system, particles are continu-ally being created and destroyed. *Hence the system cannot, as yet, be characterized by the number of particles.*

(iii) Of course as the system expands the interaction decreases and the mean free path increases. When the interaction is small enough the particles in the system at that time will become "frozen", and when the mean free path is of the order of the dimensions of the system, the system will break up into these constituent particles.

As in the Pomeranchuk model the temperature at "break-up", T_d, will be roughly of the order of the pion mass. However, the significant difference between the Landau model and the previous models is the treatment of the second stage of the collision. Since the expansion and cooling of the system takes place over a fairly long period of time, the pressure (leading to addi-tional accelerations), will give rise to significant modifications of the system which can be considered using relativistic hydrodynamics.

Although we do not propose to go into all the details of the model it is instructive to give a rough idea of its prediction for the mean multiplicity. To do this let us note, as we did after (6.4.9) that the number of particles after break-up, $\langle n \rangle$, is linearly proportional to the entropy S. Hence to determine $\langle n \rangle$ we shall have to calculate S at the time of break-up. However, the second stage of collision, the expansion, is regarded as the motion of an ideal fluid and therefore adiabatic. Thus the entropy of the system remains constant from the moment after collision, all through this second stage, and right up to the time of break-up. Consequently, it is only necessary to deter-mine S during the initial stages with the system in a state of statistical equi-librium. Such an equilibrium condition implies that the chemical potential vanishes, which is to say

$$\varepsilon - T\sigma + p = 0, \tag{6.4.12}$$

where ε is the energy density, p is the pressure, σ is the specific entropy and T is the temperature of the system in the earlier stages of the collision (given in energy units). In order to calculate σ and hence S where

$$S = \sigma V, \tag{6.4.13}$$

and V is the initial volume of the system, an equation of state is required to describe the highly compressed medium at temperatures $T \gg \mu c^2$. Landau assumed the equation

$$p = \tfrac{1}{3}\varepsilon \tag{6.4.14}$$

which is equation of state for an extremely relativistic electron-photon gas.

Combining (6.4.12) and (6.4.14) we have the conditions

$$\sigma \approx \varepsilon^{\frac{3}{4}} \quad \text{where} \quad T \approx \varepsilon^{\frac{1}{4}}. \tag{6.4.15}$$

(See the relations for black-body radiation (6.4.2), (6.4.3), (6.4.6).) From (6.4.4), (6.4.13) and (6.4.15) we find therefore

$$\langle n \rangle \propto S \propto s^{\frac{1}{4}}, \tag{6.4.16}$$

which is similar to Fermi's model; although the present model is significantly different, and since the break-up temperature is constant and low, i.e. $T \approx \mu c^2$, the produced particles in this case will be predominantly pions.

In order to compute the energy and angular distribution of particles produced in this model it is again necessary to investigate the expansion of the system. Specifically Landau used the following equation of relativistic hydrodynamics to study the motion of the substance

$$\frac{\partial T_{ij}}{\partial x_j} = 0, \quad T_{ij} = wu_iu_j + pg_{ij}, \tag{6.4.17}$$

where the specific enthalpy $w = \varepsilon + p$, and u_i represents the four velocity. The tensor g_{ij} is given by $g_{11} = g_{22} = g_{33} = 1$, $g_{44} = -1$, with $x_i = (x_1, x_2, x_3, it)$; $c = 1$. At the moment of collision the system in the c.m. frame has the form of a very flat disc whose radius in the transverse direction is of the order of μ^{-1} compared to its thickness which is of the order of $2m/(\mu\sqrt{s})$. Therefore initially the expansion of the system can be examined by solving (6.4.17) in one dimension only (Landau, 1953; Halatnikov, 1954). However, at some point the expansion of the system must become three-dimensional and the solution of (6.4.17) for this three-dimensional case must be matched on to the one-dimensional solution. This can only be done in an approximate manner but leads to an energy distribution for produced pions, which we rewrite in terms of rapidity y

$$\frac{1}{n}\frac{dn}{dy} = \frac{1}{n\sigma_{\text{inel}}}\frac{d\sigma(p + p \to \pi + X)}{dy} = \frac{\exp(-y^2/2L)}{\sqrt{(2\pi L)}} \tag{6.4.18}$$

where $L = \frac{1}{2}\log{(s/4m^2)}$. (See Gerasimova and Chernavshy (1955) for a discussion of modifications to this model to incorporate leading particle effects.)

It is further found that the hydrodynamical nature of Landau's model leads to the prediction that the transverse momentum increases slowly with energy. (For a detailed treatment of the transverse momentum distribution see Milekhin (1958).) Also it is calculated that most particles are produced at angles of about one radian. A smaller number of particles are formed at smaller angles but they possess greater energy. This is to be contrasted with the isotropic distribution suggested by Fermi's model.

For a fuller discussion of Landau's model and its relationship to other statistical models we would refer the reader to the excellent review by Feinberg (1972) and to the references of other reviews given therein. (See also Cooper and Schonberg (1973) for an analysis of production from pp collisions using a version of this model.) Clearly it is a very ambitious and detailed model which successfully incorporates many features of very high energy particle production. Because of the low decay temperature it correctly predicts the ratio of different produced particles. Also it gives a strong preference to events with only small values of transverse momenta, which together with (6.4.18) suggests the average pion energy in the c.m. system is of the order of 0·4–0·5 GeV. It is, however, essentially a *very* high energy model, and is only applicable provided $\langle n \rangle > 10$–15. (Sissakyan *et al.* 1971). For lower values of $\langle n \rangle$, i.e. at lower energies, the velocities in the system are still fairly low so that the hydrodynamical stage is not operative. In this case we return to the simple statistical picture of Pomeranchuk. Also, although Landau's choice of state equation (6.4.14) seems highly plausible it has been demonstrated by Ezawa *et al.* (1957), Namiki and Iso (1957) and Iso *et al.* (1959) that other equations of state are quite possible. Moreover, Landau's model seems deficient in that it does not take account of viscosity and heat conductivity in the initial stages of the expansion (Hamaguchi, 1953).

The most serious difficulty with the model, however, is that (as it has been written) it is a model only for central collisions whereas it would appear that peripheral reactions are dominant at least in the energy range presently accessible. In order to describe these dominant peripheral interactions (i.e. collisions with impact parameter $\neq 0$) we could perhaps consider the formation of one, two or more fireballs (nova) and apply our statistical ideas to the fireball decays. However, it is clear that the mass of the fireballs must be *very* large before the hydrodynamical picture becomes applicable, and for the most part Pomeranchuk's statistical model should be relevant. (Feinberg and Chernavsky 1951, 1953.)

Notice that in the Pomeranchuk model we had $\langle n \rangle \sim s^{\frac{1}{4}}$ so that this now means the average number of particles produced from the fireball $\langle n \rangle_M \sim M$ where M is the mass of the fireball (cf. 6.3.2). There is, therefore, a close link between these statistical ideas for peripheral collisions and the diffractive fragmentation models considered previously.

6.4c. *Hagedorn's statistical thermodynamical model.* Finally in this section, though by no means least, we come to the thermodynamical model of Hagedorn. Here the ideas of statistical equilibrium are combined with certain kinematical assumptions to provide a very satisfactory description of production phenomena both from central and peripheral collisions.

The two essentially independent ingredients of the model can be stated briefly as

H

(i) a statistical bootstrap of excited hadrons (fireballs)

(ii) a kinematical superposition of decaying fireballs.

Since they were initially proposed by Hagedorn (1965) and Hagedorn and Ranft (1968) both (i) and (ii) have been modified somewhat, though unlike most models the modifications have meant a simplification of the original model.

In order to understand (i) and (ii) as they now stand let us start back at the beginning and consider a piece of highly excited hadronic matter—a fireball. It is described as a hadronic black-body radiation enclosed in a volume $V \approx \frac{4}{3}(\pi/\mu^3)$; and kept at a temperature T, chosen such that $\langle E(T) \rangle$ equals the energy (=invariant mass) of the object. The decay of this fireball can then be described according to statistical mechanics, i.e. the one-particle momentum cecay distributions are of the Planck type

$$f_m(\mathbf{k}, T) \equiv \frac{\mathrm{d}D}{\mathrm{d}^3 \mathbf{k}} = \mathrm{const}\left[\exp\left(\frac{\sqrt{(k^2 + m^2)}}{T} \right) \pm 1 \right]^{-1}, \qquad (6.4.19)$$

where m is the mass of the produced particle and the plus sign refers to fermion production and the minus sign to boson production.

Although the temperature has been specified the mass of the fireball is arbitrary. Thus we must presumably allow a fireball to have any mass. At low masses the fireball will decay into only a few particles and will behave very much like a low mass resonance. Indeed, it is assumed there is no fundamental difference between resonances and fireballs. However, the bootstrap hypothesis for hadrons suggests that all particles and resonances are bound states or resonances of systems of such particles and resonances. Hence applying this bootstrap hypothesis to fireballs, therefore, leads Hagedorn to the postulate:

A fire ball is

—a statistical equilibrium (hadronic black-body radiation) of undetermined numbers of all kinds of fireballs, each of which in turn is considered to be

—a statistical equilibrium of undetermined number of all kinds of fireballs, each of which . . . etc.

This postulate has a very far-reaching consequence in that it determines the asymptotic form of the fireball mass spectrum $\rho(M)$. In fact, assuming $\rho(M)$ and the fireball level density $\sigma(M, V)$ are asymptotically "equal" leads to a set of solutions of the form

$$\rho(M) = cM^a \exp(M/T_0), \qquad -\tfrac{7}{2} \leqslant a \leqslant -\tfrac{5}{2} \qquad (6.4.20)$$

where c and T_0 can be approximately calculated.

Initially it appeared plausible that $a = -\tfrac{5}{2}$, though subsequent work by Frautchi (1971) excluded this possibility. Later it was argued by Frautchi and

Hamer (1971) and Nahm (1972) that a strongly convergent solution of the bootstrap hypothesis led uniquely to the value $a = -3$. We shall denote the two asymptotic forms of $\rho(M)$ that have been considered in detail by the words "weak" and "strong" such that

$$\rho(M) \to cM^{-\frac{5}{2}} \exp(M/T_0) \quad \text{(weak solution)}$$
$$\rho(M) \to cM^{-3} \exp(M/T_0) \quad \text{(strong solution)}. \tag{6.4.21}$$

To these asymptotic solutions of course we must add a lower mass form which at least up to 1 or 2 GeV can be determined experimentally. The complete expression for $\rho(M)$ (using either the weak or strong solution) according to Hagedorn takes account of the whole of strong interactions.†
Of course, certain kinematical details have still to be specified, but these can be described by the thermodynamical (black-body radiation) assumption of a system of non-interacting particles, e.g. the decay distribution (6.4.19).

The most interesting aspect of the solutions (6.4.21) is that the exponential growth of mass spectrum leads to a *universal highest temperature* $T_0 = 160$ MeV (i.e. $1.86 \times 10^{12}\,^{\circ}$K). As matter is heated to this temperature by its internal collisions it will create hadrons and hence become a black-body radiator. The temperature cannot rise above T_0 because as $T \to T_0$ the production of particles is so violent that any energy input is used to create additional particles rather than increasing the kinetic energy of those already there. Thus T_0 represents, in some respects, the boiling point of hadronic matter.

It is worth stressing that the prediction of a maximum temperature T_0 (≈ 160 MeV) is a *result* of the model and not an assumption. Furthermore, it must be regarded as a major achievement of the model that this value of T_0 when used as an upper bound on the temperature in (6.4.19) leads to the desired cut-off in the momentum distributions of decaying particles. Although the longitudinal component of the momentum of produced particles will be boosted by Lorentz transformations to the overall c.m. system, the transverse momentum distributions must still be determined by this maximum value for T.

The main phenomenological consequences of the two solutions for $\rho(M)$ in (6.4.21) can be summarized as follows. In the original "weak" solution it is found that the number of decay products in the first generation, or decay, of the fireball, is proportional to $\log M$ and the distribution of fireballs in a fireball is a Poisson distribution. In the strong solution, however, the multiplication of $M^{-\frac{1}{2}}$ to the mass distribution results in a number of decay products being approximately constant, ≈ 2.4. This means that the decay

---

† It is undoubtedly worth noting that (6.4.20) and (6.4.21) are very similar to results obtained from dual resonance models. Fubini and Veneziano (1969), Fubini *et al.* (1969) though the connection between the two models is still not understood. See however Satz (1972).

chain now looks very much simpler. At each decay step of the fireball only one light particle of mass m and kinetic energy $\approx \frac{3}{2}T$ is radiated off. (Frautchi, 1971). Since the mass of the emitted particle is small the heavy mass fireball will not undergo much recoil. The momentum spectrum, therefore, of such emitted particles should indeed be close to the Planck spectrum (6.4.19) which is isotropic in the rest frame of the fireball.

Although this statistical bootstrap of fireballs is a rather elegant idea it must be supplemented by Hagedorn's second postulate, i.e. the kinematical superposition of decaying fireballs, in order for the model to be of any detailed phenomenological use. Unlike the single- or two-fireball models, Hagedorn suggests an initial two-particle collision leads to the production of a *continuum* of fireballs corresponding to a range of impact parameters, as indicated by the arrows in Fig. 50. In the longitudinal directions there will

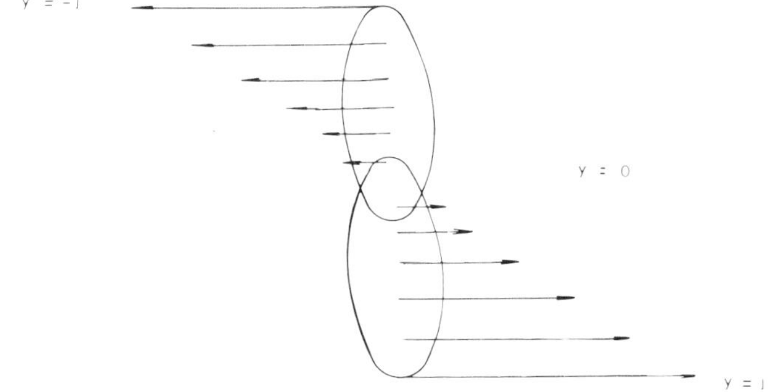

FIG. 50. The continuum of fireballs in the thermodynamical model.

be "strong collective motions" with velocities ranging from zero at the centre to almost the initial proton velocity at the outermost limits. In the rest frame of each element of the combined system it is proposed that a thermodynamical equilibrium of the hadronic matter is established appropriate to the temperature. From our preceding remarks this temperature will be somewhat lower than T_0. Now each element is considered as a "fireball" or part of a fireball, and its motion in the centre-of-mass system described by the Lorentz factor $\gamma = (1 - \beta^2)^{-\frac{1}{2}}$ where β is the local velocity of the element in units of c. In order to determine the velocity of elements, or rather the γ-distribution, we should understand the mass distributions within the colliding particles. Although these distributions will be severely Lorentz-contracted in the longitudinal directions at high energies, the transverse direction is of fundamental importance here. Rather than guessing these distributions, Hagedorn prefers to introduce the distribution $F(\lambda)$ for the entire collision, where $\lambda = \pm(\lambda - 1)/(\gamma_0 - 1)$ with γ_0 corresponding to the Lorentz factor of the colliding particles in the c.m. system. This distribution function $F(\lambda)$ (which in general may also depend on γ_0) is completely arbitrary and is chosen purely on empirical grounds. It can, therefore, incorporate aspects of diffraction or multiperipheralism as required. However, once chosen, it is assumed that there is only one function such that (Hagedorn, 1972)

(a) it is the same for all initial and produced particles

(b) it has minimal dependence on collision energy, i.e. scaling, except perhaps at $\lambda = 0$.

(Although initially a second function $F_0(\lambda)$ was introduced to account for leading particle effects, this has now been shown to be unnecessary by Letessier and Tounsi (1972).)

With these assumptions the basic formula for inclusive momentum distributions can be written as an integral over the velocity distribution, i.e.

$$\frac{\mathrm{d}^3 N}{\mathrm{d}^3 \mathbf{q}} = c \int_{-1}^{1} F(\lambda, (\gamma_0) q(\lambda) L(\lambda, \gamma_0) \{ f_m(k, T(\lambda)) \} \, \mathrm{d}\lambda, \qquad (6.4.22)$$

where $q(\lambda)$ in the strong solution represents the dependence on the available fireball mass and $L(\lambda, \gamma_0)$ is a Lorentz boost taking the isotropic Planck spectrum $f_m(k, T)$ from the λ-frame to the c.m. frame or lab. frame, whichever is desired. The temperature T is determined by assuming that in each volume element in the interaction region the loss of initial kinetic energy into heat is adiabatic, and only subsequently is the heat dissipated by radiating off particles. In order to incorporate the factorization hypothesis of limiting fragmentation the function F is split into two parts F^+ and F^- such that

$$F^+(\lambda) \neq 0 \quad \text{for } \lambda > 0$$

$$F^+(\lambda) \approx 0 \quad \text{for } \lambda < 0,$$

and similarly

$$F^-(\lambda) \neq 0 \quad \text{for } \lambda < 0$$

$$F^-(\lambda) \approx 0 \quad \text{for } \lambda > 0,$$

where $+, -$ refer to projectile or target fragments respectively.

Although the choice of $F(\lambda)$ represents a considerable degree of flexibility in the model the universality of the assumptions (a) and (b) ensure that it still retains a large degree of predictive power. Recent fits by Letessier and Tounsi, for example, seem to be quite impressive. As the model now stands it incorporates limiting fragmentation and scaling. It predicts a $\log \sqrt{s}$ dependence for the average multiplicity, due to a logarithmically increasing plateau in the rapidity variable; but at low energies this is modified by the effect of the temperature T not yet approaching its limiting value T_0. This effect is more important for heavier particles since the total production rates in the central region behave like $\exp(-m/T)$. Hence the change in this production rate for a fixed increment ΔT in the temperature T will be greater, the greater the mass m. Therefore, in a very natural way the model can also explain the strong energy dependence of the inclusive distributions for K and $\bar{p}$ production for instance near $x = 0$ observed at current accelerator energies. (Besides the original papers of Hagedorn, and Hagedorn and Ranft we would refer the reader to reviews by Feinberg (1972) and Omnés (1972) for a further *critical* summary of the model).

6.5. The gas analogy and two component models

Finally in this chapter on hadrodynamics we must discuss the possibility of there being two or more dynamical mechanisms operative in high energy production collisions. So far we have concentrated on either multiperipheralism (independent emission) or diffraction excitation; or else as in the Hagedorn model, skirted the problem by the introduction of an arbitrary (phenomenological) distribution function. As we have seen, there is good reason to believe the data cannot be described by diffractive processes alone. On the other hand, from a study of few body exclusive reactions it would appear that the diffractive mechanism is not negligible (at least for few body production). It would therefore seem very desirable to consider ways in which these two-particle production hypotheses can be combined and compared in a fairly general manner.

One way to do this is by the gas analogy initiated by Feynman and taken up by Wilson (1970) and Bander (1972). The analogue is to view the particles distributed in some momentum space variables as gas (or liquid) molecules confined within a box. Noting yet again that the average transverse momentum $\langle r \rangle$ of produced particles are severaly restricted, the distribution is essentially the one-dimensional rapidity distribution and the "box" containing the fluid is a long cylinder of length $y \propto \log s$ and radius $R \propto \langle r \rangle$. The fragmentation regions of the target and projectile correspond then to the regions near the ends of the cylinder, and any structure of the particle distributions can presumably be interpreted as boundary effects due to the viscosity of the gas (or liquid) and adhesion to the walls, etc.

It should be stressed that this is purely a mathematical analogue in order for us to visualize the many-body system at high energies. However, as such it can be very useful. For example, the analogue suggests that we interpret the normalized cross-section for observing particles with momenta $\mathbf{q}_1, \ldots, \mathbf{q}_n$ in a final state of $n + m$ particles, i.e.

$$\frac{1}{\sigma_{ab}}\frac{\mathrm{d}\sigma^{n,m}}{\mathrm{d}\phi_1\ldots\mathrm{d}\phi_n}(Y,\mathbf{q}_1,\ldots,\mathbf{q}_n)$$

where $\mathrm{d}\phi_i = \mathrm{d}y_i\,\mathrm{d}^2 r_i$, as $f^{n,m}(Y,\phi_1,\ldots,\phi_n)$ the probability distribution of finding n molecules at positions $(y_1,\mathbf{r}_1)\ldots(y_n,\mathbf{r}_n)$ in a fluid bounded by the cylinder containing $n+m$ molecules. Similarly, the n-particle inclusive cross-section

$$\frac{1}{\sigma_{ab}}\frac{\mathrm{d}\sigma}{\mathrm{d}\phi_1\ldots\mathrm{d}\phi_n}(Y,\mathbf{q}_1,,\ldots,\mathbf{q}_n)$$

can be interpreted as the probability distribution $f^n(Y,\phi_1,\ldots,\phi_n)$ for finding n molecules in a fluid with an unspecified total number of molecules in the cylinder.† If we take this analogy further we would expect that if the length of the cylinder Y is long enough the probability distributions $f^n(Y,\phi_1,\ldots,\phi_n)$ must become independent of the length. That is to say

$$f^n(Y,\phi_1,\ldots,\phi_n)\underset{Y\to\infty}{\longrightarrow}\hat{f}^n(\phi_1,\ldots\phi_n)\qquad(6.5.1)$$

which is just the scaling hypothesis (see Section 5.2). Also, if $\mathbf{q}_1$ and $\mathbf{q}_2$ are widely separated we would expect that the probability distribution $\hat{f}^2(\phi_1,\phi_2)$ should be simply the product of the two single particle distributions $\hat{f}^1(\phi_1)$ and $\hat{f}^2(\phi_2)$. That is, we can write

$$\hat{f}^2(\phi_1,\phi_2) - \hat{f}^1(\phi_1\,\hat{f}^1(\phi_2)\underset{|Y_1-Y_2|\to\infty}{\longrightarrow}0\qquad(6.5.2)$$

which is the assumption of the existence of only short range correlations. If this is valid we can deduce that the fluid density should be independent of the rapidity except within some correlation distance L from the ends of the cylinder. Thus, as we saw in the last chapter, we are led to the prediction that there will be a central plateau in the single particle rapidity distribution which is independent of both beam and target particles, except for an overall normalization. The existence of such a plateau would seem to be essential to the validity of the gas analogy, at least as presently constituted.

In the case of a classical fluid, all the thermodynamics may be obtained from either the canonical or grand canonical partition functions which are written respectively as

$$Q_n(Y) = \frac{1}{n!}\int f^{n,0}(Y,\phi_1,\ldots,\phi_n)\,\mathrm{d}\phi_1\ldots\mathrm{d}\phi_n,\qquad(6.5.3)$$

and

$$Q(z,Y) = \Sigma\,z^n Q^n(Y),\qquad(6.5.4)$$

† Notice that $f^n(Y,\phi_1,\ldots,\phi_n)$ differs from $f_{1,\ldots,n}(Y,\phi_1,\ldots,\phi_n)$ introduced in (5.1.2) by the normalization factor σ_{ab}.

where $f^{n,0}$ is the completely *exclusive* n particle distribution. Here the parameter z corresponds to the fugacity in statistical mechanics. By analogy, for particle production processes we have

$$Q(z, Y) = \sum z^n \frac{\sigma_n(Y)}{\sigma_{ab}} \tag{6.5.5}$$

where σ_n is the integrated exclusive n-body cross-section, and by definition

$$Q(1, Y) = 1$$

$$\left. \frac{\partial Q(z, Y)}{\partial z} \right|_{z=1} = \langle n \rangle = \int d\phi_1 \, f^1(Y, \phi_1) \tag{6.5.6}$$

$$\left. \frac{\partial^2 Q(z, Y)}{\partial z^2} \right|_{z=1} = \langle n(n-1) \rangle = \int d\phi_1 \, d\phi_2 \, f^2(Y, \phi_1, \phi_2)$$

where f^1 and f^2 are the one and two particle inclusive distribution functions. That is to say that $Q(z, Y)$ is a special case of the generating functional introduced in Section 5.1.

If the analogy is to be pursued we must suppose that the limit

$$\lim \log Q(z, Y) = p(z)Y + s(z) \tag{6.5.7}$$

whose existence is central to the statistical mechanical treatment (where $p(z)$ is the pressure and $s(z)$ is related to the surface tension) also exists for the particle production analogue. If this is true then we can ask—what do the models predict for the functions $p(z)$ and $s(z)$?

In the case of the multi-Regge model we obtained the result, under certain simplifying assumptions that

$$\sigma_{n+2} \propto s^{2\alpha_0 - 2} (g^2 Y)^n / n!, \tag{6.5.8}$$

where α_0, g^2 are the input trajectory and coupling constant. Hence

$$\sigma_{ab} = \sum \sigma_{n+2} = \beta(g^2) \exp\left[Y(\alpha(g^2) - 1)\right] \tag{6.5.9}$$

with

$$\alpha(g^2) = 2\alpha_0 - 1 + g^2 \tag{6.5.10}$$

and g^2 chosen so that for $g^2 = g_0^2$, $\alpha(g_0^2) = 1$. The grand canonical partition function $Q(z, Y)$ can be easily calculated in the same way as σ_{ab} and just corresponds to rewriting zg^2 for g^2. Therefore

$$\sigma_{ab}Q(z, Y) = \beta(zg^2) \exp\left[Y(\alpha(zg^2) - 1)\right] \tag{6.5.11}$$

and hence

$$\log Q = (\alpha(zg^2) - 1)Y + \log(\beta/\sigma_{ab}) \tag{6.5.12}$$

so that the limit (6.5.7) does indeed exist for this model with the "pressure" given by the output trajectory

$$p(z) = \alpha(zg_0^2) - 1. \tag{6.5.13}$$

From (6.5.10) we see that in this model the pressure is zero when $z = 1$ and monotonically increasing with increasing z.

In fact this result is rather more general than any particular multiperipheral model and depends only on the short-range correlation hypothesis. To see this let us rewrite $Q(z, Y)$ in terms of the correlation functions $\rho^n(y, \mathbf{q}_1, \ldots, \mathbf{q}_n)$ as

$$Q(z, Y) = \exp \sum \frac{(z - 1)^n}{n!} R^n(Y) \tag{6.5.14}$$

where

$$R^n(Y) = \int d\phi_1 \ldots d\phi_n \, \rho^n(Y, \mathbf{q}_1, \ldots, \mathbf{q}_n). \tag{6.5.15}$$

The $3n$-dimensional integral in (6.5.15) is over a volume proportional to Y^n so that since $\rho^n(Y, \mathbf{q}_1, \ldots, \mathbf{q}_n)$ are non-singular distributions, in general we must have

$$R^n(Y) \leqslant O(Y^n). \tag{6.5.16}$$

But the short-range correlation hypothesis implies that each $\rho^n(Y, \mathbf{q}_1, \ldots, \mathbf{q}_n)$ vanishes whenever the rapidity difference $|y_i - y_j|$ between any two produced particles is greater than some finite correlation length L. Hence (6.5.16) in this case reduces to

$$R^n(Y) = a_n Y + b_n. \tag{6.5.17}$$

Inserting (5.3.17) in (5.3.14) immediately gives the result (6.5.7) with

$$p(z) = \sum \frac{(z - 1)^n}{n!} a_n$$
$$s(z) = \sum \frac{(z - 1)^n}{n!} b_n. \tag{6.5.18}$$

Notice that from (6.5.5) we can calculate each σ_n from

$$\frac{\sigma^n}{\sigma_{ab}} = \frac{1}{n!} \frac{\partial^n}{\partial z^n} Q(z, Y) \bigg|_{z=0}, \tag{6.5.19}$$

so that in the short-range correlation picture

$$\frac{\sigma_n}{\sigma_{ab}} = \exp\left[p(0)Y\right] g_n(Y) \tag{6.5.20}$$

where $g_n(Y)$ is some polynomial of degree n. Hence each n-body cross-section will eventually fall to zero with the same power, $p(0)$ of the energy. (See Le Bellac, 1971).

Turning now to diffractive models, we must be a little careful since things are not quite so straightforward. These models predict that each n-body cross-section tends to a finite limit as the energy increases. Thus we can write

$$Q(z, Y) = \sum z^n \sigma_n \equiv q(z), \tag{6.5.21}$$

i.e. asymptotically independent of Y and hence as expected from (6.5.19) the pressure $p = 0$. However, the particular diffraction models we have so far considered, such as Hwa's model with $\sigma_n \propto 1/n^2$, do not allow us to make the analogy with a classical fluid, since $Q(z, Y)$ does not exist for $z > 1$, and they do not possess the short-range correlation condition (5.5.2). Thus the class of diffraction models we consider here therefore must have σ_n decreasing fast enough with increasing n so that $Q(z, Y)$ exists for all z, and hence give rise to *constant* average multiplicities.

Such constraints on the diffractive mechanism, of course, are by no means unrealistic. Our knowledge of diffractive phenomena, such as it is, only comes from a study of few-body exclusive reactions. It was only our wish to construct one universal model for hadrodynamics that led us to try to extrapolate this knowledge into an overall description of the data. However, as we saw in Section 6.3, this apparently led us into difficulties when examining the amount of single and double diffraction excitation.

Therefore, while the diffraction models we consider here cannot by themselves describe the data it may well be that in combination with some other mechanism they provide a very realistic description of high energy production phenomena. This brings us to the consideration that both diffractive and "multiperipheral" mechanisms may be operative, in which case the grand canonical partition function becomes

$$Q(z, Y) = q(z) + \beta(z) \exp \left[Y(\alpha(g_0^2 z) - 1) \right] \tag{6.5.22}$$

The limit (6.5.7) now depends critically on the value of z. For $z < 1$, the exponential in (6.5.22) is less than one and for $Y \to \infty$ $Q(z, Y) \to q(z)$. For $z > 1$, on the other hand, the exponential term will dominate and $\log Q(z, Y) \to Y(\alpha(g_0^2 z) - 1)$. Thus the pressure $p(z)$ is zero up to the critical point $z = 1$, after which it increases monotonically. This situation is referred to in statistical mechanics as a phase transition. Whether such an effect is actually observed in multiparticle reactions is difficult to establish, since the pressure can only be determined by extrapolating existing data to infinite energy. If the diffractive component, $q(z)$, is rather small then $Y \approx \log s$ may have to be rather large before $q(z)$ dominates over the second term in (6.5.14) for $z < 1$.

For the purpose of argument let us assume that such a phase-transition is present. The particle density is given by

$$\frac{n}{Y} = z\,\frac{\mathrm{d}p(z)}{\mathrm{d}z} \tag{6.5.23}$$

Then assuming that $p(z)$ for $z > 1$ can be determined from existing data to have a fairly linear form such that at $z = 1$ $n/Y \sim 0\cdot7$ (Bander 1972) we can make the following proposition. For n (charged particles) $> 0\cdot7\,Y$ the multiperipheral production mechanism dominates and particles in general will be fairly uniformly distributed in rapidity. For $n \ll 0\cdot7Y$ the diffractive mechanism becomes important and the particles will tend to cluster near the ends of the cylinder. In fact there is evidence to suggest that in the lower n part of the σ_n versus n plot there is a deviation from the typically Poisson distribution of the multiperipheral model (Frazer *et al.*, 1972b). However, whether this deviation is the result of a two-component picture or of an (as yet) unknown alternative dynamical mechanism giving rise to an approximately normal distribution (Kaiser, 1973; Parry and Rotelli, 1973) is not established.

In principle, the two-component description can be easily identified at sufficiently high energy by the formation of a dip in the σ_n versus n distribution. As we have mentioned, the diffractive term leads to a constant peak in this distribution since σ_n is constant for fixed n. The Poisson distribution has a peak which moves upwards with energy. Therefore at sufficient values of s these two distributions will tend to separate, giving rise to the dip. However, if the diffractive part is rather small we shall probably have to look at extremely high energies before this feature becomes apparent. In this connection it is worth remarking that the two-component picture has been used recently to account for the linear relationship between the dispersion $D = (\langle n^2\rangle - \langle n\rangle^2)^{\frac{1}{2}}$ and $\langle n\rangle$ noted by Wroblewski (1972). Assuming both components have fairly small dispersions, but one having a much larger multiplicity than the other, Van Hove (1972) has used this linearity between D and $\langle n\rangle$ to estimate the diffractive components to pp and πp cross-sections as $7\cdot7$ mb and $3\cdot4$ mb respectively. Similarly, rather small values have been obtained also by Fialkowski (1972).

It is clearly possible to elaborate on these two-component ideas and construct various types of two-component models to investigate more sublte features of the data. We shall conclude this chapter with a brief account of one such model which suggests that the current data can be explained in terms of some form of cluster production (see Section 4.2). Here, however, let us make one further observation which is independent of any particular model assumption, and which we believe is rather important: i.e., even the addition of a small diffractive component can significantly alter the predictions of the short-range correlation hypothesis. To see this let us write the n-body cross section as

$$\sigma_n = \sigma_n^D + \sigma_n^M \qquad (6.5.24)$$

where M, D refer to multiperipheral and diffractive respectively, and we may assume that σ_n^D is completely negligible for large n, so that only the low n σ_n cross-sections are modified. The partition function in this case becomes

$$Q(z, Y) = (1 - \delta)Q^M(z, Y) + \delta Q^D(z, Y) \qquad (6.5.25)$$

where

$$\delta = \sum \sigma_n^D / \sigma_{ab}$$

and from (6.5.14) we can obtain the relationship between the correlation functions $R^n(Y)$ and the separate correlation functions $R_D^n(Y)$, $R_M^n(Y)$ corresponding to the two components D and M. The first few relations are given by (Frazer *et al.*, 1972b)

$$R^1 = (1 - \delta) R_M^1 + \delta R_D^1$$

$$R^2 = \delta(1 - \delta)(R_M^1 - R_D^1)^2 + (1 - \delta)R_M^2 + \delta R_D^2 \qquad (6.5.26)$$

etc.

Now by hypothesis we have that each $R_M^n(Y)$ is proportional to Y (see 6.3.17) while each $R_D^n(Y)$ is constant. However, from (6.5.26) we find that provided δ is non-zero the two-particle correlation parameter R^2 does not grow asymptotically like Y, as suggested by R_M^2, but is proportional to $\delta(R_M^1)^2$, i.e. δY^2. Similarly, in the general case we find that R^n grows like δY^n, (notice it is still linear in δ). Hence, even though the diffractive component is small, provided it is not completely negligible we would expect a very marked energy dependence for the higher order correlation functions.

This is, therefore, a rather nice example of how a small effect in the production dynamics can interfere with some dominant mechanism to produce a very significant deviation from the naive predictions.

6.5a *Production of Particle Clusters.* In concluding this Section on hadrodynamics let us give an explicit example of a class of two component models which has received much publicity recently. Here as in (6.5.24) we write the inelastic cross-section as a sum of the terms

$$\sigma = \sigma^D + \sigma^M \qquad (6.5.27)$$

where σ^D is the diffractive part of the cross-section and σ^M corresponds to the non-diffractive contribution.

The two particle rapidity correlation density $\rho^2(y_1, y_2)$ which is defined as

$$\rho^2(y_1, y_2) = \frac{1}{\sigma}\frac{d^2\sigma}{dy_1\,dy_2} - \left(\frac{1}{\sigma}\frac{d\sigma}{dy_1}\right)\left(\frac{1}{\sigma}\frac{d\sigma}{dy_2}\right) \qquad (6.5.28)$$

can be expressed in terms of these two components as

$$\rho^2(y_1, y_2) = \alpha^M \rho_M^2(y_1, y_2) + \alpha^D \rho_D^2(y_1, y_2) + \alpha^M\alpha^D\rho_{MD}^2(y_1, y_2) \qquad (6.5.29)$$

where

$$\alpha^M = 1 - \alpha^D = \sigma^M/\sigma$$

$$\rho_M^2(y_1, y_2) = \frac{1}{\sigma^M}\frac{d^2\sigma^M}{dy_1\,dy_2} - \left(\frac{1}{\sigma^M}\frac{d\sigma^M}{dy_1}\right)\cdot\left(\frac{1}{\sigma^M}\frac{d\sigma^M}{dy_2}\right)$$

similarly for $\rho_D^2(y_1, y_2)$; and

$$\rho_{MD}^2(y_1, y_2) = \left(\frac{1}{\sigma^M}\frac{d\sigma^M}{dy_1} - \frac{1}{\sigma^D}\frac{d\sigma^D}{dy_1}\right)\left(\frac{1}{\sigma^M}\frac{d\sigma^M}{dy_2} - \frac{1}{\sigma^D}\frac{d\sigma^D}{dy_2}\right) \qquad (6.5.30)$$

From this last relation we see that, even in the central region where we might expect the diffractive component to be negligible small, the correlation function $\rho^2(y_1, y_2)$ in (6.5.29) has an extra *positive* contribution arising from the cross-product of diffractive and non-diffractive contributions. This is just a restatement of the effect given in (6.5.26). The correlation function in the central region therefore, is predicted to be composed of two parts. This first part corresponds to some short range non-diffractive component, $\rho^M(y_1, y_2)$, which in a simple multiperipheral model or independent emission model will be close to zero unless $|y_1 - y_2| < L$. The second contribution

$$\rho_{MD}^2(y_1, y_2) \approx \alpha^M\alpha^D\left(\frac{1}{\sigma^M}\frac{d\sigma^M}{dy_1}\right)\left(\frac{1}{\sigma^M}\frac{d\sigma^M}{dy_2}\right)$$

represents some long range background term which, if y_1 and y_2 are both in the central plateau region, should be independent of y_1 and y_2. Subtracting this long range correlation term from the experimental data, it is found that there is still a strong, positive, short-range correlation effect, even between particles of the same sign (see for instance Foa (1973)).

In order to account for this strong short-range correlation it has been suggested that the non-diffractive component of the two component model is given by a multi-peripheral or independent emission model, not producing single particles, but instead producing *clusters* which subsequently decay into a (small) number of particles. (See for example Pirila and Pokorski (1973), Hamer and Peierls (1973), Berger and Fox (1973), Pokorski and Van Hove (1973)). Denoting the rapidity density of clusters by $\rho(y_c)$ the average multiplicity per cluster by $\langle k \rangle$ and the one and two-particle cluster decay distributions by $D^1(y_c - y)$, $D^2(y_c - y_1, y_c - y_2)$ respectively we have

$$\int D^1(y_c - y)\, dy = \langle k \rangle$$

$$\int D^2(y_c - y_1, y_c - y_2)\, dy_1\, dy_2 = \langle k\,(k - 1)\rangle \qquad (6.5.31)$$

Inserting these quantities in Eq. (6.5.30) we find that

$$\frac{1}{\sigma^M}\frac{d\sigma^M}{dy} = \int \rho(y_c)D^1(y_c - y)\, dy_c$$

and

$$\rho_M^2(y_1, y_2) = \int \rho(y_c)D^2(y_c - y_1, y_c - y_2)\, dy_c \qquad (6.5.32)$$

where in the central region $\rho(y_c)$ will be constant because of the independent emission assumption.

Integrating the $1/\sigma^M \times d\sigma^M/dy$ and $\rho_M^2(y, y_2)$ distributions over y_1 and y_2 and comparing with (6.5.31) therefore it is possible to obtain an estimate of the mean number of particles produced per cluster. Depending on the way in which one subtracts the diffractive component, and also on the particular cluster model used, the data suggests a mean number of between two to three charged particles per cluster. Since the average charged multiplicity even at the highest present energies is found to be quite small (≈ 10) this implies a picture in which, on average, a small number of clusters are formed which subsequently each decay into only a small number of particles.

In order to determine the correlation density $\rho_M^2(y_1, y_2)$ as a function of y_1 and y_2 one has to go further and consider the shape and mass of the clusters. For example if one assumes, by analogy with the statistical bootstrap theory, that each cluster decays isotropically in its own rest frame into pions, with a distribution

$$D^1(\mathbf{q}) \propto \exp(-q^2/2\sigma^2)$$

the parameter σ relates the mass of the cluster to the number of particles per cluster (Berger and Fox, 1973) i.e.

$$M_{\text{cluster}} \approx 1{\cdot}6\,\sigma k$$

By transforming the cluster decay distribution into the overall c.m. system, and assuming some simple mass spectrum for the cluster. the model can very naturally explain the strong positive short-range correlation found in the data. In fact, quite generally one has that

$$D^1(y_c - y) \approx \frac{\langle k \rangle}{\delta \sqrt{(2\pi)}} \exp(-(y - y_c)^2/2\delta)$$

where $\delta \approx 0{\cdot}9$, independent of σ; and provided one can neglect energy-momentum constraints in the decay of the cluster, it is possible to assume that

$$D^2(y_c - y_1, y_c - y_2) = D^1(y_c - y_1)D^1(y_c - y_2)$$

It has been stressed by Pokorski and Van Hove (1974) that the (approximate) isotropic decay assumption apparently describes both the shape of this two particle rapidity correlation density and the observed damping in transverse momenta.

Clearly such models are very much oversimplified and one can think of several ways in which they could, and must be improved. However the success of even these simplified versions in describing the correlation data demands that the models be given much further consideration. In particular we await the answers to the following questions:

  (i) Are clusters merely a shorthand way of describing a much more complicated dynamical mechanism, or do they have a dynamical significance in their own right; in other words, what is a cluster?

 (ii) If clusters do exist as dynamical entities, can they be identified in the present data, and if so what are their isospin, charge, mass, and spin assignments?

(iii) What is the relationship, if any, between centrally produced clusters and clusters produced in diffractive reactions, i.e. diffraction dissociated systems?

(iv) Can two-component models, with cluster production, give a *quantitative* description, via the overlap function of the elastic scattering amplitude?

7

References

Abarbanel, H. D. I. (1971a). *Phys. Letters* **34B**, 69.

Abarbanel, H. D. I. (1971b). *Phys. Rev.* **D3**, 2227.

Abarbanel, H. D. I., Chew, G. F., Goldberger, M. L. and Saunders, L. M. (1971). *Phys. Rev. Letters* **26**, 937.

Abarbanel, H. D. I., Gribov, V. N., Kanchelli, O. V. (1972). Leningrad preprint.

Abarbanel, H. D. I. and Schwimmer, A. (1972). *Phys. Rev.* **D6**, 3018.

Abarbanel, H. D. I. and Kane, G. L. (1973), *Phys. Rev. Letters* **30**, 67.

Adair, R. K. (1968). *Phys. Rev.* **172**, 1370.

Adair, R. K. (1972). Proceedings of the Fourth International Conference on High Energy Collisions, Oxford.

Adjei, S. A., Collins, P. A., Hartley, B. J., Moore, R. W. and Moriarty, K. L. M. (1971a). *Phys. Rev.* **D3**, 150.

Adjei, S. A., Collins, P. A., Hartley, B. J., Moore, R. W. and Moriarty, K. L. M. (1971b). *Phys. Rev.* **D3**, 2425.

Albrow, M. G., Barber, D. P., Bogaerts, A., Bosnjakovic, B., Brookes, J. R., Clegg, A. B., Erne, F. C., Gee, C. N. P., Kanaris, A. D., Locke, D. H., Loebinger, F. K., Murphy, P. G., Rudge, A., Sens, J. C., Terwilliger, K. and Van der Veen, F. (1973). *Nucl. Phys.* **B51**, 388.

Alessandrini, V., Amati, D., Le Bellac, M. and Olive, D. (1971). Physics Reports **1c**, 270.

Alvarez Estrada, R. F. (1971). CERN preprint Ref. TH 1414.

Amati, D., Fubini, S. and Stanghellini, A. (1962). *Nuovo Cim.* **26**, 896.

Amado, R. D. (1967). *Phys. Rev.* **158**, 1414.

Atkinson, D., Contogouris, A. P. and Gaskell, R. (1972). *Phys. Rev.* **D6**, 1070.

Auerbach, S., Aviv, R., Sugar, R. and Blankenbeder, R. (1972). *Phys. Rev.* **D6**, 2216.

Bali, N. F., Chew, G. F. and Pignotti, A. (1967). *Phys. Rev. Letters* **19**, 614; *Phys. Rev.* **163**, 1572.

Ball, J. S., Frazer, W. R. and Nauenberg, M. (1962). *Phys. Rev.* **128**, 478.

Ball, J. S. and Marchesini, G. (1969). *Phys. Rev.* **188**, 2209; *Phys. Rev.* **188**, 2508.

Ball, J. S. and Marchesini, G. (1970). *Phys. Rev.* **D2**, 2665.

Bander, M. and Shaw, G. (1965). *Phys. Rev.* **139**, B956.

Bander, M. (1972). *Phys. Rev.* **D6**, 164.

Bardakci, K. and Mandelstam, S. (1969). *Phys. Rev.* **184**, 1640.

Bardakci, K. and Halpern, M. (1970). *Phys. Rev. Letters* **24**, 428.

Bardakci, K. and Ruegg, H. (1968). *Phys. Letters* **28B**, 342.

Barger, V. and Phillips, R. J. N. (1968). *Phys. Letters* **26B**, 730.

Bartsch, J., Bossebeck, K., Deutschmann, M., Schultz, R., Speth, R., Bottcher, H., Kaufmann, H. H., Nowak, S., Buckmann, K., Johnssen, W., Rost, M., Sternberger, K., Barnham, K., Cocconi, V. T., Dalpiaz, P. F., Ely, R., Hansen, J. D., Kittel, W., Lindblom, K. P. C., Morrison, D. R. O., Tofte, H., Brandt, S., Luth, V., Shah, T. P., Barford, N. C., Counihan, M. J., Dallman, D. P., Grammatikakis, G. A., Brochbeck, B., Gottfried, C., Markytan, M., Otter, G., Schmid, P. (1970a). *Nucl. Phys.* **B19**, 81.

Bartsch, J., Schultz, R., Steinberg, R., Gensch, V., Nowak, S., Ryseck, E., Schiller, H., Cocconi, V. T., Dalplaz, P. F., Hansen, J. D., Kellner, G., Kittel, W., Matsumoto, S., Morrison, D. R. O., Stroynowski, R., Whittaker, J. B., Counihan, M. J., Dornon, P. J., Goldsack, S. J., Otter, G., Buschbeck, B., Frohlich, A., Porth, P., Schmid, P. and Wahl, H. (1970b). *Nucl. Phys.* **B24**, 221.

Barut, A. O. and Leung, Y. C. (1965). *Phys. Rev.* **138**, B1119.

Beaupré, J. V., Deutschmann, M., Grässler, H., Kirk, H., Schulte, R., Geusch, U., Nowak, W. D., Bossen, G. J., Drevermann, H., Kanazirsky, Ch., Propach, E., Rost, M., Bockmann, K., Campbell, J. R., Cocconi, V. T., Kellner, G., Kittel, W., Morrison, D. R. O., Schiller, H., Sotirion, D. and Wahl, H. (1972). *Nucl. Phys.* **B46**, 1.

de Beer, M., Deler, B., Dalbeau, J., Neveu, M., Diem, N. T., Smadja, G. and Valladas, G. (1969). *Nucl. Phys.* **B12**, 599.

Belenkji, S. Z. and Landau, L. D. (1956). *Suppl. Nuovo Cimento* **3**, 15.

Benecke, J., Chou, T. T., Yang, C. N. and Yen, E. (1969). *Phys. Rev.* **188**, 2159.

Berger, E. L. (1968). *Phys. Rev. Letters* **21**, 701.

Berger, E. L. (1969). *Phys. Rev.* **179**, 1567.

Berger, E. L. (1971). In "Phenomenology in Particle Physics 1971", Proceedings of the Conference held at Caltech, March 25, 26, 1971.

Berger, E. L. and Fox, G. (1971). *Phys. Letters* **36B**, 389.

Berger, E. L. (1972a). Proceedings of the Fourth International Conference on High Energy Collisions, Oxford.

Berger, E. L., Jacob, M. and Slansky, R. (1972a). ANL/HEP-7211.

Berger, E. L., Oh, B. Y. and Smith, G. A. (1972b). *Phys. Rev. Letters* **29**, 675.

Berger, E. L. (1972b). *Phys. Rev. Letters* **29**, 887.

Berger, E. L., Horn, D. and Thomas, G. H. (1973). *Phys. Rev.* **D7**, 1412.

Berger, E. L. and Fox, G. C. (1973) CERN preprint TH.1700.

Bergia, S., Bonsignori, F. and Stranghellini, A. (1960). *Nuovo Cim.* **16**, 1073.

Berman, S. M. and Jacob, M. (1965). *Phys. Rev.* **139**, B1023.

Bialas, A. and Pokorski, S. (1969). *Nuc. Phys.* **B10**, 399.

Bialas, A., Dabkowski, J. and Van Hove, L. (1971). *Nucl. Phys.* **B27**, 338.

Bialas, A., Czyz, W. and Kotanski, A. (1971). Jagellonian Univ. preprint TPJU 17/71.

Bialas, A., Fialkowski, K. and Wit, R. (1972). *Nucl. Phys.* **B43**, 413.

Biebel, K. J. and Wolf, J. (1971). *Phys. Letters* **37B**, 197.

Blankenbecler, R. (1961). *Phys. Rev.* **122**, 983.

Boyling, J. D. (1963). *Ann. Phys.* (NY) **25**, 249.

Branson, D., Landshoff, P. V. and Taylor, J. G. (1963). *Phys. Rev.* **132**, 902 (*Erratum Phys. Rev.* **185** (1969), 2046).

Brau, J. E., Dao, F. T., Hodous, M. F., Pless, I. A. and Singer, R. A. (1971). *Phys. Rev. Letters* **27**, 1481.

Bros, J., Glaser, V. and Epstein, H. (1972). *Helv. Phys. Acta* **45**, 149.

Brower, R. C. and Weis, J. H. (1972). *Phys. Letters* **41B**, 631.

Brown, L. (1972). *Phys. Rev.* **D5**, 748.

Bugrij, A. I., Kenkovsky, L. L. and Kobylinsky, N. A. (1971). *Lett. Nuovo Cim.* **1**, 923 also *Nucl. Phys.* **35B**, 120.

Bugrij, A. I., Kenkovsky, L. L. and Kobylinsky, N. A. (1972). *Lett. Nuovo Cim.* **5**, 393.

Byckling, E. and Kajantie, K. (1969). *Nucl. Phys.* **B9**, 568.

Byers, N. and Yang, C. N. (1966). *Phys. Rev.* **142**, 976.

Caneschi, L. and Schwimmer, A. (1970). *Phys. Letters* **33B**, 577.

Cerulus, F. (1961). *Nuovo Cim.* **19**, 528.

Chan, H. M., Kajantie, K. and Ranft, G. (1967). *Nuovo Cim.* **49**, 157.

Chan, H. M., Loskiewicz, J. and Allison, W. M. (1968). *Nuovo Cim.* **57A**, 93.

Chan, H. M. and Paton, J. E. (1969). *Nucl. Phys.* **B10**, 516.

Chan, H. M. (1969). *Phys. Letters* **28B**, 425.

Chan, H. M. and Tsou, S. T. (1969). *Phys. Letters* **28B**, 485.

Chan, H. M., Ratio, R. O., Thomas, G. H., Törnqvist, N. A. (1970). *Nucl. Phys.* **B19**, 173.

Chan, H. M. and Hoyer, P. (1971). *Phys. Letters* **36B**, 79.

Chan, H. M., Hsue, C. S., Quigg, C. and Wang, J.-M. (1971). *Phys. Rev. Letters* **26**, 672.

Chan, H. M. and Tsou, S. T. (1971). *Phys. Rev.* **D4**, 156.

Chan, H. M., Meittinen, H. I., Roy, D. P. and Hoyer, P. (1972). *Phys. Letters* **B40**.

Chao, Y-A. (1972). *Nucl. Phys.* **B40**, 475.

Chen, M. S., Kinsey, R. R., Morris, T. W., Panvini, R. S., Wang, L. L., Wong, T. F., Stone, S. L., Ferbel, Z., Slattery, P., Werner, B., Elbert, J. W., Erwin, A. R. (1971). *Phys. Rev. Letters* **26**, 1585.

Cheng, H. and Wu, T. T. (1969). *Phys. Rev. Letters* **23**, 1311.

Chew, G. F. and Low, F. E. (1959). *Phys. Rev.* **113**, 1640.

Chew, G. F. and Pignotti, A. (1968). *Phys. Rev. Letters* **20**, 1078, *Phys. Rev.* **176**, 2112.

Chew, G. F., Goldberger, M. L. and Low, F. E. (1969). *Phys. Rev. Letters* **22**, 208.

Chew, G. F. and Snider, D. R. (1970). *Phys. Letters* **31B**, 75; *Phys. Rev.* **D1**, 3453.

Chew, G. F., Rogers, T. W. and Snider, D. R. (1970). *Phys. Rev.* **D2**, 765.

Chew, G. F. and Snider, D. R. (1971). *Phys. Rev.* **D3**, 420.

Childress, S., Franzini, P., Lee-Franzini, J., McCarthy, R. and Shamberger, R. D. Jr. (1974) *Phys. Rev. Letters* **32** 389

Chliapnikov, P. V., Nielsen, S., Ciapetti, G., Drijard, D., Dunwoodie, W., Goldschmidt-Clermont, Y., Grant, A., Henri, V. P., Muller, F., Pape, L., Sekera, Z., de Baere, W., de Wolf, W., Grard, F., Herquet, P., Peeters, P., Verbeure, F., Windmolders, R., Carney, N. C., Colley, D. C., Jobes, M., Jones, G. T., Bonnel, C., Ginestet, J., Manesse, D., Trau Ha Auh, Vignand, D. and Sene, M. (1972). *Phys. Letters* **39B**, 279.

Chou, T. T. and Yang, C. N. (1968). *Phys. Rev.* **170**, 159.

Cohen, D., Ferbel, T., Katz, W., Slattery, P., Stone, S. and Werner, B. (1972). *Phys. Letters* **28**, 1601.

Cohen-Tannoudji, G., Morel, A. and Navelet, H. (1968). *Ann. Phys.* (N.Y.), **46**, 239.

Cohen-Tannoudji, G., Drouffe, I. M., Moussa, P. and Peschanski, R. (1970). *Phys. Letters* **33B**, 183.

Cohen-Tannoudji, G., Henyey, F., Kane, G. L. and Zakrzewsky, W. J. (1971). *Phys. Rev. Letters* **26**, 112.

Coleman, T. P. (1971). *Phys. Rev.* **D4**, 1382.

Cooper, F. and Schonberg, E. (1973). *Phys. Rev. Letters* **30**, 880.

Deck, R. T. (1964). *Phys. Rev. Letters* **13**, 169.

Dehm, G. (1973). Ph.D. Thesis, Universität München.

Deler, B. and Valladas, G. (1966). *Nuovo Cim.* **45**, 559.

De Tar, C. E. (1971). *Phys. Rev.* **D3**, 128.

De Tar, C. E. and Weis, J. H. (1971). *Phys. Rev.* **D4**, 3141.

Diebold, R. E. (1972). Proceedings of XVI International Conference on High Energy Physics, Chicago—Batavia.

Drummond, I. T., Landshoff, P. V. and Zakrzewski, W. J. (1969). *Phys. Letters* **28B**, 676.

Durand, L. and Chiu, Y. T. (1965). *Phys. Rev.* **139**, B646.

Eden, R. J., Landshoff, P. V., Olive, D. I. and Polkinghorne, J. C. (1966). "The Analytic S-Matrix", Cambridge University Press.

Einhorn, M. B., Green, M. B. and Virasoro, M. A. (1971). *Phys. Letters* **37B**, 292; *Phys. Rev.* **D6**, 1675.

Einhorn, M. B. (1972). *Phys. Rev.* **D5**, 2063.

Einhorn, M. B., Ellis, J. and Finkelstein, J. (1972). SLAC-PUB-1006.

Ellis, J., Finkelstein, J., Frampton, P. H. and Jacob, M. (1971). *Phys. Letters*, **35B**, 227.

Ezawa, H., Tomozawa, Y. and Umezava, H. (1957). *Nuovo Cim.* **5**, 810.

Faddeev, L. D. (1961). *Sov. Phys. J.E.T.P.* **12**, 1014; *Sov. Phys. Doklady* **6**, 384.

Faddeev, L. D. (1963). *Sov. Phys. Doklady* **7**, 600.

Feinberg, E. L. and Chernavsky, D. S. (1951), *Dokl. Akad. Nauk. SSSR* **81**, 795.

Feinberg, E. L. and Chernavsky, D. S. (1953). *Dokl. Akad. Nauk. SSSR* **91**, 511.

Feinberg, E. L. (1972). *Physics Reports* **5c**, 237.

Ferbel, T. (1972). Proceedings III International Conference on Multiparticle Reactions, Zakopane, Poland.

Fermi, E. (1950). *Prog. Theo. Phys.* **5**, 570.

Fermi, E. (1951). *Phys. Rev.* **81**, 689.

Ferrari, E. and Selleri, F. (1962). *Suppl. Nuovo Cim.* **24**, 453.

Feynman, R. P. (1969a). *Phys. Rev. Letters* **23**, 1415.

Feynman, N. P. (1969b). "High Energy Collisions" (Gordon and Breach, New York) p. 237.

Fialkowski, K. (1972). *Phys. Letters* **41B**, 379.

Fialkowski, K. and Miettinen, H. I. (1973). *Phys. Letters* **43B**, 43.

Finkelstein, J. and Kajantie, K. (1968). *Phys. Letters* **26B**, 305.

Foa, L. (1973). Proceedings II International Conference on Elementary Particles, Aix-en-Provence, 1973.

Frampton, P. H. and Törnqvist, N. A. (1972). *Letters Nuovo Cim.* **4**, 233.

Franzen, G. (1971). *Nucl. Phys.* **B31**, 575.

Frautchi, S. C. (1971). *Phys. Rev.* **D3**, 2821.

Frautchi, S. C. and Hamer, C. J. (1971). *Phys. Rev.* **D4**, 2125.

Frazer, W. R. and Mehta, C. H. (1970). *Phys. Rev.* **D1**, 696.

Frazer, W. R., Ingber, L., Mehta, G. H., Poon, C. H., Silverman, D., Stowe, K., Ting, P. D. and Yesian, H. J. (1972a). *Rev. Mod. Phys.* **44**, 284.

Frazer, W. R., Peccei, R. D., Pinsky, S. S. and Tau, C.-I. (1972b). Univ. of California, San Diego report USCD 10P10-113.

Freedman, D. Z., Jones, C. E., Low, F. E. and Young, J. E. (1971). *Phys. Rev. Letters* **26**, 1197.

Freund, P. G. O. (1968). *Phys. Rev. Letters*, **20**, 235.

Froggatt, C. D. and Ranft, G. (1969). *Phys. Rev. Letters* **23**, 943.

Fubini, S., Gordon, D. and Veneziano, G. (1969). *Phys. Letters* **29B**, 679.

Fubini, S. and Veneziano, G. (1969). *Nuovo Cim.* **64A**, 811.

Ganguli, S. N. and Malhotra, P. K. (1972). *Phys. Letters* **42B**, 83.

Gaskell, R. and Contogouris, A. P. (1972). *Lett. Nuovo Cim.* **3**, 231.

Gault, F. D. and Walters, P. J. (1972). *Lett. Nuovo Cim.* **4**, 461.

Gerasimova, N. M. and Chernavsky, D. S. (1955). *Zh. Esksp. i Teor. Fiz.* **29**, 372.

Gilman, F. J., Pumplin, J., Schwimmer, A. and Stodolski, L. (1970). *Phys. Letters*, **31B**, 387.

Goddard, P. and White, A. R. (1971). *Nuovo Cim.* **1A**, 645; *Nuovo Cim.* **3A**, 25.

Goebel, C. (1958). *Phys. Rev. Letters* **1**, 337.

Good, M. L. and Walker, W. D. (1960). *Phys. Rev.* **120**, 1857.

Gordon, D. and Veneziano, G. (1971). *Phys. Rev.* **D3**, 2116.

Gottfried, K. and Jackson, J. D. (1964). *Nuovo Cim.* **34**, 735.

Gottfried, K. and Koefoed-Hansen, O. (1972). *Phys. Letters* **41B**, 195.

Green, M. B. and Heimann, R. L. (1969). *Phys. Letters* **30B**, 642.

Gribov, V. N. (1967). *Zh. Eksp. Teor. Fiz.* **53**, 654.

de Groot, E. H. (1972). *Nucl. Phys.* **B48**, 295.

Hagedorn, R. (1964). "Relativistic Kinematics". W. A. Benjamin, New York.

Hagedorn, R. (1965). *Nuovo Cim.* **3**, 147.

Hagedorn, R. and Ranft, J. (1968). *Suppl. Nuovo Cim.* **6**, 169.

Hagedorn, R. (1972). Proceedings III International Conference on Multiparticle Processes, Zakopane, Poland.

Halatnikov, I. M. (1954). *Zu. Eksper. Teor. Fiz.* **27**, 529.

Halliday, I. G. (1971). *Nucl. Phys.* **B33**, 285.

Hamaguchi, H. (1953). *Nuovo Cim.* **5**, 1622.

Hamer, C. J. and Peierls, R. F. (1973). *Phys. Rev.* **D8**, 1358.

Harari, H. (1968). *Phys. Rev. Letters* **20**, 1395.

Harari, H. (1969). *Phys. Rev. Letters* **22**, 562.

Henyey, F. and Zakrzewski, W. J. (1972). *Phys. Letters* **42B**, 431.

Heisenberg, W. (1949). *Nature* **164**, 65.

Heisenberg, W. (1963). "Theory of Multiple Meson Production" in Vortäge über Kosmiche Strahlung. Translated by L. V. Lindern and B. Stringfellow (Springer 1970).

Hite, G. E. (1971). *Nuovo Cim. Letters* **1**, 579.

Höhler, G. and Strauss, H. (1970). *Zeitschrift für Physik* **232**, 205.

Hopkinson, J. F. L. (1969). Daresbury Nuclear Physics Lab. preprint DNPL/P21, unpublished.

Hopkinson, J. L. and Plahte, E. (1969). *Phys. Letters* **28B**, 489.

Horn, D. and Silver, R. (1970). *Phys. Rev.* **D2**, 2082.

Hoyer, P., Petersson, B. and Törnqvist, N. A. (1970). *Nucl. Phys.* **B22**, 497.

Hoyer, P., Petersson, B., Lea, A. T., Paton, J. E. and Thomas, G. H. (1971). *Nucl. Phys.* **B32**, 285.

Humble, S. (1971). *Nucl. Phys.* **B28**, 416.

Humble, S. (1972a). *Phys. Letters* **B40**, 373.

Humble, S. (1972b). Paper presented at the III International Colloquium on Multiparticle Reactions, Zakopane, Poland.

Humble, S. (1973). *Phys. Rev.* **D7**, 1523.

Humble, S., Kaiser, G. D. and White, J. N. J. (1973). *Nucl. Phys.* **B56**, 61.

Huskins, J. (1972). *Nucl. Phys.* **B46**, 547.

Hwa, R. C. (1971). *Phys. Rev. Letters* **26**, 1143.

Hwa, R. C. (1972). Proceedings of the Fourth International Conference on High Energy Collisions, Oxford.

Iso, C., Mori, K. and Namiki, M. (1959). *Prog. Theo. Phys.* **22**, 403.

Jackson, J. D. (1964). *Nuovo Cim.* **34**, 1644.

Jackson, J. D. (1970). *Rev. Mod. Phys.* **42**, 1.

Jacob, M. and Wick, G. C. (1959). *Ann. Phys.* (N.Y.) **7**, 404.

Jacob, M. and Slansky, R. (1971). *Phys. Letters* **37B**, 408.

Jacob, M. and Slansky, R. (1972). *Phys. Rev.* **D5**, 1847.

James, F. (1970). FOWL, A General Monte Carlo Phase-Space Program CERN Program Library W 505.

Jen, C.-L., Kang, K., Shen, P. and Tan, C.-I. (1971). *Phys. Rev. Letters* **27**, 754.

Jen, C.-L., Kang, K., Shen, P. and Tan, C.-I. (1972). *Ann. of Phys.* (N.Y.) **72**, 548.

Jones, C. E., Low, F. E., Tye, S-H. H., Veneziano, G. and Young, J. E. (1972). *Phys. Rev.* **D6**, 1033.

Kaiser, G. D. (1973). *Nucl. Phys. B*. To be published.

Kane, G. L. (1972). *Acta. Phys. Pol.* **B3**, 845.

Kang, K. and Shen, P. (1972). *Phys. Rev. Letters* **29**, 1283.

Khinchin, A. I. (1949). Mathematical Foundations of Statistical Mechanics, Dover Publications, Inc.

Kibble, T. W. B. (1960). *Phys. Rev.* **117**, 1159.

Kibble, T. W. B. (1963). *Phys. Rev.* **131**, 2282.

Kittel, W., Van Hove, L., and Wojcik, W. (1970). *Comput. Phys. Commun.* **1**, 425.

Kittel, W., Ratti, S. and Van Hove, L. (1971). *Nucl. Phys.* **B30**, 333.

Kittel, W. (1972). Lectures given at the Scandinavian Conference on High Energy Physics, Spatind. CERN preprint CERN/D.Ph. II/Phys. 72-11/Rev.

Koba, Z. (1965a). *Phys. Letters* **16**, 326.

Koba, Z. (1965b). *Suppl. Progr. Theo. Phys.* Extra number 168.

Koba, Z. and Nielson, H. B. (1969). *Nucl. Phys.* **B10**, 633.

Koba, Z. (1971). *Acta. Phys. Pol.* **B2**, 111.

Koba, Z. (1972). Proceedings III International conference on Multiparticle Processes, Zakopane, Poland.

Koba, Z., Nielson, H. B. and Olesen, P. (1972). *Phys. Letters* **B38**, 25; *Nucl. Phys.* **B40**, 317.

Kopylov, G. I. (1962). *Nucl. Phys.* **36**, 425.

Krisch, A. (1971). Proceedings of the APS Division of Particles and Fields Meeting, Rochester.

Krzywicki, A. (1964). *Nuovo Cim.* **32**, 1069.

Krzywicki, A. (1965). *J. Math. Phys.* **6**, 485.

Kugler, M., Rittenberg, V. and Lipkin, H. J. (1972). *Phys. Letters* **38B**, 423.

Kwiecinski, J. (1972). *Nuovo Cim. Letters* **3**, 619.

Landau, L. D. (1953). *Izv. Akad. Nauk. SSSR Ser. Fiz.* **17**, 51.

Landshoff, P. V. and Treiman, S. B. (1961). *Nuovo Cim.* **19**, 1249.

Landshoff, P. V. and Polkinghorne, J. C. (1972). *Phys. Rep.* **5C**, 1.

Law, M. E., Kasman, J., Panvini, R. S., Sims, W. H. and Ludlam, T. (1972). "A Compilation of Data on Inclusive Reactions", Lawrence Berkeley Laboratory Report LBL-80.

Le Bellac, M. (1971). *Phys. Letters* **37B**, 413.

Leith, D. W. G. S. (1972). Proceedings of XV International Conference on High Energy Physics, Batavia, Illinois.

Letessier, J. and Tounsi, A. (1972). *Nuovo Cim.* **13A**, 557.

Lindenbaum, S. J. and Sternheimer, R. (1957). *Phys. Rev.* **105**, 1874.

Lindenbaum, S. J. and Sternheimer, R. (1961). *Phys. Rev.* **123**, 333.

Liu, F. F., Davier, M., Fries, D. C., Derado, J., Mozley, R. F., Odian, A. C., Park, J., Swanson, W. P., Villa, F. and Yount, D. (1972). *Nucl. Phys.* **B47**, 1.

Lo, S. Y. and Phua, K. K. (1972). *Phys. Letters* **38B**, 415.

Lovelace, C. (1963). In "Strong Interactions and High Energy Physics". Scottish Universities Summer School, (R. G. Moorhouse, ed.), Oliver and Boyd.

Lovelace, C. (1971). *Phys. Letters* **34B**, 500.

Lurcat, F. and Mazur, P. (1964). *Nuovo Cim.* **31**, 149.

MacFarlane, A. J. (1963). *J. Math. Phys.* **4**, 490.

Maglic, B. C., Alvarez, W. L., Rosenfeld, A. H. and Stevenson, M. L. (1961). *Phys. Rev. Letters* **7**, 178.

Mandelstam, S. (1963). *Nuovo Cim.* **30**, 1127; *Nuovo Cim.* **30**, 1148.

McKerrel, A. (1964). *Nuovo Cim.* **34**, 1289.

Miettinen, H. I. (1972). *Phys. Letters* **38B**, 431.

Miettinen, H. I. and Pirilä, P. (1972). *Phys. Letters* **40B**, 127.

Miettinen, H. I. (1973). Ph.D. Thesis, Helsinki, Finland.

Milekhin, G. A. (1958). *Zh. Eksp. i Teor. Fiz.* **35**, 978.

Moen, I. O. (1972). Cambridge preprint DAMTP 72-11.

Moen, I. O. and White, A. R. (1972). *Phys. Letters* **42B**, 75.

Moen, I. O. and Zakrzewski, W. J. (1972). Cambridge Univ. preprint DAMPT 72/45.

Morgan, D. (1968). *Phys. Rev.* **166**, 1731.

Mueller, A. H. (1970). *Phys. Rev.* **D2**, 2963.

Mueller, A. H. (1971). *Phys. Rev.* **D4**, 150.

Nahm, W. (1972). *Nucl. Phys.* **B45**, 525.

Namiki, M. and Iso, C. (1957). *Prog. Theo. Phys.* **18**, 591.

Namyslowski, J. M., Razmi, M. S. K. and Roberts, R. G. (1967). *Phys. Rev.* **157**, 1328.

Oleson, P. (1971). *Nucl. Phys.* **B32**, 609.

Olsson, M. and Yodh, G. B. (1966). *Phys. Rev.* **145**, 1309, *Phys. Rev.* **145**, 1327.

Omnés, R. (1972). *Phys. Rep.* **3C**, 1.

Parry, G. W. and Rotelli, P. (1973). *Nuovo Cim. Letters* **7**, 649.

Peccei, R. D. and Pignotti, A. (1971). *Phys. Rev. Letters* **26**, 1076.

Pene, O. and Krzywicki, A. (1969). *Nucl. Phys.* **B12**, 415.

Petersson, B. and Törnqvist, N. A. (1969). *Nucl. Phys.* **B13**. 629.

Pilkuhn, H. (1967). "The Interactions of Hadrons". John Wiley and Sons, New York.

Pirila, P. and Pokorski, S. (1973). *Phys. Letters* **B43**, 502.

Plahte, E. and Roberts, R. G. (1969). *Nuovo Cim. Letters* **1**, 187.

Pokorski, S. and Satz, H. (1970). *Nucl. Phys.* **B19**, 113.

Pokorski, S. and Van Hove, L. (1974) *Acta Physica Polonica B* (to be published).

Polkinghorne, J. C. (1972). *Nuovo Cim.* **7A**, 555.

Pomeranchuk, I. Ya. (1951). *Dokl. Akad. Nauk. SSSR* **78**, 889.

Ranft, G. (1969). *Nuovo Cim.* **60A**, 643.

Ratner, L., Ellis, R., Vannini, G., Babcock, B., Krisch, A. and Roberts, J. (1971). *Phys. Rev. Letters.* **27**, 68.

Razmi, M. S. K. (1964). *Nuovo Cim.* **34**, 150.

Roberts, R. G. and Frazer, G. M. (1967). *Phys. Rev.* **159**, 1297.

Roberts, R. G. and Roy, D. P. (1972). *Phys. Letters* **38B**, 507.

Roberts, R. G. and Roy, D. P. (1974). Rutherford Lab. preprint RL.74-022.

Rosner, J. L. (1969). *Phys. Rev. Letters* **22**, 689.

Ross, M., Henyey, F. S. and Kane, G. L. (1970). *Nucl. Phys.* **B23**, 269.

Roy, D. P. and Roberts, R. G. (1972). *Nucl. Phys.* **40B**, 141.

Ruijgrok, Th. W. (1972). Proceedings III International Conference on Multiparticle Reactions, Zakopane, Poland.

Rushbrooke, J. G. (1972). Proceedings III International Conference on Multiparticle Reactions, Zakopane, Poland.

Sanda, A. I. (1972). *Phys. Rev.* **D6**, 282.

Satz, H. (1965). *Nuovo Cim.* **37**, 1407.

Satz, H. and Schilling, K. (1970). *Nuovo Cim.* **67A**, 511.

Satz, H. and Schilling, R. (1971). Proceedings of International Colloquium on Multi-particle Reactions, Helsinki, Finland.

Satz, H. and Thomas, G. H. (1971). Univ. of Helsinki. Preprint No. 7-71.

Satz, H. (1972). Proceedings III International Conference on Multiparticle Processes, Zakopane, Poland.

Schmid, C. (1968). *Phys. Rev. Letters* **20**, 689.

Schmidt, M. G. (1971). *Letters Nuovo Cim.* **1**, 1017.

Schreiner, P. A., Stork, D. H., Ross, R. T., Clark, A. and Lyons, L. (1971). *Nucl. Phys.* **B28**, 85.

Schwarz, J. H. (1967). *Phys. Rev.* **159**, 1269.

Sens, J. C. (1972). Proceedings IV International Conference on High Energy Collisions, Oxford.

Shapiro, J. (1960). *Nuovo, Cim. Suppl.* **18**, 40.

Silverman, D. P. and Tan, C.-I. (1971). *Phys. Rev.* **D2**, 233.

Sissakyan, I. N., Feinberg, E. L. and Chernavsky, D. S. (1971). *Tr. Fiz. Inst. Akad. Nauk. SSSR* **57**, 170.

Sivers, D. and Thomas, G. H. (1972). *Phys. Rev.* **D6**, 1961.

Slater, L. J. (1966). "Generalized Hypergeometric Functions", Cambridge University Press.

Smith, D. B. (1971). Ph.D. Thesis, University of California (URCL-20632), Berkeley.

Stapp, H. P. (1957). *Phys. Rev.* **103**, 425.

Stapp, H. P. (1971). *Phys. Rev.* **D3**, 3177.

Stone, S. L., Cohen, D., Farber, M., Ferbel, T., Holmes, R., Slattery, P. and Werner, B. (1971). *Nucl. Phys.* **B32**, 19.

Suzuki, M. (1969). *Phys. Rev. Letters* **23**, 205.

Ter-Martirosyan, K. A. (1963). *Sov. Phys. J.E.T.P.* **17**, 233.

Ting, P. D. and Yesian, H. J. (1971). *Phys. Letters* **35B**, 321.

Toller, M. (1965). *Nuovo Cim.* **37**, 631.

Toller, M. (1968). *Nuovo Cim.* **57A**, 671.

Törnqvist, N. A. (1970). *Nucl. Phys.* **B18**, 530.

Treiman, S. B. and Yang, C. N. (1962). *Phys. Rev. Letters* **8**, 140.

Tye, S. H.-H. and Veneziano, G. (1972). *Phys. Letters* **38B**, 30.

Van Hove, L. (1963). *Nuovo Cim.* **28**, 798.

Van Hove, L. (1964). *Rev. Mod. Phys.* **36**, 655.

Van Hove, L. (1969). *Nucl. Phys.* **B9**, 000.

Van Hove, L. (1972). *Phys. Letters* **43B**, 89.

Veneziano, G. (1968). *Nuovo Cim.* **57A**, 190.

Veneziano, G. (1971). *Nuovo Cim. Letters* **1**, 681.

Wang, C. P. (1969). *Phys. Rev.* **180**, 1463.

Wang, J. M. and Wang, L. L. (1971). *Phys. Rev. Letters* **26**, 1287.

Watson, K. M. (1952). *Phys. Rev.* **88**, 1163.

Weis, J. H. (1971). *Phys. Rev.* **D4**, 1779.

Weis, J. H. (1972). *Phys. Letters* **B43**, 487.

White, A. R. (1972). *Nucl. Phys.* **B39**, 432 *and* 461.

Wick, G. C. (1962). *Ann. Phys.* (N.Y.) **18**, 65.

Wilson, K. G. (1970). Cornell Univ. preprint CLNS-31 (unpublished).

Wolfendale, A. W. and Wdowczyk, J. (1973). Durham Univ. preprint (unpublished).

Wroblewski, A. (1972). Warsaw Univ. preprint IFD No. 72/2.

Wu, T. T. (1961). *Phys. Rev.* **123**, 678.

Wu, T. T. and Yang, C. N. (1965). *Phys. Rev.* **137**, B708.

Yamdagni, N. K. and Gavrilas, M. (1971). Paper presented at the Amsterdam International Conference on Elementary Particles.

Zachariasen, F. (1971). Physics Reports **2C**, 1.

Zachariasen, F. and Zweig, G. (1967). *Phys. Rev.* **160**, 1322; *Phys. Rev.* **160**, 1326.

Subject Index